Subcellular Biochemistry

Volume **11**

ADVISORY EDITORIAL BOARD

J. ANDRÉ Laboratoire de Biologie Cellulaire, 4 Faculté des Sciences, 91 Orsay, France

D. L. ARNON Department of Cell Physiology, Hilgard Hall, University of California, Berkeley, California 94720, USA

J. BRACHET Laboratoire de Morphologie Animale, Faculté des Sciences, Université Libre de Bruxelles, Belgium

C. de DUVE Université de Louvain, Louvain, Belgium and The Rockefeller University, New York, NY 10021, USA

M. KLINGENBERG Institut für Physiologische Chemie und Physikalische Biochemie, Universität München, Goethestrasse 33, München 15, Germany

A. LIMA-de-FARIA Institute of Molecular Cytogenetics, Tornavagen 13, University of Lund, Lund, Sweden

O. LINDBERG The Wenner-Gren Institute, Norrtullsgatan 16, Stockholm, VA, Sweden

V. N. LUZIKOV A. N. Belozersky Laboratory for Molecular Biology and Bioorganic Chemistry, Lomonosov State University, Building A, Moscow 117234, USSR

H. R. MAHLER Chemical Laboratories, Indiana University, Bloomington, Indiana 47401, USA

M. M. K. NASS Department of Therapeutic Research, University of Pennsylvania School of Medicine, Biology Service Building, 3800 Hamilton Walk, Philadelphia, Pennsylvania 19104, USA

A. B. NOVIKOFF Department of Pathology, Albert Einstein College of Medicine, Yeshiva University, Eastchester Road and Morris Park Avenue, Bronx, NY 10461, USA

R. N. ROBERTSON Macleay Building, A12, School of Biological Sciences, The University of Sydney, Sydney, N.S.W. 2006, Australia

P. SIEKEVITZ The Rockefeller University, New York, NY 10021, USA

F. S. SJÖSTRAND Department of Zoology, University of California, Los Angeles, California 90024, USA

A. S. SPIRIN A. N. Bakh Institute of Biochemistry, Academy of Sciences of the USSR, Leninsky Prospekt 33, Moscow V-71, USSR

D. von WETTSTEIN Department of Physiology, Carlsberg Laboratory, Gl. Carlsbergvej 10, DK-2500, Copenhagen, Denmark

V. P. WHITTAKER Abteilung für Neurochemie, Max-Planck Institut für Biophysikalische Chemie, D-3400 Göttingen-Nikolausberg, Postfach 968, Germany

---

A Continuation Order Plan is available for this series. A continuation order will bring delivery of each new volume immediately upon publication. Volumes are billed only upon actual shipment. For further information please contact the publisher.

Subcellular Biochemistry

Volume 11

Series Editor
Donald B. Roodyn
University College London
London, England

PLENUM PRESS · NEW YORK AND LONDON

The Library of Congress cataloged the first volume of this title as follows:

Sub-cellular biochemistry.

   London, New York, Plenum Press.
      v. illus. 23 cm. quarterly.
     Began with Sept. 1971 issue. Cf. New serial titles.
    1. Cytochemistry – Periodicals. 2. Cell organelles – Periodicals.
QH611.S84   574.8'76   73-643479

ISBN 0-306-41959-9

This series is a continuation of the journal *Sub-Cellular Biochemistry*,
Volumes 1 to 4 of which were published quarterly from 1972 to 1975

© 1985 Plenum Press, New York
A Division of Plenum Publishing Corporation
233 Spring Street, New York, N.Y. 10013

All rights reserved

No part of this book may be reproduced, stored in a retrieval system, or transmitted, in any form or by any means, electronic, mechanical, photocopying, microfilming, recording, or otherwise, without written permission from the publisher

Printed in the United States of America

Publisher's Note

During the period of editing contributions for this volume, Dr. Roodyn was unfortunately taken ill, and we are grateful to Dr. J. B. Finean, Department of Biochemistry, University of Birmingham, U.K., for his assistance in editing the remaining contributions.

Contributors

Volkmar Braun Department of Microbiology, University of Tübingen, D-7400 Tübingen, Federal Republic of Germany

Nejat Düzgüneş Cancer Research Institute, School of Medicine, University of California, San Francisco, California 94143

Eckhard Fischer Department of Microbiology, University of Tübingen, D-7400 Tübingen, Federal Republic of Germany

Klaus Hantke Department of Microbiology, University of Tübingen, D-7400 Tübingen, Federal Republic of Germany

John H. Hartwig Hematology–Oncology Unit, Department of Medicine, Massachusetts General Hospital, Boston, Massachusetts 02114, and Harvard School of Dental Medicine, Boston, Massachusetts 02115

Knut Heller Department of Microbiology, University of Tübingen, D-7400 Tübingen, Federal Republic of Germany

A. M. Kidwai Biomembrane Laboratory, Industrial Toxicology Research Centre, Mahatma Gandhi Marg, Lucknow 226001, India

Stuart E. Lind Hematology–Oncology Unit, Department of Medicine, Massachusetts General Hospital, Boston, Massachusetts 02114, and Harvard School of Dental Medicine, Boston, Massachusetts 02115

Richard Niederman Hematology–Oncology Unit, Department of Medicine, Massachusetts General Hospital, Boston, Massachusetts 02114, and Harvard School of Dental Medicine, Boston, Massachusetts 02115

Heinz Rotering Department of Microbiology, University of Tübingen, D-7400 Tübingen, Federal Republic of Germany

Friedhelm Schroeder Department of Pharmacology, School of Medicine, University of Missouri, Columbia, Missouri 65212

Contents

Chapter 1
Cortical Actin Structures and Their Relationship to Mammalian Cell Movements
John H. Hartwig, Richard Niederman, and Stuart E. Lind

1. Introduction	1
1.1. The Structure of Cortical Cytoplasm and Organization of Actin Filaments within It	1
1.2. Structure, Assembly, and Polarity of Actin Filaments	4
2. F-Actin-Associated Proteins	6
2.1. Cross-linking of Actin Filaments	7
2.2. Regulation of Actin Filament Length	16
2.3. Stabilization of Actin Filaments	23
2.4. Retraction of Actin Filaments	24
2.5. Proteins that Interact with Actin Monomers	25
3. Movements of Cell Membranes	26
3.1. Propulsive Movements	27
3.2. Retractive Movements	30
4. Actin Structures in Mammalian Cells	31
4.1. Cortical Networks in Leukocytes	32
4.2. Platelets	35
5. Conclusions	39
6. References	39

Chapter 2
Fluorescence Probes Unravel Asymmetric Structure of Membranes
Friedhelm Schroeder

1. Introduction	51
1.1 Protein Asymmetry	52
1.2. Lipid Asymmetry	53

1.3.	Boundary Lipid	55
1.4.	Charge Asymmetry	56
2. Transbilayer Lipid Distribution		56
3. Cholesterol Transbilayer Asymmetry		60
3.1.	Fluorescent Sterol Distribution	61
3.2.	Polyene–Sterol Interactions	63
3.3.	Perturbation of Cholesterol Asymmetry	64
4. Transbilayer Structure		64
4.1.	Transbilayer Coupling	64
4.2.	Membrane Depth Gradients	65
4.3.	Transbilayer Fluidity Gradients	65
5. Phase Separations of Lipids		67
5.1.	Lateral Rearrangements	67
5.2.	Independent Monolayers	69
6. Physiological Function of Asymmetric Structure of Membranes		70
6.1.	Receptor Modulation	70
6.2.	Receptor–Adenylate Cyclase Coupling	72
6.3.	Phospholipid Methylation and Receptor Functions	74
6.4.	Role of Cholesterol Asymmetry in Secretion, Endocytosis, Transport, and Aging	75
7. Potential Artifacts		77
7.1.	Membrane Preparation	77
7.2.	Probe Location	77
7.3.	Quenching and Other Artifacts of Fluorescence Determinations	78
7.4.	Use of Filipin to Detect Cholesterol Location	82
8. Future Directions		83
9. References		84

Chapter 3
Functional Aspects of Gram-Negative Cell Surfaces
Volkmar Braun, Eckhard Fischer, Klaus Hantke, Knut Heller, and Heinz Rotering

1. Outer Membrane Proteins of *Escherichia coli*		103
1.1.	Introduction	103
1.2.	The Outer Membrane	104
1.3.	Proteins in the Outer Membrane	108
1.4.	*E. coli* K-12	109
1.5.	Lipoproteins	109
1.6.	Porins	112

1.7.	Outer Membrane Proteins That Are Part of Specific Transport Systems	113
1.8.	The Receptor Proteins of Iron Uptake Systems	116
1.9.	The BtuB Protein	116
1.10.	The OmpA Protein	117
1.11.	The TolC Protein	118
1.12.	Enzymes in the Outer Membrane	119
2. The Outer Membrane of *Neisseria*		119
3. The Outer Membrane of *Pseudomonas aeruginosa*		124
4. Export of Proteins		126
4.1.	Introduction	126
4.2.	Models for Protein Transport	126
4.3.	Signal Sequence	127
4.4.	Role of the Sequence of the Mature Protein	130
4.5.	Cellular Components of the Export System	132
4.6.	Processing	133
4.7.	Translocation into or through the Outer Membrane	135
4.8.	Concluding Remarks	136
5. Components of the Outer Membrane Involved in the Uptake of DNA		137
5.1.	Introduction	137
5.2.	Proteins	138
5.3.	Lipopolysaccharides (LPS)	142
5.4.	Lipids	144
5.5.	Divalent Cations	146
5.6.	Energy Dependence	147
5.7.	Summary	149
6. Osmoregulation		150
6.1.	Introduction	150
6.2.	Osmoregulation in the Cytoplasm	152
6.3.	Osmoregulation in the Periplasm	157
6.4.	Osmolarity-Dependent Regulation of the Protein Composition of the Outer Membrane	161
7. References		163

Chapter 4
Biochemistry of the Sarcolemma
A. M. Kidwai

1. Introduction		181
2. Isolation Techniques		182
2.1.	Isolation of Smooth Muscle Plasma Membrane	182

2.2.	Isolation of Skeletal Muscle Plasma Membrane	183
2.3.	Isolation of Cardiac Muscle Plasma Membrane	185
3.	Characterization of the Sarcolemma	186
3.1.	Morphology	187
3.2.	Enzyme Markers	187
3.3.	Chemical Composition	187
3.4.	Miscellaneous Markers	188
4.	Biochemical Properties	188
4.1.	Enzymatic Makeup	188
4.2.	Glucose Transport	189
4.3.	Calcium Transport	189
4.4.	Receptors	190
5.	Special Studies	190
5.1.	Membrane Orientation	190
5.2.	Labeling Techniques	191
6.	Conclusions	191
7.	References	192

Chapter 5
Membrane Fusion
Nejat Düzgüneş

1.	Introduction	195
2.	Fusion of Subcellular Membranes	196
2.1.	Exocytosis	196
2.2.	Endocytosis	198
2.3.	Membrane Flow	199
3.	Studies on the Mechanisms of Membrane Fusion	200
3.1.	Electron Microscopy of Secretory Processes	201
3.2.	Studies with Permeabilized Cells and Ca^{2+} Indicators	203
3.3.	Fusion of Isolated Secretory Vesicles and Other Intracellular Membranes	204
3.4.	Fusion of Erythrocytes Induced by Chemicals	210
3.5.	Fusion of Myoblasts	213
3.6.	Virus–Cell Fusion	215
3.7.	Studies with Planar and Spherical Phospholipid Bilayers	219
3.8.	Fusion of Pure Phospholipid Vesicles	223
3.9.	Fusion of Mixed Phospholipid Membranes	230

4.	Molecular Mechanisms of Membrane Fusion	234
	4.1. Close Approach of Membranes	234
	4.2. Lateral Reorganization and Destabilization of Membranes	236
	4.3. Phospholipid Specificity in Membrane Fusion	241
	4.4. Cation Specificity in Membrane Fusion: Ion Binding and Dehydration	243
	4.5. Osmotic Effects in Membrane Fusion	245
	4.6. Modulation of Membrane Fusion by Proteins, Polypeptides, and Polyamines	246
	4.7. Implications for Subcellular Membrane Fusion	253
5.	References	255
	INDEX	287

Chapter 1

Cortical Actin Structures and Their Relationship to Mammalian Cell Movements*

John H. Hartwig, Richard Niederman, and Stuart E. Lind
Hematology–Oncology Unit, Department of Medicine
Massachusetts General Hospital
Boston, Massachusetts 02114
and Harvard School of Dental Medicine
Boston, Massachusetts 02115

1. INTRODUCTION

Cytoplasmic filaments effect cell movement, organize the cytoplasm, and support the cell membrane. Three filament systems are currently recognized: actin filaments, intermediate filaments, and microtubules. In this chapter we describe the role of actin filaments in the structure and movement of metazoan cytoplasm.

In the following sections we will (1) present an overview of the biochemistry and organization of actin filaments in mammalian cells, (2) introduce the actin-associated proteins that regulate the structure of actin in these cells, and (3) provide examples of how cellular actin structures can effect the propulsion and retraction of cytoplasm.

1.1. The Structure of Cortical Cytoplasm and Organization of Actin Filaments within It

Cells display a wide assortment of membrane-bounded protrusions on their surface. These include microvilli of varying sizes and stabilities, mem-

---

*Supported by USPHS Grants HL27971, DE06321, DE06123, and HL01063, and grants from the Edwin S. Webster Foundation and Edwin W. Hiam.

brane veils, blebs, and pseudopods. These surface specializations derive their structure from the actin-filament-enriched region of cytoplasm located just beneath them. During movement, changes in cell shape appear to be generated by propulsive and retractile movements in this layer of cortical cytoplasm.

In general, cortical actin filaments are observed in electron micrographs either to be aligned with one another into bundles or to branch from one another to form networks that lack apparent order.

1.1.1. Filament Bundles

Stable microvilli decorate the luminal surface of intestinal and kidney epithelial cells. Actin filaments fill these microvilli and are precisely ordered. They lie side by side in linear aggregates, and all the filaments have the same polarity. Heavy meromyosin arrowheads point inward from the plasma membrane (Mooseker and Tilney, 1975; Matsudaria and Burgess, 1979; Begg et al., 1980). This ordered filament structure is maintained by proteins that bind, cross-link, and stabilize the bundles. The filaments are therefore spaced apart rather than being densely packed as is observed in paracrystals of purified actin filaments created by high Mg^{2+} concentrations. Similar processes are found on the surface of eggs and activated platelets but are less rigid than the brush border microvilli (Figures 1a and 1b). Although the length of filaments within the microvilli has not been investigated, individual filaments appear in electron micrographs to transverse the length of microvilli from intestine cells. However, individual filament length may vary according to the microvilli and cell type.

A different kind of actin filament bundle is found in flattened endothelial cells (Wong et al., 1983), in fibroblasts (Lazarides and Weber, 1974), in retinal rods (Burnside et al., 1983), and in the cleavage furrow of dividing cells (Schroeder, 1973). These bundles have been called stress fibers. Actin filaments in these assemblies have much less organization than those found in the bundles filling microvilli. In contrast to filaments of microvilli, a bidirectional interdigitation of filaments has been observed in electron micrographs of fibers labeled with heavy meromyosin (Begg et al., 1980, and Figures 1c and 1d). This nonuniform polarity also appears to be characteristic of filaments in the contractile ring of dividing cells. Once again whether actin filaments are long or short in stress fibers has not been determined.

1.1.2. Actin Networks

Much of the cortex of motile cells is composed of a three-dimensional net of actin filaments. Filaments in the network are short, 0.1–0.5 micrometers in length, and intersect at even shorter distances, about every 100 nm (Figure 1f). The polarity of filaments in this network appears to be random (Figure 1e).

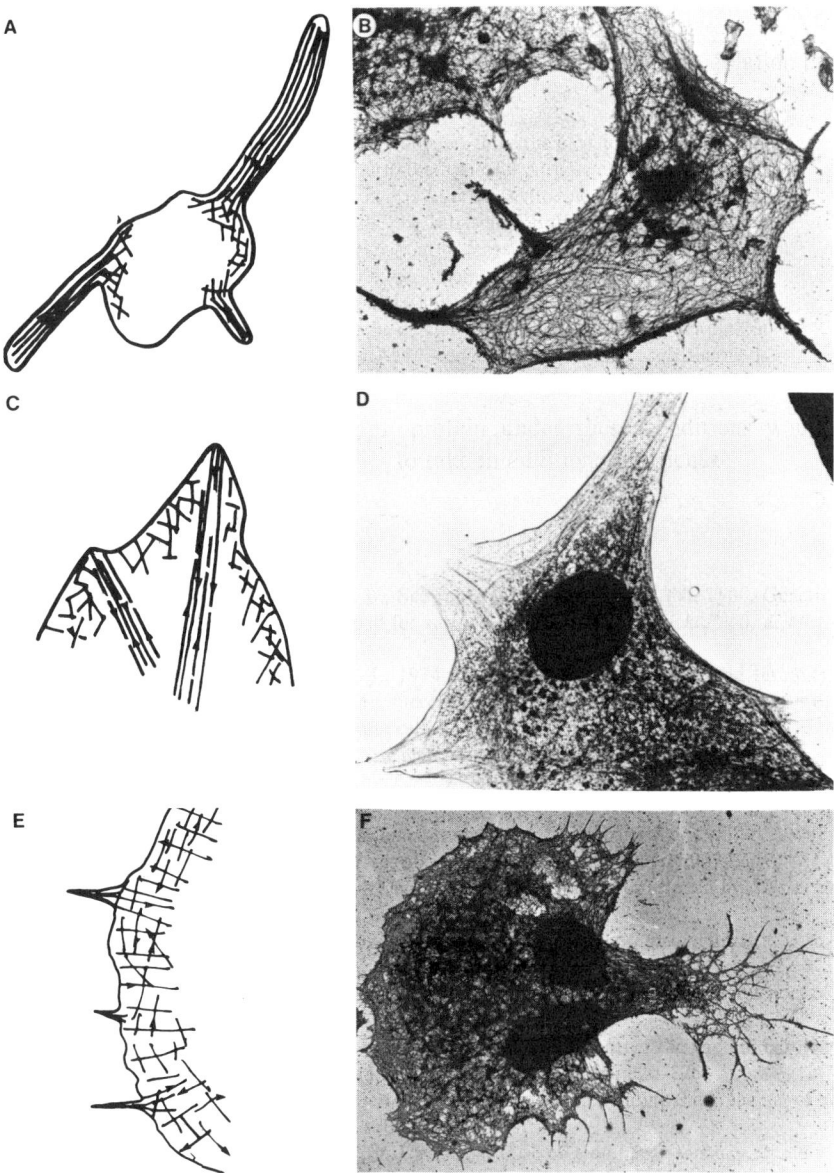

FIGURE 1. Structures formed by actin in the cortex of mammalian cells. A, C, and E illustrate the arrangement and polarity of actin filaments in different cell types. (A) Polarity of actin filaments in platelet filopodia. Filaments in these projections all pointing inward from the membrane. (C) Filaments composing stress fibers do not have a uniform polarity. (E) Filaments organized into networks are randomly oriented. B, D, and F show electron micrographs of cells and structures illustrated in A, C, and E. B) Activated platelet. D) Blood vascular endothelial cell. E) Polymorphonuclear leukocyte. Cells were permeabilized with detergent and then prepared for the electron microscope.

The network is formed because filaments intersect, forming T- and X-shaped branches (Schliwa and van Blerkom, 1981). In highly motile cells, such as leukocytes, this cortical actin filament zone is broad and fills all cellular extensions such as pseudopods, veils, or blebs. Random networks of filaments are also apparent in electron micrographs in less motile cells such as fibroblasts. These occur between the more ordered stress fibers and are found in motile regions of the cell (Henderson and Weber, 1979; Herman *et al.*, 1981).

1.2. Structure, Assembly, and Polarity of Actin Filaments

Actin is a major protein of muscle and nonmuscle cells. Although small differences in the primary structure of actins purified from different species and tissues have been demonstrated (Vanderkerckhove and Weber, 1978), all actins appear to be similar in function. Monomeric actin is a globular protein of 42,000 daltons. The actin monomers are somewhat asymmetrical in shape, with molecular dimensions of 3.5 by 6.5 nm (Aebi *et al.*, 1980).

Actin monomers can reversibly assemble into filaments. Physiological concentrations of K^+ and Mg^{2+} salts promote their assembly into long double-helical polymers (F-actin) that average 1–2 μm in length and contain 500–1000 monomers per filament (Kawamura and Maruyama, 1970; Oosawa and Asakura, 1975). In negatively stained specimens viewed in the electron microscope, the filaments are 6–8 nm in diameter and have a 37-nm axial helix repeat. In platinum-coated specimens, the same periodicity is found but the filament diameter is thickened to 10 nm by the metal deposit (Figure 2a). Within the filament, individual actin subunits are spaced every 5.5 nm, such that 13 monomers, in two strands, are contained in each axial helix repeat (Figure 2b). All monomers in the filament are oriented in the same direction, which results in the actin monomers at each end of the filament having different domains exposed.

Filament assembly occurs by at least a two-step reaction (Oosawa and Kasai, 1971). First, a small number of monomers must aggregate to form nuclei, small oligomers of two to three monomers having actin-binding sites comparable to filament ends. These nuclei serve as templates for additional actin monomers to assemble onto, a process that elongates them into filaments. Nucleation is a slow reaction, the rate of which is proportional to the third or fourth power of the actin concentration. The second is a fast reaction, the rate of which is diffusion limited. Monomers rapidly add to the oligomer ends, elongating them into filaments. Elongation can only occur when actin monomers are present in solution at concentrations greater than actin's "critical" concentration, the concentration at which the rate of assembly equals the rate of disassembly. This value, approximately 1 μM under physiological ionic conditions, is similar for all purified actins that have been tested (Gordon *et al.*, 1977).

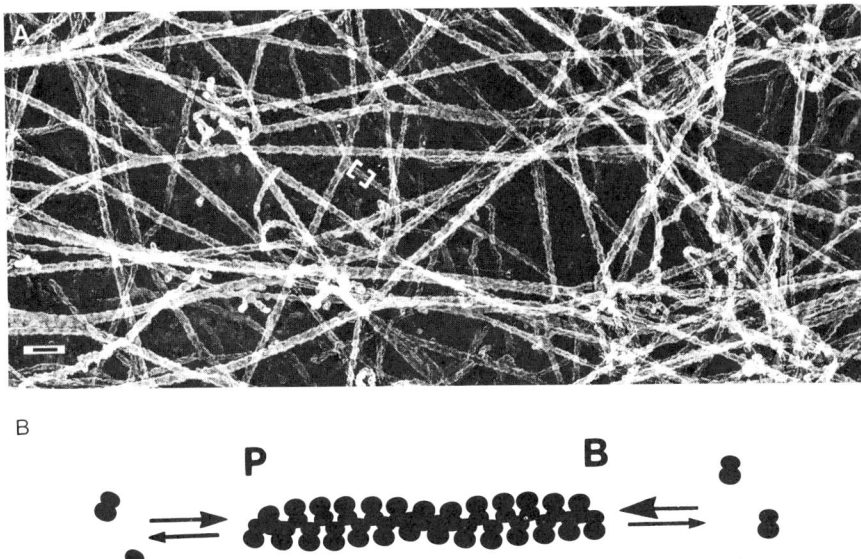

FIGURE 2. Structure of actin filaments. (A) Electron micrograph of actin filaments after fixation with glutaraldehyde, dehydration in ethanol, critical point drying in CO_2, and rotary shadowing with platinum and carbon. The bar is 100 nm. The bracketed region denotes one complete helix turn of the filament, equivalent to a length of 75 nm. (B) Model showing the arrangement of actin monomers in the filament and defining the barbed (B) and pointed (P) end of the actin filament. Actin monomers assemble most rapidly to the barbed end of the filament.

Filaments disassemble when the actin concentration is below the critical concentration. Raising the temperature favors assembly; lowering it favors disassembly (Oosawa and Kasai, 1971).

The polarity of actin was first recognized by Huxley (1969), who observed the stereospecific binding of the myosin fragment, heavy meromyosin, to actin filaments in the electron microscope. Myosin fragments bind to actin filaments at an angle that forms repeating arrowhead patterns. A "pointed" and "barbed" end of each filament can therefore be defined (Figure 2b). The affinity of monomers is greater at the barbed filament end and therefore the ends of the actin filament have been found to elongate at different rates (Woodrum et al., 1976; Kondo and Ishiwata, 1976). The rate of addition of actin monomers at the barbed end is about five to six times faster than at the pointed end (Pollard and Mooseker, 1981). Polar assembly occurs because actin monomers are asymmetric and all orient in the same direction when assembled into filaments. Therefore, the ends of filaments have unique regions of their subunits exposed.

Filament polarity has important implications in cell movement. Actin filaments attach to membranes selectively at their barbed ends (Begg *et al.*, 1978; Mooseker and Tilney, 1975). This orientation dictates that the force produced by membrane-attached filaments and cytoplasmic myosin will have an inward direction. This is because myosin filaments pull actin filaments in the "pointed" direction. The kinetics of filament assembly indicate that the assembly of filaments toward the cell membrane would be the favored event since filaments prefer to assemble in the barbed direction. However, the mechanism whereby filaments become attached to the membrane remains unknown. Filaments may elongate inward from nucleating sites attached to the cytoplasmic face of the plasma membrane, although this process is less kinetically favored. A more likely possiblity is that filaments elongate outward from cytoplasmic nucleating sites and then dock on filament end binding proteins attached to the cytoplasmic face of the membrane.

2. F-Actin-Associated Proteins

Cells have actin filaments arranged in their cytoplasm in varying degrees of order. Order, or as we shall see, disorder, is imparted onto actin by additional proteins. In solution, purified actin filaments, because of their extreme length, tend to align randomly (Zaner and Stossel, 1982; Niederman *et al.*, 1983). Therefore, actin-associated proteins must be present in cytoplasm to facilitate side by side alignments and bundling of actin or facilitate network formation by promoting filament branching and shortening, thereby inhibiting these parallel (anisotropic) alignments.

We are just beginning to understand how different actin-based structures are assembled within cells. Extensive effort has been applied to isolating actin-associated proteins from different cells, and a large number of proteins have been purified and characterized with respect to their interaction with actin. Because of the size and complexity of this list, this review considers only proteins of mammalian cells. Proteins of similar function, although often with quite different molecular properties, have been isolated from cells of many lower vertebrates and invertebrates.

The structure and movement of actin is orchestrated by five classes of associated proteins, proteins that:

1. Cross-link actin filaments.
2. Regulate actin filaments' length.
3. Stabilize actin filaments.

4. Retract actin filaments.
5. Interact with actin monomers.

2.1. Cross-linking of Actin Filaments

A large number of actin cross-linking proteins have been isolated from mammalian cells (Table I). These differ in size, molecular properties, calcium sensitivity, and the kind of structures they produce.

2.1.1. Actin Networks

Cross-linker proteins of this class are thought to control the perpendicular (orthogonal) branching of actin filaments within networks and perhaps the association of filaments with the cell membrane.

2.1.1.1. Actin-Binding Protein and Filamin. Actin is cross-linked into orthogonal (isotropic) networks by a family of large, flexible proteins. The first of this class of proteins was isolated from leukocytes and called actin-binding protein (Hartwig and Stossel, 1975; Stossel and Hartwig, 1975, 1976; Boxer and Stossel, 1976; Brotschi *et al.*, 1978; Hartwig and Stossel, 1981; Hartwig *et al.*, 1980). This protein subsequently has been isolated from platelets (Schollmeyer *et al.*, 1978; Rosenberg *et al.*, 1981; 1982) and BHK cells (Schloss and Goldman, 1979). A related and very similar protein, filamin, has been purified from smooth muscle tissue (Wang *et al.*, 1975; Shizuta *et al.*, 1976; Wang, 1977; Wallach *et al.*, 1978), and IgG antibodies prepared against guinea pig vas deferans filamin have been reported to cross-react with the leukocyte protein (Wallach *et al.*, 1978). However, functional assays have revealed differences that suggest that actin-binding protein and filamin are not identical (Brotschi *et al.*, 1978). Both of these proteins are homodimers in physiological, 0.1 M KCl buffers. Subunits have molecular weights of 270,000 daltons, and in the electron microscope are extended molecules 80 nm long and 3 nm wide (Figure 3; Tyler *et al.*, 1980; Hartwig *et al.*, 1980). Each chain is very flexible and contains a calcium-insensitive actin-binding site and a self-association site. These sites are located on opposite ends of the monomer. These subunits associate to generate a bipolar dimer 160 nm in length with its actin-binding sites near its free ends. Consequently, each molecule can bind and cross-link two actin filaments.

The importance of actin-binding protein and filamin as cytoplasmic actin filament cross-linkers has been established by the following criteria. First, both molecules are localized within cortical cytoplasm. Fluorescent antibodies against actin-binding protein are concentrated in the motile regions of moving leukocytes (Stendahl *et al.*, 1981; Valerius *et al.*, 1981). Antibodies against filamin localize strongly in stress fibers and in the cortical cytoplasm of fibro-

Table I
Mammalian Actin Filament Cross-Linking Proteins

Class	Cell	Structure	Function	References
Isotropic				
Actin-binding protein	Leukocytes (macrophages, polymorphonuclear leukocytes, lymphocytes)	2 × 270K	Cortical actin networks 1% of cell protein Ca^{2+} insensitive 3–5X more potent than smooth muscle filamin 75% of actin-gelling activity in leukocytes	Hartwig and Stossel, 1975; 1981; Stossel and Hartwig, 1975; 1976; Brotschi et al., 1978; Stendahl et al., 1980; Boxer and Stossel, 1976; Valerius et al., 1981 Thortesson et al., 1981
	Platelets	2 × 270K	Cortical networks ? Microspikes 7% of cell protein	Rosenberg and Strachter, 1980, 1981; Scholmeyer et al., 1978
	Amphibian oocytes	2 × 270K	All the actin-gelling activity in these cells	Corwin and Hartwig, 1983
	BHK-21 cells	? × 250K	Stress fibers	Schloss and Goldman, 1979
(Filamin)	Smooth muscle fibroblasts	2 × 250K	Stress fibers	Wang et al., 1975; Wang and Singer, 1977; Wang, 1977; Shizuta et al., 1976; Wallach et al., 1978; Davies et al., 1977, 1982
Nonerythrocyte spectrin	Intestinal epithelial cells	500K 1 × 260K 1 × 240K	Cross-links actin bundles together in terminal web	Glenney et al., 1981, 1982a,b
	Brain	500K 1 × 250K 1 × 240K	Axoplasm Ca^{2+} sensitive	Levine and Willard, 1981

Protein	Source	MW	Properties	References
	3T3	500K	Cortical cytoplasm	Burridge et al., 1982
	Skeletal muscle	500K		Nelson and Lazarides, 1983
Anisotropic α-actinin	Skeletal muscle	200K 2 × 100K	Z-lines; Ca²⁺-insensitive; Effect maximal at 4°, much less at 25 and 37°C	Maruyama and Ebasbi, 1965; Maruyama et al., 1977; Singh et al., 1981
	Ehrlich ascites	2 × 115K	Ca²⁺-sensitive; 5X less potent than filamin in gelling actin (called actinogelin by cited authors)	Mimura and Asano, 1979, 1982; Maruyama et al., 1981
	Kidney	?		Maruyama et al., 1981; Kuo et al., 1982
	3T3 cells	2 × 105K	Same as above	Burridge and Feramisco, 1981
	Platelets	2 × 105K		Rosenberg et al., 1981
	Macrophages	2 × 105K		Bennet et al., 1984
	Amoeba	2 × 85 – 95K	Ca²⁺-sensitive	Brier et al., 1983; Condeelis and Vahley, 1982; Pollard, 1981
Villin	Intestinal epithelial cells	95K	Characterized by a number of labs; Cross-links to form bundles in the absence of Ca²⁺	Bretscher & Weber, 1978; Glenney et al., 1981; Craig and Powell, 1980; Mooseker et al. 1980
Fimbrin	Same as above	68K	Bundles actin independent of Ca²⁺; Localized in a number of different cell types	Bretscher and Weber, 1981; Glenney et al., 1981
Vinculin	130K		Bundles; Localizes in attachment plaques	Jockusch and Isenberg, 1981; Isenberg et al., 1981

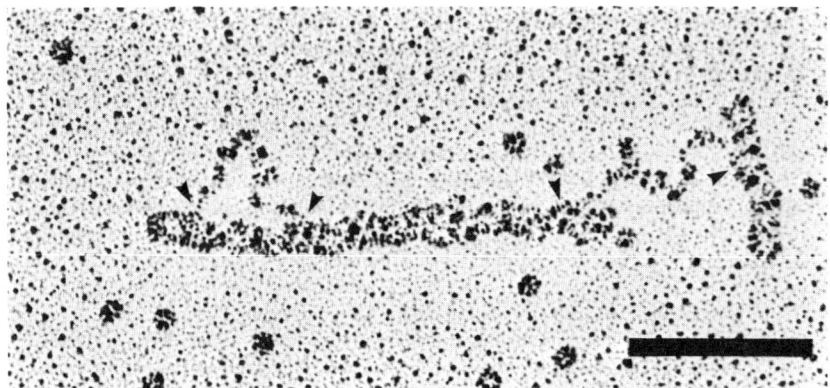

FIGURE 3. Electron micrograph showing the structure of individual molecules of leukocyte actin-binding protein cross-linking actin filaments. The proteins were sprayed onto mica, dried under vacuum, and the coated with platinum at 5° while rotating. The bar is 100 nm. The arrowheads point to the ends of the actin-binding protein molecules cross-linking the actin filaments.

blasts (Wang et al., 1975). Second, they appear to be the most potent actin filament cross-linkers of mammalian cells. In leukocytes, where actin filament cross-linking activity has been quantitated, actin-binding protein accounts for 75% of this activity. These proteins are major components of mammalian cells and generally account for about 1–2% of the total cellular protein, although in certain cells (i.e., platelets) it accounts for as much as 7% of the total cell protein. Third, the vast majority of the actin-binding protein is associated with the detergent-insoluble cell cytoskeleton. Finally, leukocyte actin-binding protein links actin filaments into networks with high efficiency. Only 1.5 molecules of actin-binding protein have been demonstrated to be required per filament to form actin gels (Brotschi et al., 1978). This is near the one molecule per filament ratio predicted to be the minimum required to gel actin filaments for a cross-linking molecule containing two actin-binding sites (Flory, 1953; Yin et al., 1980). Filamin, on the other hand, is one-third as potent as actin-binding protein in gelling actin.

There are two striking features of actin networks formed with actin-binding protein *in vitro* and in cells. The filaments are short and they are highly branched. The angle of filament branching is usually between 60 and 90° (Schliwa and van Blerkom, 1980; Heuser and Kirschner, 1980). Actin assembled in the presence of actin-binding protein, at molar ratios that exist within the cell, forms a very similar orthogonal network (Figure 4). This structure forms *in vitro* because actin-binding protein nucleates filament assembly (Hartwig et al., 1980). Since each molecule of actin-binding protein contains two

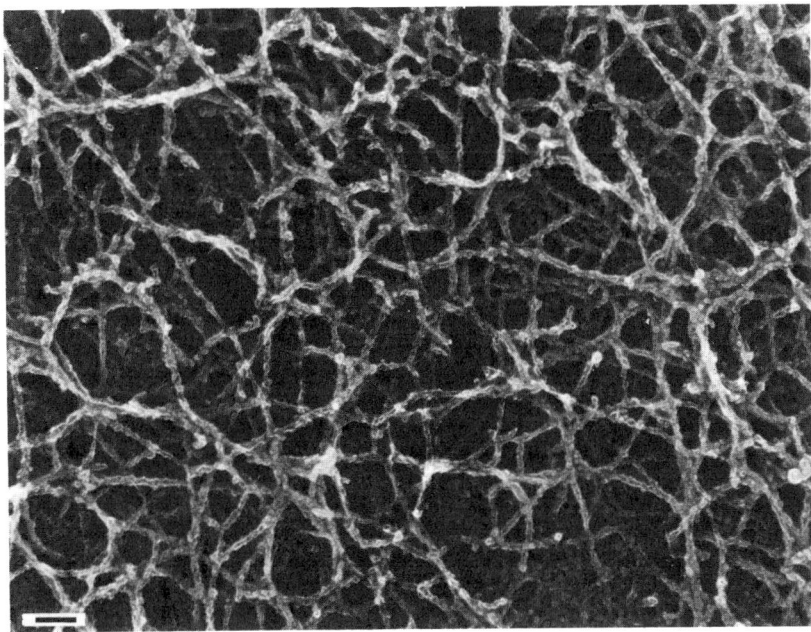

FIGURE 4. Electron micrograph showing the structure of actin assembled in the presence of actin-binding protein at a molar ratio of 1 actin-binding protein molecule to 50 actin monomers. Actin-binding protein induces the formation of a orthogonal network composed of short actin filaments (compare with Figure 1). The specimen was prepared as described in Figure 1. The bar is 100 nm.

binding sites for actin, two templates are present from which actin monomers elongate. The spatial orientation of these sites in solution determines the angle at which the elongating filaments branch and this must be near perpendicular if an orthogonal network is to form. Likewise, filamin has been reported to nucleate actin filament assembly (Koteliansky *et al.,* 1981). One aspect of the assembly of actin from actin-binding protein nuclei is noteworthy. Filaments growing off actin-binding protein nuclei do so most rapidly in the "barbed" direction. The cell may use this mechanism to assemble filaments from the cortical actin network toward the cell membrane. Filaments arriving at the membrane would have the polarity that has been observed in the electron microscope.

2.1.1.2. Spectrin and Related Proteins (Fodrin, Intestinal TW 260/240, Nonerythrocyte Spectrin). Spectrin, a 460,000-dalton protein, associates with actin to form a two-dimensional, submembranous lamina in red blood cells. Like actin-binding protein, spectrin is an elongated molecule. Unlike actin-binding protein, it is a heterodimer composed of subunits of 240,000 (α) and 220,000 (β) daltons. In electron micrographs, dimers are 100 nm in length with the subunits aligned side by side (Shotton et al., 1979; Tyler et al., 1980). Dimers have one site for binding actin and one for self-association. These are located at opposite ends of the molecule. In physiological buffers, spectrin dimers reversibly associate into bipolar tetramers 200 nm in length (Ungewickell et al., 1979; Shotton et al., 1979). This association is necessary to create a structure having two binding sites for actin, the minimum required to crosslink filaments (Brenner and Korn, 1979; Fowler and Taylor, 1980). Relative to dimers of mammalian actin-binding protein, spectrin dimers appear more rigid in the electron microscope, where the subunit chains are generally found to be interwoven, not dangling free, as is observed in actin-binding protein molecules. This increased rigidity is reflected in actin filament cross-linking. Filaments cross-linked by spectrin acquire 200-nm spacings. Therefore, actin assembled in its presence forms dense anastomosing networks similar in structure to the submembranous network in erythrocytes (Cohen et al., 1980).

Spectrin, in addition to binding and cross-linking actin, binds to the plasma membrane protein ankryin (Bennett and Stenbuck, 1979). Ankryin is in turn bound to the integral anion transport protein in the membrane (Bennett and Stenbuck, 1979).

Until recently it was believed that spectrin was a specialized product of erythrocytes. However, in 1981 Levine and Willard identified a high-molecular-weight protein in brain that they called fodrin and that was associated with cellular actin structures. Similar molecules also called intenstinal TW 260/240 or nonerythrocyte spectrin were later purified from this and other cell types (Glenney et al., 1982a,b,c; Burridge et al., 1982; Collier and Wang, 1982a,b; Bennet et al., 1982) and antigenically demonstrated to be related to erythrocyte spectrin (Goodman et al., 1981; Burridge et al., 1982; Glenney et al., 1982; Nelson and Lazarides, 1983). Like spectrin, fodrin is also an asymmetric heterodimer composed of subunits of 260,000 and 240,000 daltons. The 240,000 (β) chain cross-reacts with IgG antibodies prepared against the β subunit of erythrocyte spectrin (Glenney et al., 1982b; Burridge et al., 1982). This protein tetramerizes to cross-link actin. In electron micrographs the tetramers are 260 nm long and fairly straight. The binding sites for actin are near the ends of fodrin and molecules observed cross-linking actin filaments are bound end on to the filament forming T-shaped fodrin-actin filament branches.

The role of this protein in the cytoplasm of cells remains to be determined. Its remarkable similarities with red cell spectrin suggest that it may be involved in actin–membrane attachments. This protein also contains binding site(s) for the ubiquitous cellular protein calmodulin. Calmodulin has been demonstrated to confer calcium sensitivity onto a number of cellular processes, and the cross-linking of actin by fodrin has been reported to be affected by changes in calcium concentrations (Glenney *el al.,* 1982b).

2.1.2. Actin Filament Bundles

Proteins that align actin filaments into parallel aggregates differ greatly in structure from those inhibiting these alignments. They are much smaller proteins and fall into two categories: the rigid rods and the globular proteins. These proteins have been isolated from muscle and nonmuscle cells, localized in the cell cortex, and demonstrated to interact with actin filaments *in vitro* but have no known role in effecting cell movement. Nevertheless, in Section 4, we will present models for cell movements that incorporate some of these proteins.

2.1.2.1. Actinin. α-Actinin was first purified from skeletal muscle in 1965 (Maruyama and Ebasbi). Eleven years later, antigens related to α-actinin were demonstrated to be components of nonmuscle cells (Lazarides, 1976). Recently, they have been isolated and characterized in different cells (Mimura and Asano, 1979, 1982; Kuo *et al.,* 1982; Burridge and Feramisco, 1981; Rosenberg *et al.,* 1981). About 1% of the cellular protein is α-actinin (a molar ratio of α-actinin to actin of 1 to 50). While purified α-actinins have been clearly demonstrated to bind, cross-link, and induce the side-to-side alignment of filaments, for unknown reasons these effects are maximally manifested at low temperatures. Increasing the temperature markedly diminishes the effects of this protein on actin viscosity (Singh *et al.,* 1981; Jockusch and Isenberg, 1981). Moreover, the gelation of actin, a property of this protein at low temperature, does not occur at temperatures in the physiological range for mammalian cells, that is, 37–40°C (Bennett *et al.,* 1984).

In electron micrographs, α-actinin is a small rod, 4 nm in diameter and 40 nm in length (Figure 5). The orientation of individual subunits in these rods is unknown. The binding sites for actin appear to be located near the ends of these rods and rods can be observed between filaments within loose bundles formed by α-actinin (Jockusch and Isenberg, 1981).

Similarities and differences exist in α-actinins purified from muscle tissue and nonmuscle cells. Both proteins are homodimers in physiological solutions. However, subunit size appears to vary with each cell type. The muscle protein consists of polypeptides with molecular weights of 105,000 daltons each. Subunits in the nonmuscle α-actinins differ in molecular weight from 95,000 to 115,000 daltons, depending on the individual cell. Functionally, both cross-link actin. However, only the nonmuscle protein is calcium sensitive being inhibited by μM calcium concentrations. The explanation for this inhibition is that cal-

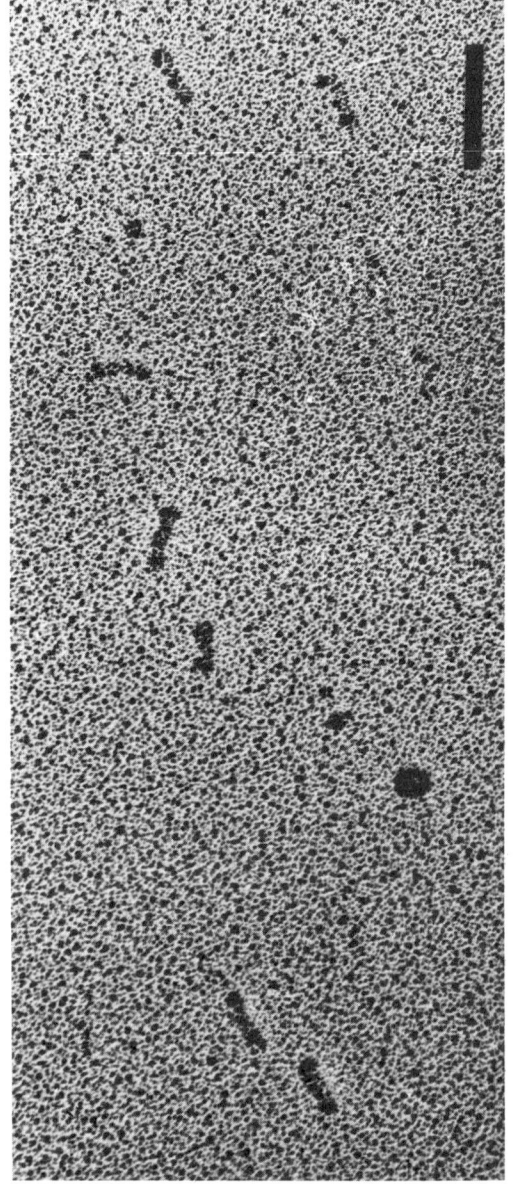

FIGURE 5. Electron micrographs of leukocyte alpha-actinin molecules. The molecules are visualized as described in Figure 3. The bar is 50 nm.

cium has been reported to greatly diminish the binding of α-actinin to actin (Burridge and Feramisco, 1981; Maruyama *et al.,* 1981).

In summary, the cellular role of α-actinin is unclear. It was initially thought to be involved in the linking of actin filaments to membranes, but this idea has not been supported by much experimental evidence.

2.1.2.2. Fimbrin. Fimbrin is a protein of the actin filament bundles of microvilli found on the surface of intestinal epithelial cells (Bretscher and Weber, 1980; Bretscher, 1981; Glenney *et al.,* 1981). It is composed of a single 68,000 dalton subunit. It binds to actin with high capacity, 1 molecule of fimbrin binds for every 2–3 actin monomers in a filament, and forms tightly packed actin filament bundles in which all the filaments have the same polarity. The filament-to-filament spacing in these bundles is about 9 nm (Matsudaria *et al.,* 1983). This is less than the spacing of filaments within bundles of the microvillus core.

Antigens cross-reactive with IgG antibodies prepared against intestinal fimbrin localize in the cortical cytoplasm near the plasma membrane of a variety of motile cells (Brestcher and Weber, 1980a). Curiously, this region of cortical cytoplasm lacks actin bundles and is instead composed of actin networks. Its role in aligning filaments into bundles in intestinal cells must therefore be altered in these different cell types.

2.1.2.3. Villin. Villin is a specialized actin cross-linking and severing protein found in intestinal and kidney epithelial cells (Bretscher and Weber, 1979, 1980; Matsudaria and Burgess, 1979, 1982; Mooseker *et al.,* 1980; Craig and Powell, 1980). It has also been demonstrated recently in oocytes and eggs (Corwin and Hartwig, 1983). Villin is globular protein composed of 95,000 dalton subunits (Bretscher and Weber, 1979; 1980; Glenney *et al.,* 1981).

Villin is a bifunctional protein. In the presence of submicromolecular concentrations of calcium ions (or in buffers containing EGTA) it binds to actin filaments and promotes the formation of bundles (Figure 6A). In the presence of micromolar calcium, it changes character: It cuts actin filaments (Figure 6B). After cutting the filament, it remains bound on the "barbed" end of newly formed filaments. The mechanism of this calcium-mediated effect is identical to that of gelsolin (discussed below in Section 3.2.2.1.). The structure of villin–actin bundles is different from that of fimbrin–actin aggregates. Filaments in the villin–actin bundles pack more loosely, having a center-to-center filament spacing of about 12 nm as opposed to the 9-nm spacing of the fimbrin-induced bundles (Matsudaria *et al.,* 1983). The polarity of actin filaments in villin-induced bundles has not been reported.

By immunofluorescent staining, it localizes only in the apical core filaments of microvilli on the luminal surface of intestinal and kidney cells.

2.1.2.4. Vinculin. Vinculin is concentrated in the cortical cytoplasm of fibroblasts, particularly in adhesion plaques, regions of the cell in contact with the substratum (Geiger *et al.*, 1979; Geiger, 1979, 1982; Burridge and Feramisco, 1980; Feramisco *et al.*, 1982). These plaques are common in cultured cells, and electron micrographs reveal them to be points where filaments in stress fibers end and, it is postulated, insert into the plasma membrane.

Vinculin has been purified from smooth muscle tissue and nonmuscle cells and its interaction with actin characterized. It is a protein of 130,000 daltons that decreases the viscosity of solutions of actin filaments (Jockusch and Isenberg, 1981; Isenbert *et al.*, 1982; Wilkins and Lin, 1982). However, the mechanism of this effect is controversial, as two very different mechanisms have been proposed. One group of investigators has reported that it binds selectively to the "barbed" end of actin filaments (Wilkins and Lin, 1982; how binding of this sort can alter actin's viscosity is discussed in Section 2.2). Another group, however, has reported that the decrease in viscosity is due to filament-bundling activity (Jockusch and Isenberg, 1981; Isenberg *et al.*, 1982) and has presented electron micrographs showing that actin assembled in the presence of vinculin organized in long, ordered filament bundles. Resolution of this question is needed before a role for vinculin can be formulated.

2.2. Regulation of Actin Filament Length

A number of proteins that regulate the length of actin filaments have been purified from mammalian cells (Table II). These molecules accomplish this task by binding to and capping the ends of actin filaments.

Alterations in the length of actin filaments can be very important in regulating the integrity of actin in networks and in bundles. Gels form when the filaments become cross-linked together by one or many of the proteins described previously. This gelation occurs when the amount of added cross-linker is sufficient to join all the polymers in the solution together, forming in essence a giant network. When the concentration of added cross-linker reaches this point, defined as the gel point (V_c), the physical properties of the solution abruptly change from liquid to solid (gel). Since all known mammalian cross-linking proteins are bivalent for actin, each molecule added can bind two actin polymers and link one additional polymer into the network. Therefore, the minimal number of added cross-linker necessary for incipient gelation of an actin solution is one per actin filament. Independent of cross-linkers, the gel point is also influenced by the concentration of actin in the solution and the length of the actin polymers. V_c is inversely related to the weight-average length of the actin polymers, X_w, and directly related to the actin polymer concentration c as defined in the following relationship (Flory, 1953):

$$V_c = c/X_w$$

Table II
Mammalian Proteins that Regulate the Length of Actin Filaments

Class	Cell	Size	Function	References
Barbed end Ca^{2+}-sensitive				
Gelsolin	Leukocytes	91,000	Regulate cortical actin network	Yin and Stossel, 1979, 1980; Yin et al., 1980, 1981
	Platelets	91,000	Same	Wang and Bryan, 1981; Lind et al., 1982; Im et al., 1982 Yin et al., 1981
	Other mammalian cells			
Villin	Intestinal epithelial cells	95,000	Cross-links actin filaments in microvilli in EGTA; cuts in uM calcium; calcium regulated; function unknown	Bretscher and Weber, 1979, 1980; Glenney et al., 1981; Glenney and Weber, 1981; Nunnally et al., 1981; Mooseker et al., 1980
Ca^{2+}-insensitive				
Vinculin	Smooth muscle Other cells	130,000	?	Wilkins and Lin, 1982; Geiger et al., 1980
Pointed end Ca^{2+}-insensitive				
Acumentin	Leukocytes	65,000	Prevents unwanted filament assembly	Southwick and Stossel, 1981; Southwick et al., 1982; Southwick and Hartwig, 1982
β-actinin	Muscle	71,000 (1 × 37,000 1 × 34,000)	?	Maruyama et al., 1977
	Other mammaliam cells	?		Maruyama et al., 1981

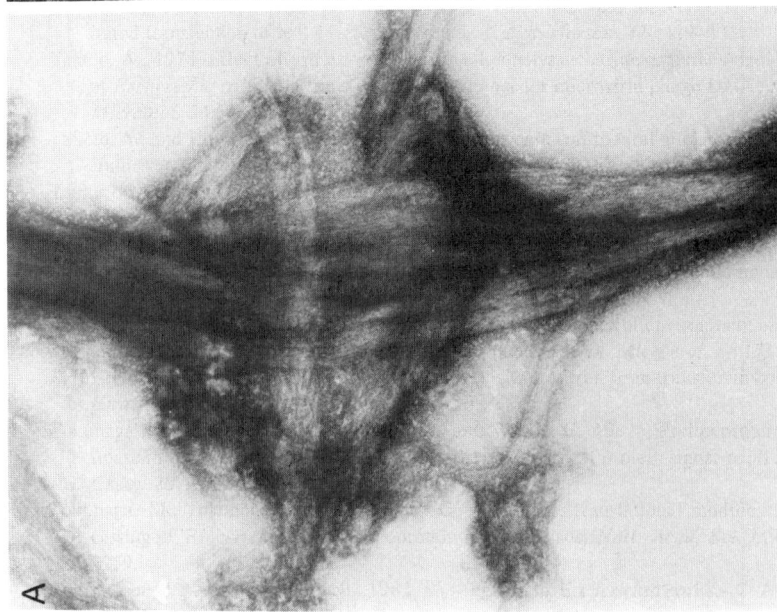
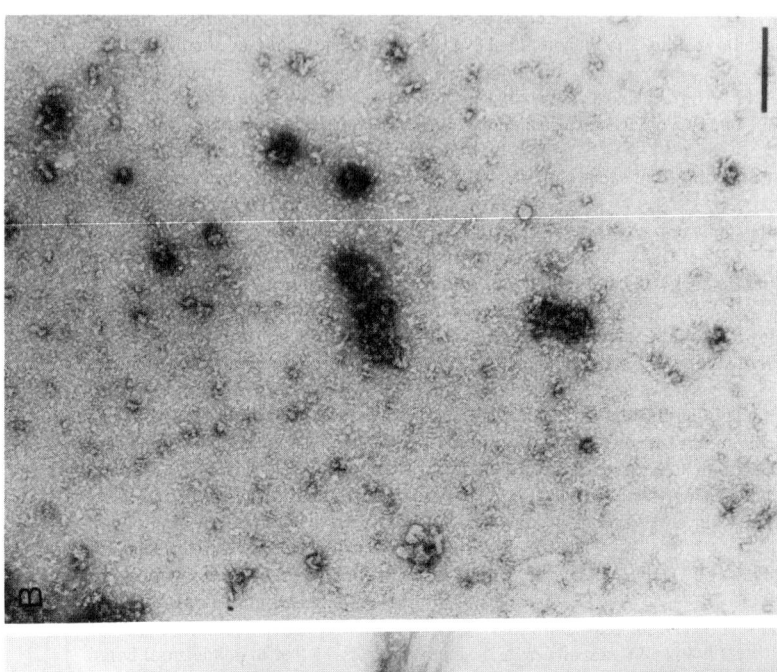

FIGURE 6. Electron micrographs demonstrating the effect of oocyte villin on actin structure in the presence (B) or absence (A) of calcium. (A) Bundles of filaments are apparent when actin is assembled in the presence of villin in the low-calcium (EGTA-containing) buffers. The molar ratio of villin to actin was 1 to 4. (B) In the presence of calcium, villin completely fragments actin filaments. The bar in part b is 200 nm.

Since V_c is dependent on the weight-average actin filament length, small alterations in actin filament length caused by actin regulatory proteins can have large effects on the network or gel state of the actin polymers in solution and in the cortical cytoplasm of cells.

2.2.1. Cytochalasins

Much of what has been learned about the cytoplasmic proteins that bind to the ends of actin filaments is derived from earlier work examining the mechanism of actin of the cytochalasins, agents that disrupt cell movement (Wessells *et al.*, 1971; Yahara *et al.*, 1982). The cytochalasins bind to actin filaments with high affinity, with a K_a of 10 nM^{-1} for cytochalasin B (Hartwig and Stossel, 1979; Brown and Spudich, 1981; Flanagan and Lin, 1980). Only a few molecules, possibly as few as one, are bound per filament. At least one of these sites is the "barbed" end (Brenner and Korn, 1979; Maclean-Fletcher and Pollard, 1980). Cytochalasin B binding to the barbed end allows actin depolymerization to outpace polymerization. The net effect of this is shorter filaments. Filament shortening will disrupt the integrity of actin assemblies.

The cytochalasins affect the length of both preformed actin filaments and filaments assembled in their presence *in vitro*. When added to actin filament solutions, cytochalasins decrease the viscosity of the solution (Spudich and Lin, 1972; Low and Dancker, 1976). This effect is attributed to an increase in the number of filaments and a decrease in the length distribution of these filaments (Hartwig and Stossel, 1979; Maruyama *et al.*, 1980). When added to actin monomers prior to filament assembly, the cytochalasins can both stimulate the onset of actin polymerization (Low and Dancker, 1976) and decrease the rate of filament elongation (Brenner and Korn, 1980a,b; Brown and Spudich, 1979, 1981; Flanagan and Lin, 1980; MacLean-Fletcher and Pollard, 1980). All of these effects can be attributed to their binding to the "barbed" filament end. These effects occur at drug concentrations lower than those required to disrupt cell movements. This result is expected for an agent that enters the cell by passive diffusion through the membrane where a higher drug concentration on the outside of the cell is the driving force of the reaction.

Direct evidence for cytochalasin's binding to the end of actin was provided in electron micrographs recording the effect of the drug on the directional assembly of actin from filament nuclei (MacLean-Fletcher and Pollard, 1980). These micrographs demonstrated that it inhibits addition of monomers to the preferred or "barbed" end of actin filaments serving as nuclei (Maclean-Fletcher and Pollard, 1980; Pollard and Mooseker, 1981).

2.2.2. Proteins that Bind to the "Barbed" End of Actin Filaments

As indicated earlier in the work on the cytochalasins, a powerful regulation of actin filament length and cell structure can be mediated by agents that bind and cap actin's "barbed" end. Proteins that bind and cap the barbed end of actin filaments have been identified and furthermore are sensitive to calcium levels.

2.2.2.1. Gelsolin. All mammalian cells examined, with the exception of erythrocytes, contain a 91,000-dalton protein that severs actin filaments in the presence of micromolar calcium concentrations. This protein, gelsolin, was first isolated from macrophages (Yin and Stossel, 1979; 1980; Yin et al., 1980) and has now been identified and purified from platelets (Grumet and Lin, 1980; Wang and Bryan, 1981; Lind et al., 1982; Im et al., 1982), brain (Bamburg et al., 1980; Nishida et al., 1981), plasma (Norberg et al., 1980; Harris and Gooch, 1981; Harris and Schwartz, 1981; Chaponnier et al., 1979), and the adrenal medulla (Grumet and Lin, 1981). IgG antibodies to macrophage gelsolin cross-react with antigens in a large variety of mammalian cells, plasma, and muscle tissue (Yin et al., 1981). Other names given to what seem to be very similar molecules are brevin and actin depolymerizing factor (ADF). Gelsolin composes about 1% of the total cellular protein leading to estimates of about 1 gelsolin molecule to 50 actin monomers being present in cytoplasm. In leukocytes, it has been shown to account for the bulk of the calcium-dependent actin filament severing activity measured in cell extracts.

Gelsolin in the presence of calcium binds to and shortens actin filaments. Each molecule of gelsolin binds two moles of calcium with high affinity, with a K_a of 1 μM^{-1}, and severs actin only when calcium is bound to it. After severing the filament, gelsolin remains bound to the barbed end of one of the newly formed filament fragments (Yin et al., 1981, and Figures 7 and 8) even if the calcium level falls (Bryan and Kurth, 1984, Janmey et al., 1985). Decreases in filament length could also result from some depolymerization of actin filaments when gelsolin is bound to the filament end. However, most of its effect on actin is attributed to fragmentation because of the rapidity with which filaments become short and because gelsolin only slightly increases the critical concentration of actin solutions.

Gelsolin, in the presence of calcium, also binds to actin monomers. This binding is sufficient to stablize nuclei and stimulate the overall rate of actin filament assembly. Although the onset of filament assembly (the rate-limiting step in actin assembly) is stimulated, the rate of actin filament elongation is greatly reduced because gelsolin is bound to the "barbed" end, diminishing the number of ends for actin additon by one-half and blocking the end of the filament having the fastest growth rate.

2.2.2.2. Villin. As previously noted, villin has at least two binding sites

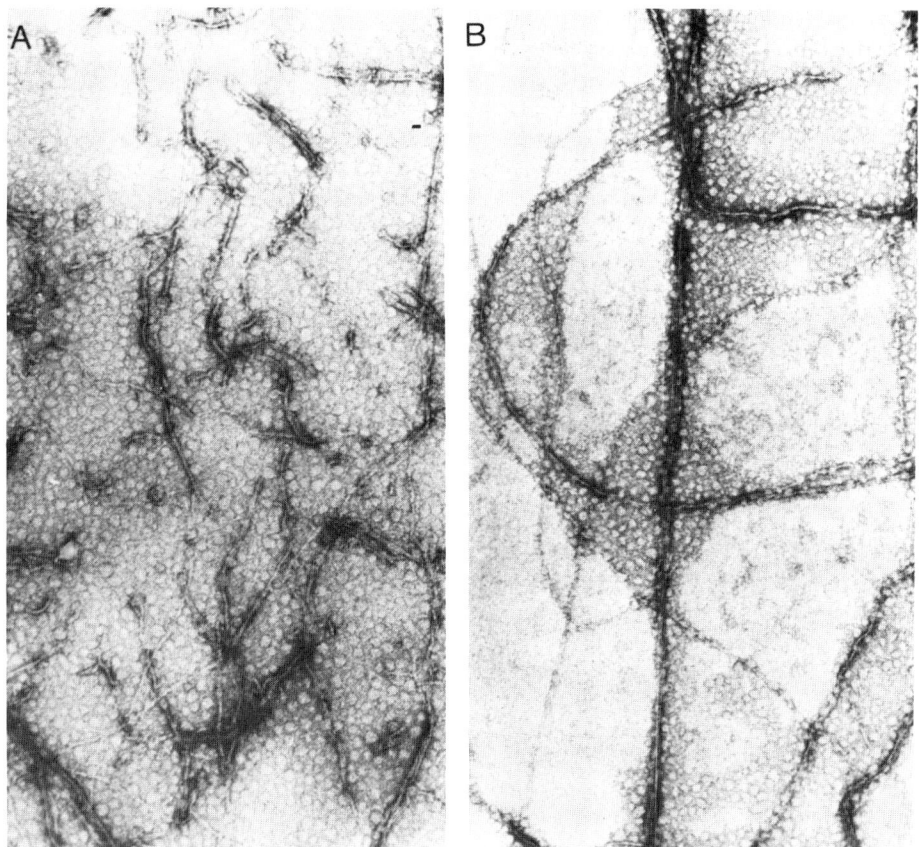

FIGURE 7. Electron micrographs of actin filaments fragmented by gelsolin. Effect of leukocyte gelsolin on filament length in the presence (A) and absence (B) of calcium.

for actin. In low calcium, it uses one of these sites to bundle filaments (Figure 6). The binding of calcium to villin activates a second actin binding site. The result of binding is filament severing (Figure 6).

Although remarkably similar in structure and function, villin and gelsolin are products of different genes (Mooseker and Yin, unpublished). Villin has a slightly larger molecular weight than gelsolin (Bretscher and Weber, 1979; 1980). Like gelsolin, villin severs actin filaments in the presence of micromolar calcium concentrations (Glenney *et al.*, 1981) via a mechanism identical to that used by gelsolin (Nunnally *et al.*, 1981; Glenney *et al.*, 1981). It binds to monomeric actin in the presence of calcium and to actin polymers in both the presence and absence of calcium. In binding to actin, it forms stable structures composed of three monomers and one villin molecule (Glenney *et al.*, 1981).

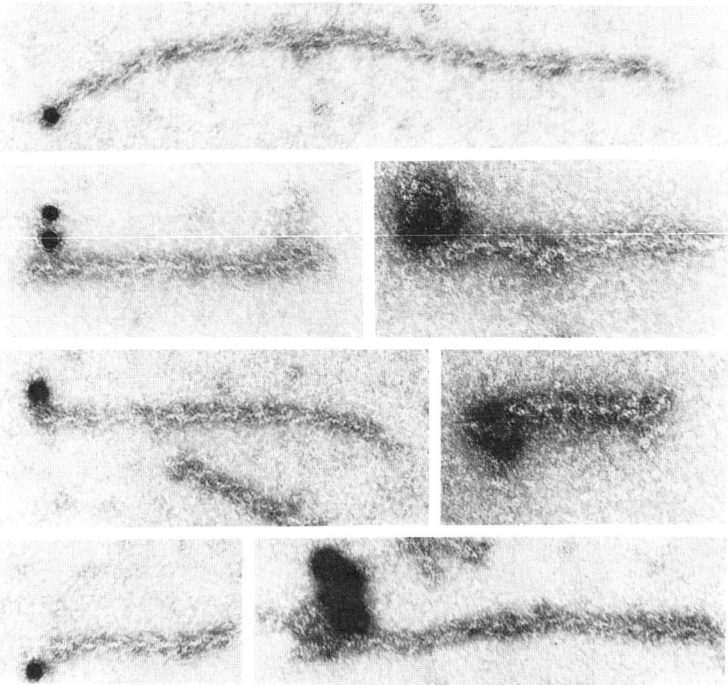

FIGURE 8. Electron micrographs demonstrating directly that gelsolin binds to the "barbed" end of actin filaments in the presence of calcium. Gelsolin was visualized by binding gold particles coated with Staphalococus aureus protein A to actin filaments. The filaments were previously incubated with antileukocyte gelsolin IgG. The polarity of filaments is visualized with heavy meromyosin labeling.

These structures also nucleate filament assembly (Craig and Powell, 1980; Mooseker et al., 1981). Binding of villin to actin also occurs at the "barbed" end of filaments in the same manner as gelsolin (Bonder and Mooseker, 1983; Glenney et al., 1981).

2.2.3. Proteins that Bind to the Pointed End of Actin Filaments

Two proteins have been demonstrated to bind to the "pointed" end of actin filaments. Calcium does not influence this binding, and no evidence is available that the binding is otherwise regulated directly. These proteins shorten actin filaments and therefore lower the viscosity of actin filament solutions.

2.2.3.1. β-Actinin. β-Actinin, a skeletal muscle protein, is a 71,000-dalton heterodimer composed of 37,000- and 34,000-dalton subunits (Maruyama, 1971; Maruyama *et al.,* 1977). It binds to actin filaments with low stoichometry and inhibits the reannealing of mechanically fragmented actin filament. β-Actinin is thought to cap the pointed end of actin filaments. When the reannealing of mixed populations of unlabeled actin filaments and heavy meromyosin-labeled filament incubated with β-actinin was studied in the electron microscope, filament annealing was inhibited selectively at the pointed end, suggesting that β-actinin binds to this end, thereby blocking filament-to-filament attachments. Proteins of similar function but different molecular size have been reported in nonmuscle cells (Maruyama *et al.,* 1981).

2.2.3.2. Acumentin. A 65,000-dalton leukocyte protein has been demonstrated to inhibit elongation at the "pointed" end of actin filaments (Southwick and Stossel, 1980; Southwick and Hartwig, 1982; Southwick *et al.,* 1982). Like β-actinin, this protein shortens actin filaments and lowers the viscosity of actin solutions without increasing the critical concentration of actin in the solution, a property predicted for agents that cap the "pointed" end of actin filaments.

When compared to gelsolin, however, acumentin is less potent on a mole-for-mole basis in shortening actin filament length, most likely because of a lower affinity of acumentin for actin than gelsolin for actin. Leukocytes compensate for this low affinity for producing large quantities of this protein. Five percent of the total protein in leukocytes is acumentin, a molar ratio to actin of one acumentin to eight actin monomers. These high concentrations suggest that acumentin functions to buffer the pointed ends of actin filaments, inhibiting spontaneous filament growth from this end of the filament. The low affinity of acumentin for actin filament ends may allow it to the displaced by proteins possessing higher affinities for this end.

2.3. Stabilization of Actin Filaments

Tropomyosin can protect filaments from fragmentation by the abundant cellular proteins that bind to and sever actin filaments in cytoplasm, particularly the calcium-insensitive proteins.

2.3.1. Tropomyosin

Nonmuscle tropomyosins have been purified from platelets (Cohen and Cohen, 1972; der Terrossian *et al.,* 1981; Côté and Smillie, 1981a,b; Côté *et al.,* 1978a,b), brain (Fine *et al.,* 1973), fibroblasts (Masaki, 1975; Talbot and MacLeod, 1983), BHK-21 cells (Schloss and Goldman, 1980), and macrophages (Fattoum *et al.,* 1983). Although a component of the actin and myosin regulatory system in skeletal muscle, its role in nonmuscle cells is not clear.

Nonmuscle tropomyosins are similar in structure to skeletal muscle tropomyosin, although smaller in mass. They are heterodimers of about 60,000 daltons composed of nonidentical subunits of about 30,000 daltons, whereas the muscle molecule has a mass of about 70,000 daltons. This mass difference is apparent in electron micrographs (der Terrossian, 1981; Fattoum *et al.*, 1983) where the nonmuscle protein is a linear strand 33 nm in length, 4–5 nm shorter than its muscle counterpart. Tropomyosin molecules are thought to bind to actin filament in the grooves of the double helix forming end-to-end assemblies along the length of the filament (Huxley, 1969). Its binding to actin filaments is cooperative. Molecules prefer to align end to end even when low molar rations of tropomyosin to actin are present rather than to disperse evenly along the entire filament (Yang *et al.*, 1979). This is important because the amount of tropomyosin in most nonmuscle mammalian cells is low and may be sufficient to cover only part of the actin filaments present in cells.

Tropomyosins, either muscle or nonmuscle types, have been demonstrated to protect actin filaments from the fragmenting activity of gelsolin (Fattoum *et al.*, 1983; Bernstein and Bamburg, 1982) and cytochalasin B (Spudich, 1972) and to decrease spontaneous filament breakage (Wegner, 1982). The inhibitory effect on filament severing, at least in the case of gelsolin, is due to decreased binding of gelsolin in the presence of tropomyosin (Fattoum *et al.*, 1983). Tropomyosin has also been demonstrated to decrease the binding of actin-binding proteins to actin filaments (Maruyama and Ohashi, 1978) and to modulate the enzymatic interaction of nonmuscle myosins and actin (Hartshorne and Siemankowski, 1981). Therefore, tropomyosin may function in the cell to regulate selectively the binding of various actin-associated proteins, thereby determining the type of actin structure formed.

2.4. Retraction of Actin Filaments

Most eukaryotic cells contain the protein myosin, which functions to pull on actin filaments (Pollard and Weihing, 1974). The structure and function of this protein has been defined in muscle (Huxley, 1969; Taylor, 1972; Eisenberg and Greene, 1980). Myosin is a large, asymmetical hexamer of 470,000 daltons composed of two 200,000-dalton heavy chains and four light chains in the 15,000–25,000 molecular weight range. Its cellular function is the conversion of chemical energy, in the form of ATP, into mechanical work, in the form of actin filament movement. Actin activates the Mg^{2+}-ATPase of myosin and promotes this transduction as well as being involved in its mechanics. The movement of actin filaments by myosin (Figure 9) requires the assembly of molecules into bipolar filaments composed of a minimum of two molecules. Myosin molecules are polarized structures having the actin-binding and enzymatic (ATPase) sites on one end of the molecule and their self-association sites on

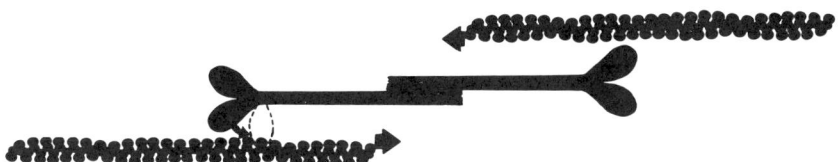

FIGURE 9. Diagram illustrating the interaction of actin with myosin. Myosin is shown assembled into bipolar filaments of two molecules, the minimum number required for actin filament cross-bridging and contraction. The arrows define both filament polarity and the direction the actin filaments move relative to myosin.

the other end. Myosin molecules and filaments bind to actin filaments in a stereospecific fashion such that the movement of the myosin along the actin filament occurs from the pointed to the barbed filament end; actin filaments are pulled in the direction pointed by bound heavy meromyosin. Therefore, the contraction of actin filaments by a bipolar myosin filament can lead only to the inward movement of the actin filaments past the myosin filaments.

The function of myosins appears to be similar in nonmuscle cells although specific modifications in structure and regulation have been demonstrated. Nonmuscle myosins assemble into smaller filaments than muscle myosins *in vitro*. The muscle filaments are many micrometers in length containing hundreds of myosin molecules. The nonmuscle myosin assembles only into 300-nm-long minifilaments composed of from 28 to 30 molecules (Niederman and Pollard, 1976; Kendrick-Jones and Scholey, 1981). However, the existence of such minifilaments has yet to be demonstrated in the cytoplasm of nonmuscle cells *in situ*.

Recent evidence indicates that both filament assembly and the enzymatic activitation of myosin Mg^{2+}-ATPase activity by actin is regulated by phosphorylation of the 20,000-dalton light chains (Adelstein, 1978; Scholey *et al.*, 1981). Both of these events are potentiated when the light chains are phosphorylated on a specific serine residue by myosin light chain kinase. The kinase is in turn activated by the binding of calcium–calmodulin complex.

2.5. Proteins that Interact with Actin Monomers

Not all the actin in cytoplasm is assembled into long filaments. A large percentage, approximtely 50%, of cellular actin may be stored as either monomers or small oligomers. Oligomers could exist if both ends were capped by proteins such as acumentin and gelsolin. Monomeric actin can be present in cytoplasm at high concentrations because it can complex to a protein called profilin.

2.5.1. Profilin

A small protein of molecular weight 15,000 binds selectively to actin monomers in a one-to-one stoichometry (Carlsson *et al.*, 1977; Blikstad *et al.*, 1980; Harris and Weeds, 1978; Markey *et al.*, 1978). This protein has been named profilin and the actin–profilin complex has been termed *profilactin* (Tilney, 1976).

The interaction of profilin with actin has two effects on the assembly reaction (Figure 10). First, it inhibits actin nucleation. Since nucleation is the rate-limiting step in actin filament assembly, large pools of profilactin are stable and filament formation does not occur. Second, if exogenous nuclei are added to profilactin, the assembly of monomers onto them is modified by profilin. The affinity of profilin for actin monomers is greater than that for the "pointed" end of actin filaments but less than that of the "barbed" end (Figure 10b). Therefore, actin nuclei capped at the "barbed" end will not assemble in the presence of profilin (Markey *et al.*, 1982). However, nuclei capped at the "pointed" end will generate filaments. Likewise, filaments capped at the "barbed" end will disassemble in the presence of profilin. Those capped at the "pointed" end will not. Therefore, the addition or activation of a cellular "barbed" end capping protein to filaments in the presence of profilin can provide a mechanism for filament dissassembly.

3. MOVEMENTS OF CELL MEMBRANES

Cells both push and pull their membranes. The outward flowing of cytoplasm to form large pseudopodia in leukocytes or the rapid extension of microspikes from the surface of activated platelets and eggs are examples of propulsive movements of cortical cytoplasm. Propulsive outward movements must

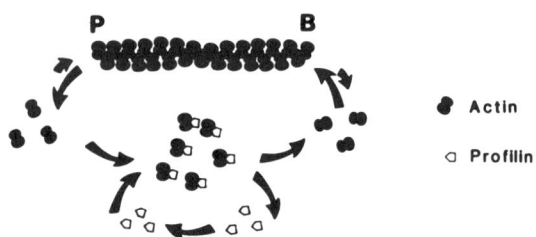

FIGURE 10. Diagram illustrating the effect of profilin on actin assembly. Profilin binds to monomeric actin, preventing filament nucleation in the absence of added nuclei. If filaments are present as indicated in this diagram, profilin will draw monomers off from the pointed end of the actin filaments. If the barbed end of the filament is free, it will, in turn, draw the monomers from profilactin.

occur in conjunction with retractive movements, since cell volume is conserved (Weiss and Garber, 1952; Albrecht-Buehler, 1982; Chen, 1981). Each outward movement of cytoplasm is accompanied by the cell retracting a equal volume of cytoplasm somewhere from its surface.

Biochemical and morphological evidence suggest that retractive and propulsive movements occur through different mechanisms. The most likely possibility is that retractive movements are mediated by myosin and propulsive movements are mediated by actin assembly. Biochemical evidence strongly suggests that both of these processes are controlled by the calcium concentration of cytoplasm.

In the following paragraphs we first integrate the assembled information of actin and its associated proteins to provide hypothetical mechanisms of membrane movements. We then focus on the mammalian leukocyte and platelet as cellular examples of these movements.

3.1. Propulsive Movements

It has been established that the rapid assembly of actin into filaments can provide the force required to push cytoplasm and the cell membrane outward in the acrosomal reaction of certain marine sperm (Tilney, 1973, 1976). In this reaction, actin, apparently bound to profilin, and present at extremely high concentrations, >100 mg/ml, rapidly assembles into filaments that push the membrane outward. Elongation occurs on a specialized structure, the actomere (Tilney, 1978). The top surface of this structure contains multiple nucleating sites for actin equivalent in conformation to the "barbed" end of actin filaments. When the acrosomal reaction is activated by contact of the sperm with the egg, these sites become exposed, initiating filament elongation, thereby forming a spike 50 μm in length from the sperm. As predicted for "barbed"-end filament growth (Tilney and Kallenbach, 1979), this process is inhibited by cytochalasin (Tilney and Inoue, 1982).

How do cells other than sperm effect propulsion? The answer is not yet clear but a number of facts allow us to speculate that actin polymerization may propel cells forward. Most cells maintain a large fraction of their actin in small oligomers or monomers. Recent experiments have demonstrated that this pool is dynamic, in that it decreases when the cells are activated to move (Pribluda *et al.,* 1981; Krishna and Varani, 1982; Casella *et al.,* 1982). This finding is consistent with the idea that filament assembly may be a driving force for cell movement. As discussed earlier, pools of monomers can exist in conjugation with profilin as profilactin. Profilactin prevents the monomers themselves from nucleating filament assembly. However, the abundance of cytoplasmic proteins that can nucleate assembly would overcome this nucleation block were it not for two cellular controls. First, only the free "barbed" ends of the filaments can compete with profilin for actin and therefore dissociate monomers. Second, cells contain proteins to cap these ends. Thus, proteins that cap the "barbed" end must dissociate to allow assembly initiation. Furthermore, submicromolar free calcium concentrations inhibit the binding of some of the "barbed" end capping proteins to actin. Therefore, if a cell can regionally drop its calcium concentration, it can expose the necessary nuclei onto which actin monomers can assemble after they dissociate from profilactin (Figure 11). Proteins that constitutively bind to the ends of actin filaments such as the "pointed" end proteins, acumentin and β-actinin, cannot function in this capacity since they will always be bound to actin.

Another point is important to recognize. Filament assembly does not need to proceed from an organized structure such as found in the sperm actomere to generate cytoplasmic extensions. The assembly of actin onto nucleating sites attached to an actin network could lead to formation of large, blunt pseudopodia (Figure 12b). Although filament assembly would be equal in all directions from the network, intracellular organelles, the nucleus, the dense perinuclear matrix of intermediate filaments, and cell–substrate attachments would act as a barrier to inward expansion and drive the pseudopods out (Figure 12b). Microspikes, cored by actin bundles of lower organizational order than those found in sperm, could also be formed by a modification of this pro-

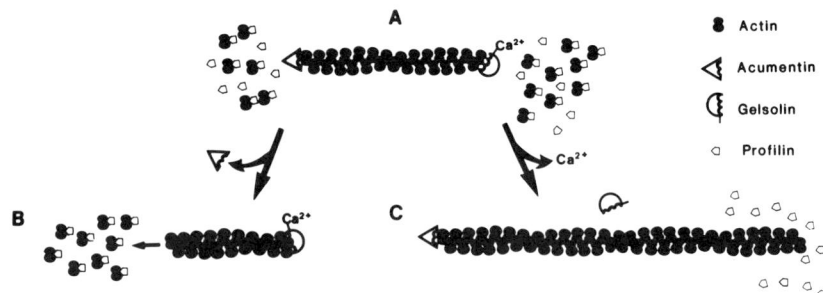

FIGURE 11. Diagram illustrating the control of actin assembly and disassembly by leukocyte proteins that cap filament ends and profilin. (A) Both ends of the filament capped. Profilin will not change the length of actin filaments if acumentin, gelsolin, and calcium are present. Acumentin and gelsolin are shown bound to their respective filament end. Capping of the ends blocks both the dissociation of monomers from the filament and their addition. (B) Filament disassembly. When filaments are severed by gelsolin, (calcium increases) free "pointed" ends are formed. Filament fragments lacking acumentin will disassemble in the presence of profilin. (C) Filament assembly. Gelsolin binding is reversed. As the calcium concentration decreases into the submicromolar range a mechanism exists to dissociate gelsolin off the "barbed" end of the filament. The presence of free "barbed" ends dissociates actin from profilin, leading to filament elongation. This also results in an increase of free molecules of profilin (as long as the "pointed" ends of filaments are capped by acumentin).

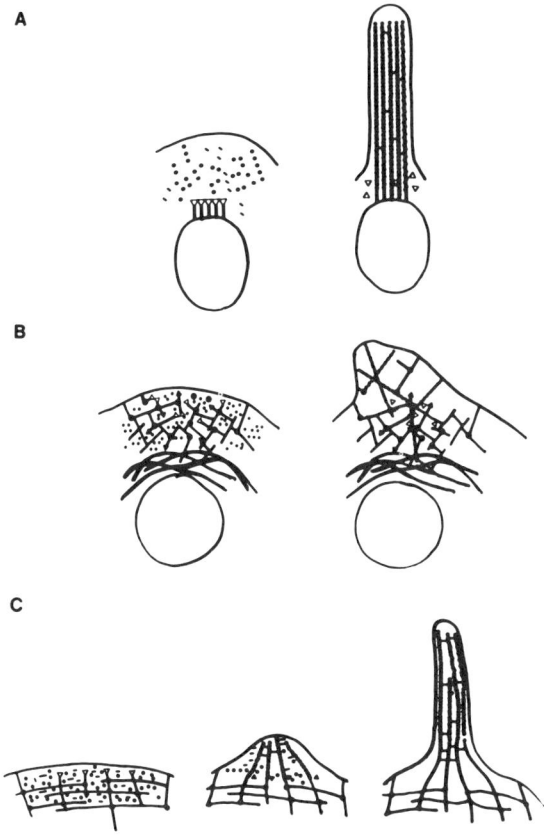

FIGURE 12. Examples illustrating the extension of membranes by actin assembly. (A) The acrosomal reaction of certain marine sperm (from Tilney, 1978). Activation leads to the dissociation of cellular proteins (▽) from the "barbed" end of the acromere. Filaments elongate by dissociating actin from profilin. Bundles form because the acromere functions to align the filaments. They remain bundled because cross-linking proteins (—) stabilize them. (B) Hypothetical mechanism for network assembly. Filament assembly begins with the exposure of free "barbed" filament ends. In this case, the barbed ends are attached to a preexisting three-dimensional actin network and X- and T-pieces of branched filaments by cross-linking molecules (●). Polymerization and perpendicular cross-linking results in an enlarged network that fills the cytoplasm, pushing the membrane outward. Capped "barbed" and "pointed" filament ends are indicated by (▽) and (▲), respectively. (C) Hypothetical mechanism for filament bundling in the absence of a specialized nucleating structure. The stimulus for assembly both frees the "barbed" end of nuclei of capping proteins (▽) and simultaneously activates filament cross-linking proteins (/). Elongation proceeds from actin dissociating from profilin. The newly formed filaments are zipped up by the cross-linking protein(s).

cess. The presence of free cross-linking proteins during assembly could zip up the filaments as they form (Figure 12c).

It should be noted that propulsive movements do not require actin filaments to be attached to the plasma membrane (Mooseker et al., 1982). Filaments that are elongating are able to push out against the membrane, thus changing its shape, without being firmly anchored by a cytoskeleton–membrane linkage.

3.2. Retractive Movements

Cytoplasmic retractions, generated by the interaction of actin and myosin, require actin filaments to interdigitate with myosin filaments with opposing polarities (Figure 9). Clear examples of this have been demonstrated to occur in stress fibers (Begg et al., 1978), in the contractile ring of dividing cells (Schroeder, 1973), and in retinal rods (Burnside et al., 1983). The simplest scheme for the formation of a pocket on the cell surface is by a mechanism analogous to that occurring in the muscle sarcomere. A minimum of two actin filaments, attached to the plasma membrane by their "barbed" end, would interact in the cytoplasm with a myosin filament (Figure 13a). Contraction by myosin slides these filaments past one another, invaginating the plasma membrane.

This mechanism, however, would require linear filaments to transverse long stretches of cytoplasm. This may be true of the filaments composing the cores of intestinal microvilli (Mooseker and Tilney, 1975; Matsudaira and Burgess, 1979) but is rarely observed in the branching networks found in the cortical cytoplasm of high motile cells. Instead, short filaments dominate cyto-

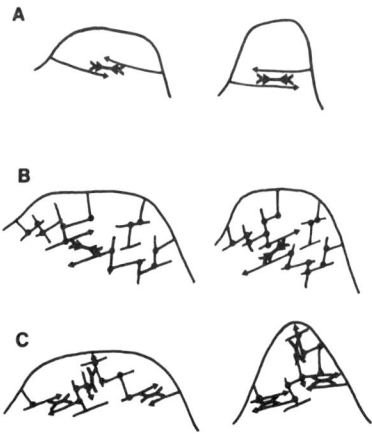

FIGURE 13. Illustrations of mechanisms for myosin-based cellular movements. (A) Model based on the arrangement of actin and myosin in the muscle sacromere. It requries the presence of long actin filaments that run between the cell membrane and cytoplasmic myosin filaments. (B) Model for contraction of cross-linked actin filaments (gel networks). In this model, short actin filaments are linked together by cross-linking proteins, allowing force to be transmitted through larger regions of cytoplasm. (C) Model based on (b) except that myosin molecules are dispersed in smaller units throughout the cytoplasm.

plasm. The cross-linking of filaments together by specific proteins may remedy this problem. Branching filament networks, in which the vertices are cross-linked, are equivalent to long filaments in three-dimensional space (Figure 13b). Similarly, short filaments, if present in stress fibers, can be effectively turned into long filaments by cross-linking proteins. This cross-linking would allow force to be transmitted thorughout large volumes of cytoplasm. Myosin may also be dispersed in small units throughout the cortical cytoplasm, allowing it to function both in filament cross-linking and contraction (Figure 13c). Dispersion of myosin in these smaller units may be mandated by the low ratio of myosin to actin in cells.

Myosin-based retraction may be involved in the formation of stress fibers in cells and in the sliding of filaments within these fibers. These fibers may be used in the movement and maintenance of cell shape, that is, in the cleavage furrow of dividing cells (Schroeder, 1973) and in endothelial cells lining the blood vascular system (Wong et al., 1983). The nonuniform polarity of stress fibers suggests that they are not dependent on a carefully coordinated assembly system and may be the product of a process that occurs after filament assembly has taken place. As diagrammed in Figure 13, tension exerted on the cytoplasm within a cell may collapse an actin filament network down upon itself, leading to stress fibers pointing in the direction of tension. The studies of Harris and co-workers (Harris et al., 1980) have shown that fibroblasts are able to "pucker" latex films on which they are plated. Since this is a randomly generated actin network in the cytosol, at least with respect to the polarity of the filaments, stress fibers would not be expected to contain the uniform polarity seen in actin bundles, as has been shown by Begg (Begg et al., 1978).

As opposed to propulsive movements, retraction of membrane requires that actin filaments be attached to the membrane. Two proteins have been suggested to mediate these attachments: vinculin and nonerythrocyte spectrin. Vinculin has been localized at points thought to correspond to membrane–substrate attachment. The suggestion that the nonerythrocyte spectrin may function in this capacity is based on the known role of its erythrocyte homologue in linking actin to the membrane of erythrocytes.

4. ACTIN STRUCTURES IN MAMMALIAN CELLS

Mammalian cells share in common a large number of actin-associated proteins. How is it, then, that cells produce such individualized shapes and surface structures?

There are at least two possibilities: (1) each cell type has specific control proteins, that is, each cell type functionally modifies structurally similar proteins; and (2) each cell type varies the amount of actin-associated proteins. The latter approach appears to be the principal choice of cells. Use of the first appears most applicable to specialized cells such as erythrocytes and certain epithelial cells such as kidney and intestinal cells. The second option is certainly the simplest and reports demonstrating difference in protein concentration are widespread in the literature.

4.1. Cortical Networks in Leukocytes

Motile cells such as leukocytes form broad pseudopods when they move. A branching actin lattice is the principal structure observed in these moving regions of cytoplasm (Figure 14).

Two proteins, actin-binding protein and α-actinin, participate in the formation of the cortical network seen in leukocytes. The most important cross-linker of leukocytes extracts is actin-binding protein, which accounts for 75% of the cross-linking activity in such extracts. The network is characterized by a striking perpendicularity of branching; the majority of filaments intersect at close to a right angle. The network's boundaries can be defined in the light microscope as the clear, organelle-free region of hyaline cytoplasm between the plasma membrane and the lysosomes.

Studies of both polymorphonuclear leukocytes (Valerius *et al.*, 1981) and pulmonary macrophages (Stendahl *et al.*, 1980) with immunofluorescent techniques have shown that actin-binding protein is concentrated in regions of moving cytoplasm such as pseudopods formed to engulf extracellular particles. The assembly of actin into orthogonal filament networks by actin-binding protein may be important in the formation of lamellipods at the leading edge of the cell. Actin-binding protein nucleated assembly would form a network composed of short filaments by recruiting monomers that would otherwise polymerize into long, floppy filaments (Figure 15). These long actin filaments alone might lack the rigidity needed to push the plasma membrane outward if they were not cross-linked into bundles or networks. The use of networks instead of bundles makes efficient use of cellular actin by dispersing a minimal amount of protein into a structure having maximal extension. Further, in these highly motile cells, rapid cycles of filament assembly–disassembly may be, in part, responsible for the rapidity of cell movement. Since actin is assembling from profilactin (assembles only onto the barbed end), and gelsolin would bind this end in the presence of calcium, filaments can form only if the calcium concentration in cytoplasm is submicromolar. The distribution of actin into a network where the cross-linker is restricted to the vertices of the network would allow gelsolin (see below) to fragment the network easily into short filaments, initiating filament disassembly.

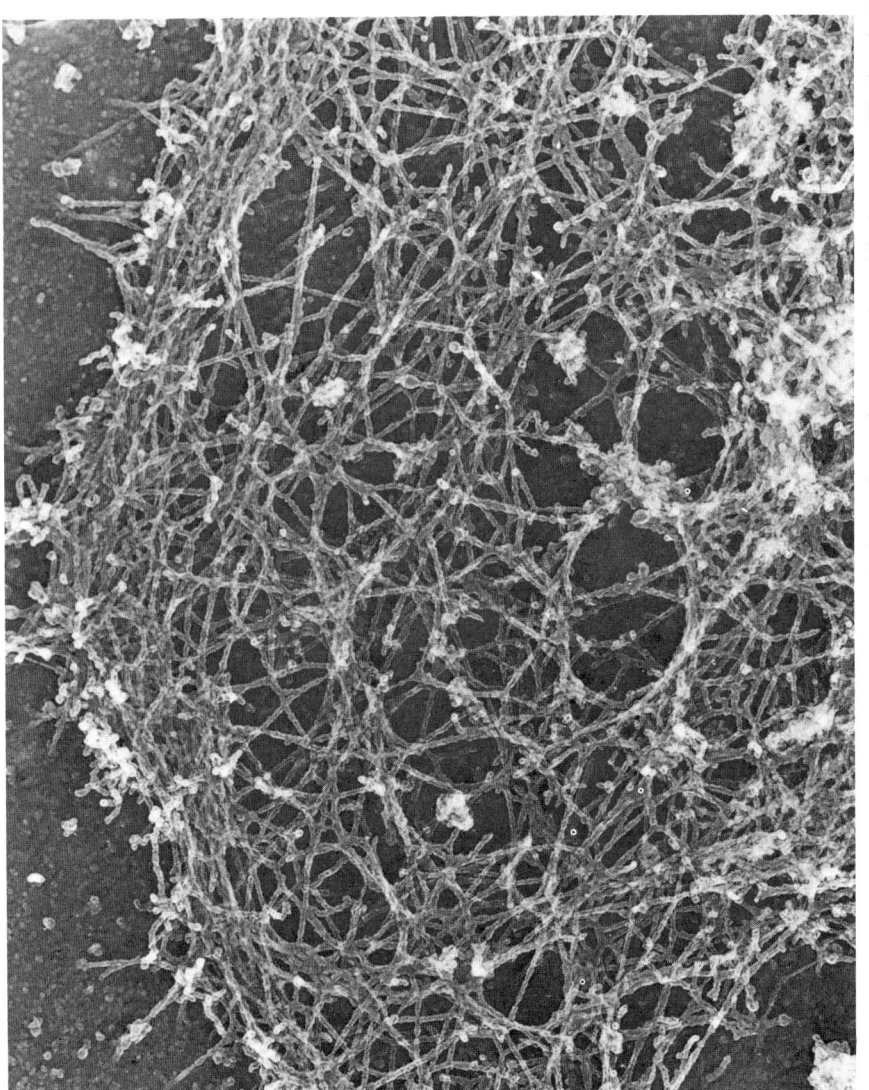

FIGURE 14. Electron micrograph of macrophage cortical cytoplasm after extraction with detergent. The bar is 100 nm.

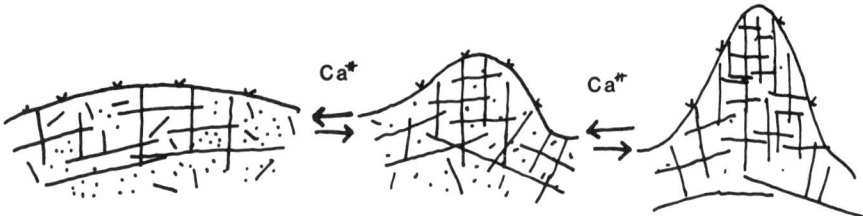

FIGURE 15. Model for pseudopod formation in leukocytes. A branching network is formed by the assembly of actin (from profilactin) onto branched nuclei. Nuclei are composed of two actin oligomers cross-linked by actin-binding protein molecules. Gelsolin is bound to these nuclei in micromolar calcium, preventing assembly. Decreased calcium results in the dissociation of gelsolin and initiation of the branching assembly process.

The cortical network is attenuated at the site of formation of phagocytic vacuoles (phagosomes) and appears progressively to "dissolve" as the phagosome travels inward to fuse with primary lysosomes. The exact mechanism responsible for the local disassembly of the network remains uncertain. The actin-filament-fragmenting protein, gelsolin, may be responsible for this phenomenon. We postulate that a signal is generated at the pit of the forming phagosome that causes a decrease in the amount of calcium being pumped out of the cytosol at that point. The resulting local increase in the free cytosolic calcium could allow gelsolin to bind and cut actin filaments, fracturing the network. This would remove the filament barrier present between lysosomes and the plasma membrane and allow free access of the granules to the forming phagosome. Filament clearing would be facilitated by myosin-based filament retraction. Increased calcium would lead to myosin activation via myosin light chain kinase and filaments would be pulled out of the pit of the phagosome toward regions of cytoplasm where the calcium concentration is low (Stendahl and Stossel, 1980). As the phagosome enters the cell, the plasma membrane over the area of ingestion would once again become capable of extruding calcium. The local cytoplasmic calcium concentration would then fall below the level needed to activate gelsolin.

The mechanism whereby phagolysosomes move toward the center of the cell also remains to be elucidated. Gelsolin may participate in this event if granules are able to increase the free calcium concentration in cytoplasm, perhaps by leaking calcium taken up from the surrounding medium during particle ingestion into the cytosol. Such local increases would activate gelsolin to cut actin filaments near the vacuole, thereby helping to clear a path for the granules as they move toward the cell center.

4.2. Platelets

Blood platelets are anucleate cells that circulate in the form of a disk. Activation of the cells by any one of a number of stimuli (e.g., thrombin, ADP, collagen, arachidonic acid and some of its derivatives, to name a few) causes a change in the shape of the cells, secretion of the contents of the cell's granules, and aggregation of the platelets to each other.

Activated platelets undergo sequential alterations in morphology. The rapid extension of filopods from the cell surface is the earliest event (Allen *et al.*, 1979). Extension rates up to 7.5 μm/min have been recorded for these fingerlike projections. These filopodia contain bundles of actin filaments with uniform polarity in which heavy meromyosin points toward the cell body (Nachmias and Asch, 1976). The platelet then "fills in" the area between these projections with an actin network resembling that found in leukocyte cortical cytoplasm. While these propulsive movements occur at the edge of the cell, cytoplasmic granules aggregate into the center of cell, where they secrete their contents into an invaginated membrane system that is contiguous with the surrounding medium (White, 1974). The retraction of cytoplasmic projections and aggregation of platelets to each other occurs in the last phase of platelet activation. When compared to the rate of filopod extension, the rate of retraction is a much slower 1.9 μm/min.

Morphological and biochemical studies have demonstrated that changes in the degree of actin polymerization accompany alterations in the shape of platelets. Nachmias (1980) explored the organization of filamentous structures in the cytoplasm of activated and unactivated platelets in the electron microscope after extraction of the cell with detergent. This method is now commonly used to reveal the cellular "cytoskeleton." The cytoskeletons of unactivated platelets were composed of an amorphous granular material lacking distinct microfilaments. After activation, cytoskeletons were abundantly filled with microfilaments (Figure 16). Filament bundles developed and these were connected to a dense network of filaments, similar to that seen in macrophages (compare Figure 16 with Figure 14). The majority of these filaments have been identified as actin filaments by labeling studies employing myosin fragments. Biochemical studies support this conclusion. Carlsson (Carlsson *et al.*, 1979) measured the amount of monomeric actin in platelets before and after activation using the DNase inhibition assay. DNase binds to monomeric but not polymeric actin, thereby losing its ability to cleave DNA, a process that can be assayed spectrophotometrically. Carlsson *et al.* demonstrated that the majority of the unactivated platelet's actin is monomeric. In activated cells this actin rapidly assembles into filaments. Their findings have been supported by the

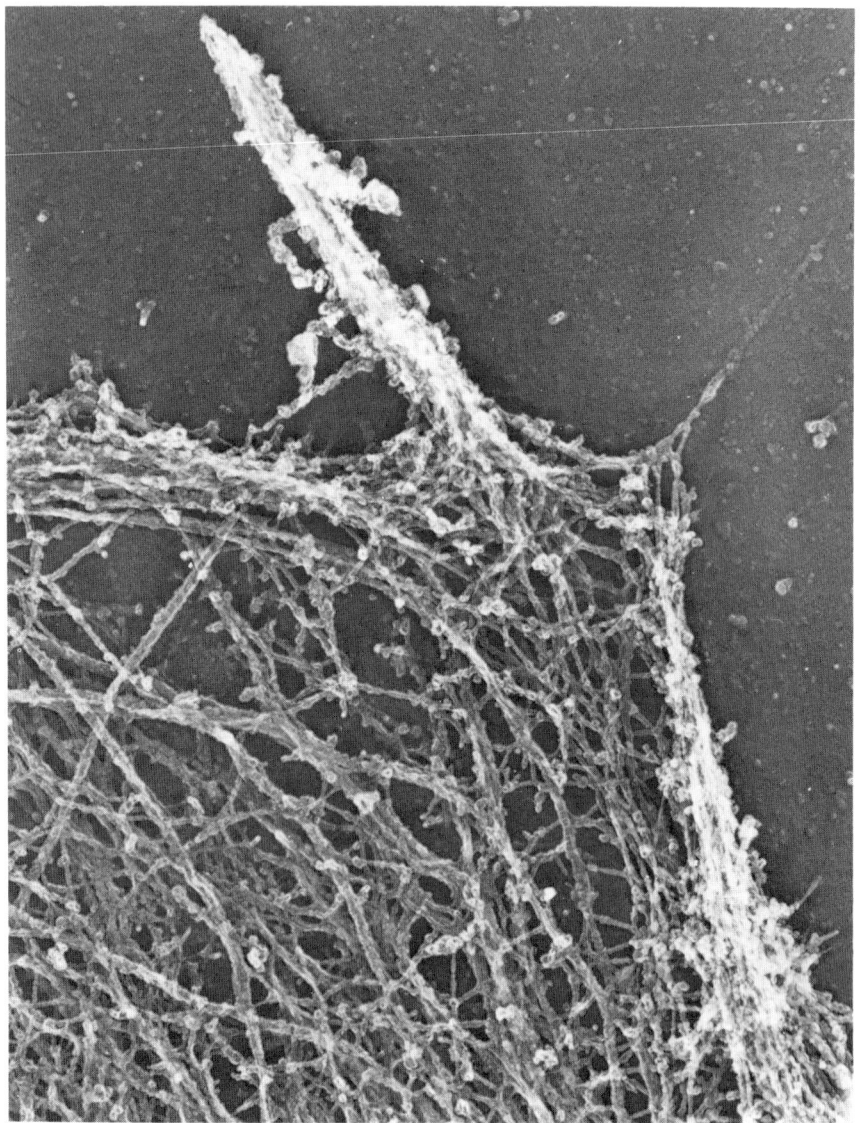

FIGURE 16. Electron micrograph of the cortical cytoplasm of activated platelet after detergent extraction. The bar is 100 nm.

work of others (Markey et al., 1981; Pribluda et al., 1981; Jennings et al., 1981). Similar conclusions have been reached by using cytochalasin D to inhibit actin filament elongation upon stimulation of platelets (Casella et al., 1982).

Enough data are now available to formulate a series of hypotheses about the molecular events accompanying platelet activation that are responsible for the morphological changes seen in the microfilament network of the cell. Similar formulations have been made previously (Adelstein and Pollard, 1978; Cohen, 1979; Lind and Stossel, 1982) and will continue to evolve as more data become available.

Circulating platelets in the discoid form have the majority of their actin as monomers. Small oligomers of actin are also likely to be present. The distinction is important theoretically, although data are not available to document what proportions of the nonfilamentous actin are in these two pools. Monomeric actin cannot be free in the cytosol as it would nucleate and assemble into filaments. Profilin has been suggested to serve a reservoir function in platelets, maintaining the monomers in a pool that is available for polymerizing filaments when free barbed ends of suitable nuclei are available. The finding that filaments do not assemble from actin oligomers indicates that the "barbed" ends of nuclei are capped in the resting platelet.

Platelet cytoplasm also contains a number of addition proteins that would be predicted to nucleate actin. These include the proteins that cross-link actin filaments, α-actinin and actin-binding protein. Since nucleation does not occur in the resting cell these proteins also must be inactive in their capacity to nucleate. High calcium would inactivate α-actinin, but not the other two proteins. One possibility is that these proteins do indeed nucleate filament assembly but gelsolin immediately caps the "barbed" ends of the growing polymers. This would account for the presence of oligomers in the first place. Another possibility could be positive modulation by phosphorylation. The phosphorylation of actin-binding protein increases upon activation (Carroll and Gerrard, 1982).

Activation of the platelet changes the intracellular conditions, permitting the rapid extension of filopods with their polarized bundle of actin filaments. An increase in local calcium concentration leads to filament assembly, primarily at the "barbed" end of actin filaments. Bundles could grow from a nucleating structure similar to the acromere of certain sperm if present in the cytoplasm. However, no structure of similar morphology or function has been identified in the cytoplasm of platelets. If cell activation is limited to only one region of membrane (via the binding of ligands to membrane receptors), alterations in the concentration of actin in the cell would be restricted to cytoplasm immediately beneath engaged receptors. Gelsolin must become inactivated in some way for filaments to rapidly elongate when driven by the large profilactin pool in the cell. The presence of α-actinin and the high concentration of tropomyosin would lead to filament bundles despite this disordered start in filament growth. Tropomyosin would bind to the elongating filaments, structurally strengthen them, and block molecules of actin-binding protein from binding to them. Such blockage would have the important effect of preventing filaments from branching off the sides of the elongating filaments. Diminished calcium would also activate α-actinin molecules in the stimulated region of cytoplasm. They could cross-link the filaments together and effectively zip up the filaments into a bundle.

Filopod elongation is followed by the formation of broad pseudopods that fill in between the filopod projections. Since initial filament growth will deplete the cytoplasm of free tropomyosin molecules, actin-binding protein can now bind to actin and cross-link it. A branching assembly of the remaining profilactin from these nuclei would be expected to result. This would fill the cytoplasm between the filopods and result in the network obversed to fill this cytoplasmic space in electron micrographs.

As actin filaments in the bundles of the filopods and in the branching network elongate, they would be expected to exert a force equally withn the cytoplasm. In platelets, the circumferential ring of microtubules and the intracellular granules are the only structures present in the cytoplasm to resist this force (i.e., for the expanding network to push against). Since activation has been reported to result in the dissolution of the microtubules, only the granules may oppose network expansion and these elements will be pushed, as actin assembles, into the center of the cell. Such aggregates of filamentous material and granules are observed in the center of activated platelets in electron micrographs (White, 1974; Figure 2a).

Retraction of filopods and pseuodpods may indicate that the concentration of calcium in the center of the cell is rising, at least transiently. Though some work suggests that there is a rise in intracellular calcium with activation (Feinstein, 1980), more recent work (Brass and Shattil, 1982) suggests that there may simply be greater rate of flux of calcium through the cytosol. Transient fluxes may still be adequate for certain calcium-dependent processes to take place, however. An increase in intracellular calcium may activate the calmodulin-dependent myosin phosphorylation system (Adelstein *et al.*, 1978; Daniel *et al.*, 1981), thereby assembling myosin into filaments and allowing actin to activate myosin's ATPase activity. This would cause a myosin-based contraction of the bundles and network into the center of the cell. Increased calcium would also cause gelsolin to cut actin filaments within the network. Therefore, as actin filaments are pulled into the cell center, a mechanism exists for their disassembly. Myosin based contraction might also help collapse and centralize the granules. The high calcium in this region would also promote granule release (VanderMeulen and Grinstein, 1982).

5. CONCLUSIONS

1. Actin filaments are the predominant structural elements in cortical cytoplasm. These filaments can be arranged into a wide variety of three-dimensional arrays. Their order can vary from branching networks to almost crystalline bundles.

2. The formation of actin assemblies is regulated by other proteins that bind to and interact with actin filaments. Elongated, flexible proteins nucleate actin growth and cross-link actin filaments into branched networks. Small, rod-shaped proteins effect actin bundling. Therefore, these proteins can function in concert with actin to provide cytoplasm with both extensibility and rigidity.

3. The stability of ctyoplasmic actin assemblies is regulated by another class of proteins that bind to, sever, then cap the ends of filaments. Separate proteins exist to cap opposite ends of the filament, but only the proteins that bind to the barbed end of actin are regulated by calcium. By severing cross-linked actin structures these proteins provide cytoplasm with fluidity.

4. The transient formation of three-dimensional actin assemblies depends on intrinsic filament assembly and disassembly, processes also regulated by proteins that cap the ends of actin filaments. Filaments can assemble from storage pools of monomeric actin complexed to profilin. Assembly of filaments from this pool is promoted only by free barbed ends of filaments. Depolymerization of actin filaments into the monomer pool is facilitated by capping this barbed filament end. Similarly, proteins that cap the pointed end of filaments inhibit disassembly. Rapid actin assembly provides cytoplasm with propulsive force.

5. The contractility of cytoplasm is powered by myosin.

6. REFERENCES

Aebi, U., Smith, P. R., Isenberg, G., and Pollard, T. D., 1980, Structure of crystalline actin sheets, *Nature* **288**:296–298.

Adelstein, R. S., 1978, Myosin phosphorylation, cell motility, and smooth muscle contraction, *Trends Biochem. Sci.* **3**:27–30.

Adelstein, R. S., Conti, M. A., and Barylko, B., 1978, The role of myosin phosphorylation in regulating actin-myosin interaction in human blood platelets, *Thromb. Haemost.* **40**:241–244.

Adelstein, R. S., and Pollard, T. D., 1978. Platelet contractile proteins, *Prog. Hemostasis and Thromb.* **4**:37–58.

Albrecht-Buehler, G., and Bushnell, A., 1982, Reversible compression of cytoplasm. *Exp. Cell Res.* **140**:173–189.

Allen, R. D., Zacharski, L. R., Widirstky, S. T., Rosenstein, R., Zaitlin, L. M., and Burgess, D. R., 1979: Transformation and motility of human platelets. *J. Cell Biol.* **83**:126–142.

Bamburg, J. R., Harris, H. E., and Weeds, A. G., 1980. Partial purification and characterization of an actin depolymerizing factor from brain, *FEBS Letters* **121**:178–182.

Begg, D. A., Rodewald, R., and Rebhun, Ll. 1978, The visualization of actin filament polarity in thin sections. Evidence for the uniform polarity of membrane-associated filaments, *J. Cell Biol.* **79**:846–852.

Bennett, V., and Stenbuck, P. J., 1979, The membrane attachment protein for spectrin is associated with band 3 in human erythrocyte membranes, *Nature* **280**:468–473.

Bennett, V., Davis, J., and Fowler, W. E., 1982, Brain spectrin, a membrane-associated protein related in structure and function to erythrocyte spectrin, *Nature* **299**:126–131.

Bennett, J., Zaner, K. S. and Stossel, T. P., 1984, Purification, structure, and function of macrophage a-actinin. *Biochem.* **23**: 5081-5086.

Bernstein, B. W., and Bamburg, J. R., 1982. Tropomyosin binding to f-actin protects the f-actin from disassembly by brain actin-depolymerizing factor (ADF), *Cell Motility*, **2**:1–8.

Blikstad, I., Sundkvist, I., and Eriksson, S., 1980, Isolation and characterization of profilactin and profilin from calf thymus and brain, *Eru. J. Biochem.* **105**:425–433.

Blikstad, I., Eriksson, S., and Carlsson, L., 1980. Alpha-actinin promotes polymerization of actin from profilactin. *Eur. J. Biochem.* **109**:317–323.

Bonder, E. M., and Mooseker, M. S., 1983, Direct electron microscopic visualization of barbed end capping and filament cutting by intestinal microvillar 95-k dalton protein (villin): a new actin assembly assay using the limulus acrosomal process, *J. Cell Biol.* **96**:1097–1107.

Boxer, L. A., and Stossel, T. P., 1976, Isolation and properties of actin, myosin, and a new actin-binding protein of chronic myelogenous leukemia leukocytes, *J. Clin. Invest.* **57**:5696–5705.

Brass, L. F., and Shattil, S. J., 1982. Changes in surface-bound and exchangeable calcium during platelet activation. *J. Biol. Chem.* **257**:14000–14005.

Brenner, S. L., and Korn, E. D., 1979, Spectrin-actin interaction. Phosphorylated and dephosphorylated spectrin tetramer cross-link actin. *J. Biol. Chem.* **254**:8620–8627.

Brenner, S. L., and Korn, E. D., 1980a, Substoichiometric concentrations of cytochalasin D inhibit actin polymerization. Additional evidence for an F-actin treadmill. *J. Biol. Chem.* **254**:9982–9985.

Brenner, S. L., and Korn, E. D., 1980b, The effects of cytochalasins on actin polymerization and actin ATPase provide insights into the mechanism of polymerization, *J. Biol. Chem.* **255**:841–844.

Bretscher, A., and Weber, K., 1979, Villin: The major microfilament-associated protein of the intestinal microvillus, *Proc. Natl. Acad. Sci. USA* **76**:2321–2335.

Bretscher, A., and Weber, K., 1980a, Fimbrin, a new microfilament-associated protein present in microfilli and other cell surface structures. *J. Cell Biol.* **86**:335–340.

Bretscher, A., and Weber, K., 1980b, Villin is a major protein of the microvillus cytoskeleton which binds both G and F actin in a calcium-dependent manner, *Cell* **20**:839–847.

Bretscher, A., 1981, Fimbrin is a cytoskeletal protein that crosslinks F-actin in vitro, *Proc. Natl. Acad. Sci. USA* **78**:6849–6853.
Brier, J., Fechheimer, M., Swanson, J., and Talyor, D. L., 1983, Abundance, relative gelation activity, and distribution of the 95,000-dalton actin-binding protein from dictyostelium discoideum. *J. Cell Biol.* **97**:178–185.
Brotschi, E. A., Hartwig, J. H., and Stossel, T. P., 1978, The gelation of actin by actin-binding protein, *J. Biol. Chem.* **253**:8988–8993.
Brown, S. S., and Spudich, J. A., 1979, Cytochalasin inhibits the rate of elongation of actin filament fragments, *J. Cell Biol.,* **83**:657–662.
Brown, S. S., and Spudich, J. A., 1981, Mechanism of action of cytochalasin: Evidence that it binds to actin filament ends, *J. Cell Biol.* **88**:487–491.
Bryan, J., Kurth, M., 1984, Actin-gelsolin interactions: Evidence for two actin-binding sites. *J. Biol. Chem.* **269**: 7480–7487.
Burnside, B., Smith, B., Nagata, M., and Porrello, K., 1982, Reactivation of contraction in detergent-lysed telepst retinal cones, *J. Cell Biol.* **92**:199–206.
Burridge, K., and Feramisco, J. R., 1980, Microinjection and localization of a 130K protein in living fibroblasts: a relationship to actin and fibronectin, *Cell* **19**:587–595.
Burridge, K., and Feramisco, J. R., 1981, Non-muscle a-actinins are calcium-sensitive actin-binding proteins, *Nature* **294**:565–567.
Burridge, K., Kelly, T., and Mangeat, P. 1982, Nonerythrocyte spectrins: actin-membrane attachment proteins occurring in many cell types, *J. Cell Biol.* **95**:478–486.
Carlsson, L., Markey, F., Blikstad, I., Persson, T., and Lindberg, U., 1979, Reorganization of actin in platelets stimulated by thrombin as measured by the DNase I inhibition assay, *Proc. Natl. Acad Sci. USA* **76**:6376–6380.
Carlsson, L., Nystrom, E., Sundkvist, I., Markey, F., and Lindberg, U., 1977, Actin polymerizability is influenced by profilin, a low molecular weight protein in non-muscle cells. *J. Mol. Biol.* **115**:465–483.
Carroll, R. C., and Gerrard, J. M., 1982, Phosphorylation of platelet actin-binding protein during platelet activation, *Blood* **59**:466–471.
Casella, J. F., Flanagan, M. D., and Lin, S., 1982, Cytochalasin D inhibits actin polymerization and induces depolymerization of actin filaments formed during platelet shape change, *Nature* **293**:302–305.
Chaponnier, C., Borgia, R., Rungger-Brandle, E., Weil, R., and Gabbiani, G., 1979, An actin-destabilizing factor is present in human plasma, *Experientia* **35**:1039–1041.
Chen, W. T., 1981. Mechanism of retraction of the trailing edge during fibroblast movement, *J. Cell Biol.,* **90**:187–200.
Collier, N. C., and Wang, K., 1982a, Purification and properties of human platelet P235. *J. Biol. Chem.* **257**:6937–6943.
Collier, N. C., and Wang, K., 1982b, Human platelet P235: A high Mr protein which restricts the length of actin filaments, *FEBS Letters* **143**:205–210.
Cohen, C. M., Tyler, J. M., and Branton, D., 1980, Spectrin-actin associations studied by electron microscopy of shadowed preparations, *Cell* **21**:875–883.
Cohen, I., 1979, The contractile system of blood platelets and its function, *Meth. Achiev. Exp. Pathol.* **9**:40–86.
Cohen, I., and Cohen, C., 1972, A tropomyosin-like protein from human platelets, *J. Mol. Biol.* **68**:383–387.
Coodeelis, J., and Vahley, M., 1982, A calcium- and pH-regulated protein from dictyostelium discoideum that cross-links actin filaments. *J. Cell Biol.* **94**:466–471.

Corwin, H., and Hartwig, J. H., 1983, The isolation of actin-binding protein and villin from toad oocytes. *Devel. Biol.* **99**:61–74.

Côté, G. O., and Smillie, L. B., 1981, The interaction of equine platelet tropomyosin with skeletal muscle actin. *J. Biol. Chem.* **256**:7257–7261.

Côté, G. P., and Smillie, L. B., 1981, The effects of platelet tropomyosin on the ATPase activities of muscle actomyosin subfragment-1 in the absence and presence of troponin, its components and calmodulin. *J. Biol. Chem.* **256**:11999–12004.

Côté, G., Lewis, W. G., and Smillie, L. B., 1978, Non-polymerizability of platelet tropomyosin and its NH_2- and COOH- terminal sequences, *FEBS Letters* **91**:237–241.

Côté, G. P., Lewis, W. G., Pato, M. D., and Smillie, L. B., 1978, Platelet tropomyosin: lack of binding to skeletal muscle troponin and correlation with sequence, *FEBS Letters* **94**:131–135.

Craig, S. W., and Powell, L. D., 1980, Regulation of actin polymerization by villin, a 95,000-dalton cytoskeleton component of intestinal brush borders, *Cell* **22**:739–746.

Daniel J. L., Molish, I. R., and Holmsen, H., 1981, Myosin phosphorylation in intact platelets, *J. Biol. Chem.* **256**:7510–7514.

der Terrossian, E., Fuller, S. D., Stewart, M., and Weeds, A. G., 1981, Porcine platelet tropomyosin. Purification, characterization, and paracrystal formation. *J. Mol. Biol.* **153**:147–167.

Davies, P., Shizuta, Y., Olden, K., Gallo, M., and Pastan, I., 1977, Phosphorylation of filamin and other proteins in cultured fibroblasts. *Biochem. Biophys. Res. Comm.* **74**:495–502.

Davies, P., Shizuta, Y., and Pastan, I., 1982, Purification and properties of avian filamin and mammalian filamins, *Methods in Enzymology* **85B**:322–328.

Eisenberg, E., and Greene, L. E., 1980, The relation of muscle biochemistry to muscle physiology, *Ann. Rev. Physiol.* **42**:293–309.

Fattoum, L., Hartwig, J. H., and Stossel, T. P., 1983, Isolation and some structural properties of macrophage tropomyosin, *Biochem.* **22**:1187–1193.

Feinstein, M. B., 1980, Release of intracellular membrane-bound calcium precedes the onset of stimulus-induced exocytosis in platelets, *Biochem. Biophys. Res. Comm* **93**:593–596.

Feramisco, J. R., Smart, J. E., Burridge, K., Helfman, D. M., and Thomas, G. P., 1982. Coexistence of vinculin and a vinculin-like protein of higher molecular weight in smooth muscle, *J. Biol. Chem.* **257**:11024–11031.

Fine, R. E., and Blitz, A. L., 1975, A chemical comparison of tropomyosins from muscle and non-muscle tissues, *J. Mol. Biol.* **95**:447–454.

Flanagan, M. D., and Lin, S., 1980, Cytochalasins block actin filament elongation by binding to high affinity sites associated with F-actin, *J. Biol. Chem.* **255**:835–838.

Flory, P., 1953, *Principles of Polymer Chemistry*, Cornell University Press, Ithaca, New York.

Fowler, V., and Talyor, D. L., 1980, Spectrin plus band 4.1 cross-link actin. Regulation by micromolar calcium, *J. Cell Biol.* **85**:361–376.

Geiger, B., 1979, A 130K protein from chicken gizzard: its localization at the termini of microfilament bundles in cultured chicken cells. *Cell* **18**:193–205.

Geiger, B., Tokuyasu, K. T., Dutton, A. H., and Singer, S. J., 1979, Vinculin, an intracellular protein localized at specialized sites where microfilament bundles terminate at cell membranes, *Proc. Natl. Acad. Sci. USA*. **77**:4127–4130.

Geiger, B., Tokuyasu, K. T., Dutton, A. H., and Singer, S. J., 1980, Vinculin, an intracellular protein localized at specialized sites where microfilament bundles terminate at cell membranes, *Proc. Natl. Acad. Sci. USA* **77**:4127–4131.

Geiger, B., 1982, Microheterogeneity of avian and mammalian vinculin, *J. Mol. Biol.* **159**:685–701.

Glenney, J. R., and Weber, K., 1981, Calcium control of microfilaments: Uncoupling of the F-actin-severing and -bundling activity of villin by limited proteolysis in vitro, *Proc. Natl. Acad. Sci. USA* **78**:2810–2814.

Glenney, J. R., Kaulfus, P., and Weber, K., 1981, F-actin assembly modulated by villin: $Ca2+$-dependent nucleation and capping of the barbed end, *Cell* **24**:471–480.

Glenney, J. R., Kaulfus, P., Matsudaira, P., and Weber, K., 1981, F-actin binding and bundling properties of fimbrin, a major cytoskeletal protein of microvillus core filaments, *J. Biol. Chem.*, **256**:9283–9288.

Glenney, J. R., Glenney, P., Osborn, M., and Weber, K., 1982a, A high molecular weight f-actin and calmodulin binding protein with spectrin-related morphology is a major consituent of the terminal web microfilament organization of isolated intestinal brush borders, *Cell* **28**:843–854.

Glenney, J. R., Glenney, P., and Weber, K., 1982b. Erythroid spectrin, brain fodrin, and intestinal brush border proteins (TW-260/240) are related molecules containing a common calmodulin-binding subunit bound to a variant cell type specific subunit, *Proc. Natl. Acad. Sci. USA* **79**:4002–4005.

Glenney, J. R., Glenney, P., and Weber, K., 1982c. F-actin binding and crosslinking properties of porcine brain fodrin, a spectrin-related molecule, *J. Biol. Chem.* **257**:9781–9787.

Goodman, S. R., Zagon, I. S., and Kulikowsky, R. R., 1981. Identification of a spectrin-like protein in non-erythyroid cells, *Proc. Natl. Acad. Sci. USA* **78**:7570–7574.

Gordon, D. J., Boyer, J. L., and Korn, E. D., 1977, Comparative biochemistry of non-muscle actins, *J. Biol. Chem.* **252**:8300–8309.

Grumet, M., and Lin, S., 1980, A platelet inhibitor protein with cytochalasin-like activity against actin polymerization in vitro, *Cell* **21**:439–444.

Grumet, M., and Lin, S., 1981, Purification and characterization of an inhibitor protein with cytochalasin-like activity from bovine adrenal medulla, *Biochim. Biophys. Acta.* **678**:381–387.

Harker, L. A., Malpass, T. W., Branson, H. E., Hessel, E. A., and Slichter, S. J., 1980, Mechanism of abnormal bleeding in patients undergoing cardiopulmonary bypass: acquired transient platelet dysfunction associated with selective alpha-granule release, *Blood* **56**:824–834.

Harris, A. K., Wild, P., and Stopak, D., 1980, Silicone rubber substrata: A new wrinkle in the study of cell locomotion, *Science*, **208**:177–179.

Harris, D. A., and Schwartz, J. H., 1981, Characterization of brevin, a serum protein that shortens actin filaments. *Proc. Natl. Acad. Sci. USA* **78**:6798–6802.

Harris, H. E., and Weeds, A. B., 1978, Platelet actin: sub-cellular distribution and association with profilin, *FEBS Letters* **90**:84–88.

Harris, H. E., and Gooch, J., 1981, An actin depolymerizing protein from pig plasma, *FEBS Letters* **123**:49–53.

Hartshorne, D. J., and Siemankowski, R. F., 1981, Regulation of smooth muscle actomyosin. *Ann. Rev. Physiol.* **43**:519–530.

Hartwig, J. H., and Stossel, T. P., 1975, Isolation and properties of actin, myosin, and a new actin-binding protein in rabbit alveolar macrophages, *J. Biol. Chem.* **250**:5696–5705.

Hartwig, J. H., and Stossel, T. P., 1979, Cytochalasin B and the structure of actin gels, *J. Mol. Biol.* **134**:539–553.

Hartwig, J. H., Tyler, J., and Stossel, T. P., 1980. Actin-binding protein promotes the bipolar and perpendicular branching of actin filaments. *J. Cell Biol.* **87**:841–848.

Hartwig, J. H., and Stossel, T. P., 1981, The structure of actin-binding protein molecules in solution and interacting with actin filament. *J. Mol. Biol.* **145**:563–581.

Henderson, D., and Weber, K., 1979, Three-dimensional organization of microfilaments and microtubules in the cytoskeleton. Immunoperoxidase labelling and stereo-electron microscopy of detergent-extracted cells. *Exp. Cell Res.* **124**:301–316.

Hermans, I. M., Crisona, N. J., and Pollard, T. D., 1981, Relation between cell activity and the distribution of cytoplasmic actin and myosin. *J. Cell Biol.* **90**:84–91.

Heuser, J. E., and Kirschner, M. W., 1980. Filament organization revealed in platinum replicas of freeze-dried cytoskeletons, *J. Cell Biol.* **86**:212–234.

Huxley, H. E., 1969. The mechanism of muscular contraction. *Science* **164**:1356–1366.

Im, T., Kamitani, T., Tatsumi, N., Okuda, K., and Kusunose, M., 1982, Purification of calcium-sensitive regulatory protein of platelets which inhibits the gelation of actin. *Biochem. Biophys. Res. Comm.* **107**:173–180.

Isenberg, G., Leonard, K., and Jockusch, B. M., 1982, Structural aspects of vinculin-actin interactions. *J. Mol. Biol.* **158**:231–249.

Janmey, P. J., Chapponnier, C, Lind, S. E. et al., 1985, Interactions of gelsolin and gelsolin: actin complexes with actin biochemistry (in press).

Jennings, L. K., Fox, J., Edwards, H., and Phillips, D. R., 1981, Changes in the cytoskeletal structure of human platelets following thrombin activation. *J. Biol. Chem.* **256**:6927–6932.

Jockusch, B. M., and Isenberg, G., 1981, Interaction of a-actinin and vinculin with actin: Opposite effects on filament network formation. *Proc. Natl. Acad. Sci. USA.* **78**:3005–3009.

Kawamura, M., and Maruyama, K., 1970, Electron microscopic particle lengtyh of F-actin polymerized in vitro. *J. Biochem.* **67**:437–457.

Kendrick-Jones, J., and Scholey, J. M., 1981, Myosin-linked regulatory systems, *J. Muscle Res. and Cell Motility* **2**:347–372.

Kondo, H., and Ishiwata, S., 1976, Uni-directional growth of F-actin. *J. Biochem.* **79**:159–171.

Korn, E. D., 1981, The polymerization of actin, in: *International Cell Biology 1980–1981*, (H. G. Schweiger, ed.), p. 336–345, Springer-Verlag, Berlin.

Koteliansky, V. E., Shirinsky, V. P., Gneushev, G. N., and Smirnov, V. N., 1981. Filamin, a high relative molecular mass actin-binding protein from smooth muscles, promotes actin polymerization, *FEBS Letters,* **136**:98–100.

Krishna, K. M., and Varani, J., 1982, Actin polymerization induced by chemotactic peptide and concanavalin A in rat neutrophils, *J. Immun.* **129**:1605–1607.

Kuo, P. F., Minura, N., and Asano, A., 1982, Purification and characterization of actinogelin, a calcium-sensitive actin-accessory protein, from rat liver, *Eur. J. Biochem.* **125**:277–282.

Lazarides, E., and Weber, K., 1974, Actin antibody: the specific visualization of actin filaments in non-muscle cells. *Proc. Natl. Acad. Sci. USA* **71**:2268–2272.

Lazarides, E., 1976, Actin, Alpha-actinin, and tropomyosin interaction in the structural organization of actin filaments in non-muscle cells, *J. Cell Biol.* **68**:202–219.

Levine, J., and Willard, M., 1981, Fodrin: axonally transported polypeptides associated with the internal periphery of many cells, *J. Cell Biol.* **90**:631–643.

Lind, S. E., and Stossel, T. P., 1982, The microfilament network of the platelet, *Prog. Hemostasis Thromb.* **6**:63–84.

Lind, S., Yin, H., and Stossel, T. P., 1981, Human platelet gelsolin, a regulator of actin filament length, *J. Clin. Invest.* **69**:1384–1387.

Low, I., and Dancker, P., 1976, Effect of cytochalasin B on formation and properties of muscle F-actin. *Biochim. Biophys. Acta.* **430**:366–374.

MacLean-Fletcher, S., and Pollard, T. D., 1980, Mechanism of actin of cytochalasin B on actin, *Cell* **20**:329–341.

Markey, F., Lindberg, U., and Eriksson, L., 1978, Human platelets contain profilin, a potential regulator of actin polymerisability. *FEBS Letters* **88**:75–79.

Markey, F., Persson, T., and Lindberg, U., 1981, Characterization of platelet extracts before and after stimulation with respect to the possible role of profilactin as microfilament precursor, *Cell* **23**:145-153.

Markey, F., Larsson, H., Weber, K., and Lindberg, U., 1982, Nucleation of actin polymerization from profilactin. Opposite effects of different nuclei. *Biochem. Biophys Acta* **704**:43-51.

Maruyama, K., and Ebashi, S., 1965, a-Actinin, a new structural protein from striated muscle. II. Actin on actin. *J. Biochem.* **58**:13-19.

Maruyama, K., 1971, A study of beta-actinin, myofibrillar protein from rabbit skeletal muscle. *J. Biochem.* **69**:369-386.

Maruyama, K., Kimura, S., Ishii, T., Kuroda, M., Ohashi, K., and Muramatsu, S., 1977, a-actinin, a regulatory protein of muscle. *J. Biochem.* **81**:215-232.

Maruyama, K., Hartwig, J. H., and Stossel, T. P., 1980, Cytochalasin B and the structure of actin gels. II. Further evidence for the splitting of F-actin by cytochalasin B. *Biochim. Biophys. Acta.* **626**:494-500.

Maruyama, K., and Ohashi, O., 1978, Tropomyosin inhibits the interaction of F-actin and filamin. *J. Biochem.* **84**:1017-1019.

Maruyama, K., and Sakai, H., 1981, Cell beta-actinin, an accelerator of actin polymerization, isolated from rat kidney cytosol. *J. Biochem.* **89**:1337-1340.

Maruyama, K., Mimura, N., and Asano, A., 1981, Rheological studies on calcium-sensitive gelation of actin filaments by actinogelin. *J. Biochem.* **89**:317-319.

Masaki, T., 1975. Tropomyosin-like protein in chick embryo fibroblasts, *J. Biochem.* **77**:901-904.

Matsudaria, P. T., and Burgess, D. R., 1979, Identification and organization of the components of the isolated microvillus cytoskeleton. *J. Cell biol.* **83**:667-673.

Matsudaria, P. T., and Burgess, D. R., 1982, Partial reconstruction of the microvillus core bundle: characterization of villin as a Ca++-dependent, actin bundling/depolymerizing protein. *J. Cell Biol.* **92**:648-656.

Matsudaria, P. T., Mandelkow, E., Renner, W., Hesterberg, L. K., and Weber, K., 1983, Role of fimbrin and villin in determining the interfilament distances of actin bundles, *Nature* **301**:209-214.

Mimura, N., and Asano, A., 1979, Ca^{2+}-sensitive gelation of actin filaments by a new protein factor, *Nature* **282**:44-48.

Mimura, N., and Asano, A., 1982, Characterization and localization of actinogelin, a Ca2+-sensitive actin accessory protein, in nonmuscle cells. *J. Cell Biol.* **93**:899-909.

Mooseker, M. S., and Tilney, L. G., 1975. Organization of an actin filament-membrane complex. Filament polarity and membrane attachment in the microvilli of intestinal epithelial cells. *J. Cell Biol.* **67**:725-743.

Mooseker, M. S., Graves, T. A., Wharton, K. A., Falco, N., and Howe, C. L., 1980, Regulation of microvillus structure: Calcium-dependent solation and crosslinking of actin filaments in the microvilli of intestinal epithelial cells. *J. Cell Biol.* **87**:809-822.

Mooseker, M. S., Pollard, T. D., and Wharton, K. A., 1982, Nucleated polymerization of actin from the membrane-associated ends of microvillar filaments in the intestinal brush border. *J. Cell Biol.* **95**:223-233.

Nachmias, V. T., 1980, Cytoskeleton of human platelets at rest and after spreading, *J. Cell Biol.* **86**:795-802.

Nachmias, V. T., and Asch, A., 1976. Regulation and polarity: results with myxomycete plasmodium and with human platelets, in: *Cell Motility* (R. Goldman, T. Pollard, J. Rosenbaum, eds.), *Cold Spring Harbor Conference on Cell Proliferation* **3**:771-783.

Nelson, W. J., and Lazarides, E., 1983, Expression of the β subunit of spectrin in nonerythroid cells. *Proc. Natl. Acad. Sci. USA* **80**:363-367.

Niederman, R., and Pollard, T. D., 1975, Human platelet myosin. II. In vitro assembly and structure of myosin filaments. *J. Cell Biol.* **67**:72–92.

Niederman, R., Amrein, P., and Hartwig, J. H., 1983, The three dimensional structure of actin filaments in solution and an actin gel made with actin-binding protein. *J. Cell Biol.* **96**:1400–1413.

Nishida, E., Kuwaki, T., Maekawa, S., and Sakai, H., 1981, A new regulatory protein that affects the state of actin polymerization, *J. Biochem.* **89**:1655–1658.

Nunnally, M. H., Powell, L. D., and Craig, S. W., 1981. Reconstitution and regulation of actin gel-sol transformation with purified filamin and villin, *J. Biol. Chem.* **256**:2083–2086.

Norberg, R., Thorstensson, R., Utter, G., Fagraeus, A., 1979, F-actin-depolymerizing activity of human serum, *Eur. J. Biochem.* **100**:575–582.

Oosawa, F., and Asakura, S., 1975, *Thermodynamics of the polymerization of protein.* Academic Press, London, New York.

Oosawa, F., and Kasai, M., 1971, 1 Actin. in: *Subunits in Biological Systems,* Chapter 6, pp. 261–321, Marcel Dekker, New York.

Pollard, T. D., 1981, Purification of a calcium-sensitive actin gelation protein from acanthamoeba. *J. Biol. Chem.* **256**:7666–7670.

Pollard, T. D., and Mooseker, M. S., 1981, Direct measurement of actin polymerization rate constants by electron microscopy of actin filaments nucleated by isolated microvillus cores. *J. Cell Biol.* **88**:654–659.

Pollard, T. D., and Weihing, R. R., 1974, Actin and myosin and cell movement. *CRC Crit. Rev. Biochem.* **2**:1–65.

Pribluda, V., Laub, F., and Rotman, A., 1981, The state of actin in activated human platelets, *Eur. J. Biochem.* **116**:293–296.

Rosenberg, S., Stracher, A., and Lucas, R. C., 1981, Isolation and characterization of actin and actin-binding protein from human platelets, *J. Cell Biol.* **91**:201–211.

Rosenberg, et al., 1981, Effect of actin-binding protein on the sedimentation properties of actin. *J. Cell Biol.* **94**:51–55.

Rosenberg, S., Stracher, A., and Burridge, K., 1981, Isolation and characterization of a calcium-sensitive a-actinin-like protein from human platelet cytoskeletons, *J. Biol. Chem.* **256**:12986–12991.

Schliwa, M., and van Blerkom, J., 1981. Structural interaction of cytoskeletal components, *J. Cell Biol.,* **90**:222–235.

Schloss, J. A., and Goldman, R. D., 1979, Isolation of a high molecular weight actin-binding protein from baby hamster kidney (BHK-21) cells. *Proc. Natl. Acad. Sci. USA* **76**:4484–4488.

Schloss, J. A., and Goldman, R. D., 1980. Microfilaments and tropomyosin of cultured mammalian cells: Isolation and characterization, *J. Cell Biol.* **87**:633–642.

Scholey, J. M., Taylor, K. A., and Kendrick-Jones, J., 1981, The role of myosin light chains in regulating actin-myosin interaction, *Biochimie* **63**:255–271.

Schollmeyer, J. V., Rao, G., and White, J. G., 1978, An actin-binding protein in human platelets, *Am. J. Pathol.* **93**:433–446.

Schroeder, T. E., 1973, Actin in dividing cells: contractile ring filaments bind heavy meromyosin, *Proc. Natl. Acad. Sci. USA* **70**:1688–1692.

Shizuta, Y., Shizuta, H., Gallo, M., Davies, P., and Pastan, I., 1976, Purification and properties of filamin, an actin-binding protein from chicken gizzard, *J. Biol. Chem.* **251**:6562–6567.

Shotton, D. M., Burke, B. E., and Branton, D., 1979, The molecular structure of human erythrocyte spectrin. Biophysical and electron microscopical studies. *J. Mol. Biol.* **131**:303–329.

Singh, I., Goll, D. E., Robson, R. M., and Stromer, M. H., 1981, Effect of -actinin on actin structure. Viscosity studies. *Biochim. Biophys. Acta.* **669**:1-6.
Southwick, F. S., and Stossel, T. P., 1981, Isolation of an inhibitor of actin polymerization from human polymorphonuclear leukocytes. *J. Biol. Chem.* **256**:3030-3036.
Southwick, F. S., and Hartwig, J. H., 1982, Acumentin, a protein in macrophages which caps the "pointed" end of actin filaments, *Nature* **297**:303-307.
Southwick, F. S., Tatsumi, N., and Stossel, T. P., 1982, Acumentin, an actin modulating protein of rabbit pulmonary machrophages. *Biochem.*
Spudich, J. A., 1972, Effects of cytochalasin B on actin filaments. *Cold Spring Harbor Symp Quant Biol.* **37**:585-593.
Spudich, J. A., and Lin, S., 1972, Cytochalasin B, its interaction with actin and actomyosin from muscle. *Proc. Natl. Acad. Sci. USA* **69**:442-446.
Stendahl, O. I., Hartwig, J. H., Brotschi, E. A., and Stossel, T. P., 1980, Distribution of actin-binding protein and myosin in macrophages during spreading and phagocytosis, *J. Cell Biol.* **84**:215-224.
Stendahl, O., and Stossel, T. P., 1980, Actin-binding protein amplifies actomyosin contraction, and gelsolin confers calcium control on the direction of contraction. *Biochem. Biophys. Res. Comm.* **92**:675-681.
Stossel, T. P., and Hartwig, J. H., 1975, Interactions between actin, myosin, and a new actin-binding protein of rabbit alveolar macrophages. Macrophage myosin Mg++-adenosine trophosphatase requires a cofactor for activation. *J. Biol. Chem.* **250**:5708-5712.
Stossel, T. P., and Hartwig, J. H., 1976, Interaction of actin, myosin, and a new actin-binding protein of rabbit pulmonary macrophages. II. Role in cytoplasmic movement and phagocytosis. *J. Cell Biol.* **68**:602-614.
Stossel, T. P., Hartwig, J. H., Yin, H. L., Zaner, K. S., and Stendahl, O. I., 1982, Actin gelation and the structure of cortical cytoplasm, *Cold Spring Harbor Symp. Quant. Biol.* **46**:569-578.
Talbot, K., and MacLeod, A. R., 1983. Novel form of non-muscle tropomyosin in human fibroblasts, *J. Mol. Biol.* **164**:159-174.
Taylor, E. W., 1972, Chemistry of muscle contraction, *Ann. Rev. Biochem.* **41**:577.
Thorstensson, R., Utter, G., Norberg, R., Fagraeus, A., Hartwig, J. H., Yin, H. L., and Stossel, T. P., 1982, Distribution of actin, myosin, actin-binding protein and gelsolin in cultured lymphoid cells, *Exp. Cell Res.* **140**:395-400.
Tilney, L. G., Hatano, S., Ishikawa, H., and Mooseker, M. S., 1973, The polymerization of actin: Its role in the generation of the acrosomal process of certain echinoderm sperm. *J. Cell Biol.* **59**:109-126.
Tilney, L. G., 1976, The polymerization of actin. II. How non-filamentous actin becomes nonrandomly distributed in sperm: Evidence for the association of this actin with membranes. *J. Cell Biol.,* **69**:51-72.
Tilney, L. G., 1976, The polymerization of actin. III. Aggregates of nonfilamentous actin and its associated proteins: A storage form of actin, *J. Cell Biol.* **69**:73-89.
Tilney, L. G., 1978, Polymerization of actin. V. A new organelle, the actomere, that initiates the assembly of actin filaments in thyone sperm. *J. Cell Biol.* **77**:551-564.
Tilney, L. G., and Kallenbach, N., 1979, Polymerization of actin. VI. The polarity of the actin filaments in the acrosomal process and how it might be determined, *J. Cell Biol.* **81**:608-623.
Tilney, L. G., and Inoue, S., 1982, Acrosomal reaction of thyone sperm. II. The kinetics and possible mechanism of acrosomal process elongation. *J. Cell Biol.* **93**:820-827.
Tyler, J. M., Anderson, J. M., and Branton, D., 1980. Structural comparison of several actin-binding molecules, *J. Cell Biol.* **85**:489-495.

Ungewickell, E., Bennett, P. M., Calvert, R., Ohanian, V., and Gratzer, W. B., 1979, In vitro formation of a complex between cytoskeletal proteins of the human erythrocyte, *Nature* **280**:811–814.

Valerius, N. H., Stendahl, O. I., Hartwig, J. H., and Stossel, T. P., 1981, Distribution of actin-binding protein and myosin in polymorphonuclear leukocytes during locomotion and phagocytosis. *Cell* **24**:195–202.

Vanderkerckhove, J., and Weber, K., 1978, Actin amino-acid sequences. Comparison of actins from calf thymus, bovine brain, and SV40-transformed mouse 3T3 cells with rabbit skeletal muscle actin. *Eur. J. Biochem.* **90**:451–462.

VanderMeulen, J., and Grinstein, S., 1982, Calcium-induced lysis of platelets secretory granules, *J. Biol. Chem.* **257**:5190–5195.

Wallach, D., Davies, P. J. A., and Pastan, I., 1978, Purification of mammalian filamin. Similarity to high molecular weight actin-binding protein in macrophages, platelets, fibroblasts and other tissues, *J. Biol. Chem.* **253**:3328–3335.

Wang, K., Ash, J. F., and Singer, S. J., 1975, Filamin, a new high-molecular weight protein found in smooth muscle and nonmuscle cells. *Proc. Natl. Acad. Sci. USA* **72**:4483–4486.

Wang, K., and Singer, S. J., 1977, Interaction of filamin with F-actin in solution, *Proc. Natl. Acad. Sci. USA* **74**:2021–2025.

Wang, K., 1977, Filamin, a new high-molecular-weight protein found in smooth muscle and nonmuscle cells. Purification and properties of chicken gizzard filamin, *Biochem.* **16**:1857–1865.

Wang, L. L., and Bryan, J., 1981, Isolation of calcium-dependent platelet proteins that interact with actin, *Cell* **25**:637–649.

Wessells, N. K., Spooner, B. S., Ash, J. F., Bradley, M. O., Ludvena, M. A., Taylor, E. L., Wrenn, J. T., and Yamada, K. M., 1971, Microfilaments in cellular and development processes, *Science,* **171**:135–143.

Wegner, A., 1982, Treadmilling of actin at physiological salt concentrations, *J. Mol. Biol.* **161**:607–615.

Weiss, P., and Garber, B., 1952, Shape and movement of mesenchyme cells as functions of the physical structure of the medium. Contributions to a quantitative morphology, *Proc. Natl. Acad. Sci. USA* **38**:264–280.

White, J. G., 1974, Electron microscopic studies of platelet secretion, *Prog. Hemostasis Thromb.* **2**:49–98.

Wilkins, J. A., and Lin, S., 1982, High affinity interaction of vinculin with actin filaments in vitro. *Cell* **28**:83–90.

Wong, A. J., Pollard, T. D., and Hermans, I., 1983, Actin filament stress fibers in vascular endothelial cells in vivo. *Science.* **219**:867–869.

Woodrum, D. T., Rich, S. A., and Pollard, T. D., 1975, Evidence for biased bidirectional polymerization of actin filaments using heavy meromyosin prepared by an improved method, *J. Cell. Biol.* **67**:231–237.

Yahara, I., Harada, F., Sekita, S., Yoshihira, K., and Natori, S., 1982, Correlation between effects of 24 different cytochalasins on cellular structures and cellular events and those on actin in vitro. *J. Cell Biol.* **92**:69–78.

Yang, Y., Korn, E. D., and Eisenberg, E., 1979, Cooperative binding of tropomyosin to muscle and acanthamoeba actin, *J. Biol. Chem.* **254**:7137–7140.

Yin, H. L., and Stossel, T. P., 1979, Control of cytoplasmic actin gel-sol transformation by gelsolin, a calcium-dependent regulatory protein, *Nature* **281**:583–586.

Yin, H. L., and Stossel, T. P., 1980. Purification and structural properties of gelsolin, a Ca^{2+}-activated regulatory protein of macrophages, *J. Biol. Chem.* **255**:9490–9493.

Yin, H. L., Zaner, K., and Stossel, T. P., 1980, Ca^{2+}-control of actin gelation. Interaction of gelsolin with actin filaments, and regulation of actin gelation. *J. Biol. Chem.* **255**:9494–9500.

Yin, H. L., Hartwig, J. H., Maruyama, K., and Stossel, T. P., 1981. Ca^{2+}-control of actin polymerization. Interaction of macrophage gelsolin with actin monomers and effects on actin polymerization, *J. Biol. Chem.* **256**:9693–9697.

Yin, H. L., Albrecht, J. H., and Fattoum, A., 1981, Identification of gelsolin, a Ca^{2+}-dependent regulatory protein of actin gel-sol transformation in a variety of cells and tissues and its intracellular distribution in phagocytic cells. *J. Cell. Biol.* **91**:901–906.

Zaner, K. S., and Stossel, T. P., 1982, Some perspectives on the viscosity of actin filaments. *J. Cell Biol.* **93**:987–991.

Chapter 2

Fluorescence Probes Unravel Asymmetric Structure of Membranes

Friedhelm Schroeder
Department of Pharmacology
School of Medicine
University of Missouri
Columbia, MO 65212

1. INTRODUCTION

In the past decade fluorescence probe molecules have become increasingly useful in unraveling the asymmetric distribution of proteins, lipids, and sterols in biological membranes. The intrinsic sensitivity and the ability to discriminate between extremely short-lived membrane events provide fluorescence methodology with distinct advantages over NMR, ESR, X-ray scattering, infrared spectroscopy, calorimetry, and optical rotary dispersion–circular dichroism (Badley, 1976). Advances in fluorescence microscopy such as fluorescence photobleaching recovery (Peters *et al.*, 1974; Edidin and Fambrough, 1973; Axelrod *et al.*, 1976), fluorescence lifetime determination by phase and modulation or by photon counting (Ware, 1971; Weber, 1981; Isenberg, 1975), and computer-centered spectrofluorimetry (Holland *et al.*, 1973, 1977; Christman *et al.*, 1980, 1981; Wampler, 1976) have greatly enhanced the utility of fluorescence probe molecules in determination of biological membrane asymmetry. An evaluation of the asymmetric structure of membranes requires recognition of two possible asymmetric distributions of components either in the plane of the membrane (lateral, horizontal) or normal to the plane (transbilayer, vertical). Both of these asymmetric aspects of membrane structure may have important physiological consequences as detailed herein.

1.1. Protein Asymmetry

The use of fluorescence probes in measuring the asymmetric orientation and distribution of proteins was recently reviewed (Etemadi, 1980; Cherry, 1979). Proteins appear to have a fixed transbilayer orientation in membranes, as they do not undergo flip-flop. However, proteins do diffuse laterally by a process that may involve extracellular receptors and intracellular linkage to the cytoskeleton. Several factors important to protein movement and agglutination of cells by lectins have emerged: plasma membrane fluidity, cell surface charge, and the cytoskeleton (McCaleb and Donner, 1981). Membrane lipids such as fatty acids (Horwitz *et al.*, 1974; Rule *et al.*, 1979; Rittenhouse and Fox, 1974; Hampton *et al.*, 1980; Tombaccini *et al.*, 1980; Ruggini and Fallani, 1973; Rittenhouse *et al.*, 1974a), sterol–phospholipid ratio (Marshall *et al.*, 1979), lysophospholipids (Weltzien, 1975), and phospholipid polar head group composition all modulate membrane fluidity (Vaughan and Keough, 1974; Blume and Ackerman, 1974) and concanavalin A agglutinability of cells (Schroeder, 1982c; Hampton *et al.*, 1980). These results were consistent with the possibility that lateral membrane fluidity was the rate-limiting step affecting diffusion of lectin receptors (Rule *et al.*, 1979). However, alteration of phospholipid polar head groups in LM fibroblasts did not alter the fluidity of the plasma membrane but drastically modified Con A agglutinability and hemagglutination (Schroeder, 1982c). In addition, the use of fluorescence photobleaching recovery methods to directly measure membrane protein lateral mobility demonstrated that membrane proteins on nearly all tissues investigated were not entirely free to diffuse in the cell membrane (Cherry, 1979; Jacobson *et al.*, 1982; Webb *et al.*, 1981) but could be released from constraints by certain chemical treatments such that the lateral diffusion rate of proteins increased from $D < 10^{-10}$–10^{-12} cm^2/sec to $D > 10^{-9}$ cm^2/sec, close to the value for membrane lipid diffusion (Webb *et al.*, 1981; Tank *et al.*, 1982). Thus, other membrane properties such as cytoskeleton–protein interactions or saccharide–saccharide interactions may be more important determinants of membrane glycoprotein diffusion than lipid fluidity alone (Wolf *et al.*, 1980; Golan and Veatch, 1980; Low *et al.*, 1982; Cherry *et al.*, 1980; Sheetz *et al.*, 1982). However, cytoskeletal constraint does not exclusively account for the lipid mediated effects unless membrane lipid composition, membrane lipid transbilayer asymmetry, lateral phase separations, or boundary lipid or other lipid properties also dramatically affect glycoprotein–cytoskeleton interactions. Some evidence consistent with this possibility comes from the observation that phagocytosis, which requires cytoskeleton–membrane protein interactions (Silverstein *et al.*, 1977; Sherman, 1979; Kavet and Brain, 1980), is also dramatically affected by alterations in membrane fatty acid unsaturation (Roberts and Quastel, 1963; Mahoney *et al.*, 1977; Schroit and

Gallily, 1979), cholesterol content (Heininger and Marshal, 1979), and phospholipid composition (Schroeder, 1981, 1982a; Kier and Schroeder, 1983a; Schroeder and Kinden, 1983).

1.2. Lipid Asymmetry

The general area of membrane lipid asymmetry is still a field in its infancy. Much information of a phenomenological nature has accumulated but almost nothing is known of the origin, regulation, and function of membrane lipid asymmetry in biological membranes. Over the past eight years the literature was extensively reviewed (Rothman and Lenard, 1977; Bergelson and Barsukov, 1977; Op den Kamp, 1979; Etemadi, 1980). With few exceptions (Sessions and Horwitz, 1981) a general picture of the transbilayer distribution of lipids in mammalian plasma membranes has emerged (Figure 1). The net neutral charged zwitterionic phospholipids (phosphatidylcholine and sphingomyelin) are enriched in the outer monolayer, while the anionic phospholipids (phosphatidylserine, phosphatidylethanolamine, and phosphatidylinositol) are enriched in the inner monolayer. These conclusions are based on data for plasma membranes of erythrocytes (data prior to 1979 reviewed by Op den

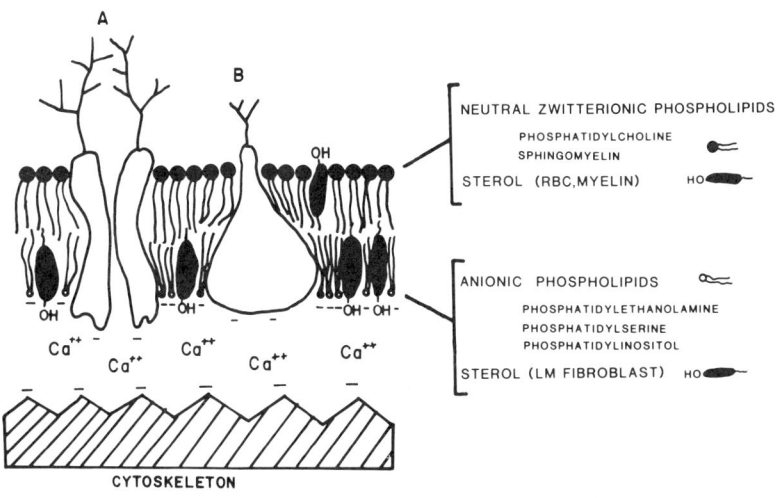

FIGURE 1. Asymmetric distribution of lipids in mammalian plasma membranes.

Kamp, 1979; Jokinen and Gahmberg, 1979; Renooij and Van Golde, 1979; Van Deenen, 1979; Van Meer et al., 1980; Gupta and Mishra, 1981; Marinetti and Gattieu, 1982; Shukla and Hanahan, 1982), of LM fibroblasts (Sandra and Pagano, 1978; Fontaine and Schroeder, 1979; Schroeder, 1980a,b, 1981a,b, 1982a; Schroeder et al., 1981, 1982a,b), of neurons (Fontaine et al., 1979, 1980), of B-16 melanoma variants (Schroeder, 1984), and of LM metastatic cell lines (Kier and Schroeder, 1985). The origin, regulation, and physiological significance of this asymmetric distribution of phospholipids is not known at this time. However, it has been shown that the aminophospholipid asymmetry is generated *de novo* by the cell and is not an artifact generated by exchange processes (Sandra and Pagano, 1978; Fontaine and Schroeder, 1979; Fontaine et al., 1979, 1980). Certainly the outer monolayer of the cell membrane as well as plasma lipoproteins was shown to be enriched in choline-containing phospholipids (sphingomyelin and phosphatidylcholine), and a direct relationship between the two exists (Figure 2). LM cell and synaptosomal plasma membranes are not exposed to plasma but are cultured in chemically defined media or exposed to CSF, which is very low in lipoproteins (Davson, 1967). However, they still have an asymmetric distribution of lipids, thereby providing convincing evidence for the *de novo* origin of lipid asymmetry in mammalian plasma membranes from transformed as well as normal cells. The

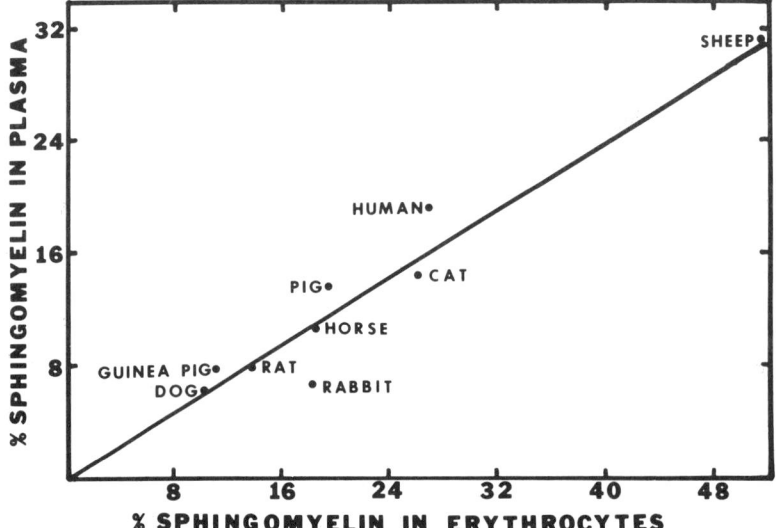

FIGURE 2. Relationship between plasma and red blood cell surface membrane sphingomyelin content. Values were obtained from Nelson (1972, 1973), Rouser et al. (1968), Skipski et al. (1967), and White (1973).

aminophospholipid asymmetry is stable to membrane fluidizing agents, depolarization, energy inhibitors (NaN$_3$ and KCN), ATP synthesis inhibitors, and microtubular and microfilament disrupting agents (Schroeder et al., 1981).

Another aspect of membrane lipid asymmetry concerns the transbilayer distribution of fatty acids. In the red cell the outer monolayer of the plasma membrane was enriched in unsaturated fatty acids (Renooij and Van Golde, 1979; Marinetti and Cattieu, 1982). In contrast, the inner monolayer of plasma membranes of other red blood cell investigations (Emmelot and Van Hoeven, 1975; Renooij and Van Golde, 1977), of LM fibroblasts (Fontaine and Schroeder, 1979), synaptosomes (Fontaine et al., 1979, 1980), and of platelets (Perret et al., 1979) was enriched with unsaturated fatty acids. Virus and mycoplasma membranes also have an asymmetric distribution of unsaturated fatty acids (Op den Kamp, 1979).

The determination of the transbilayer distribution of cholesterol has proved elusive—primarily because of a lack of easily available methodology. Cholesterol appears to be enriched in the outer monolayer of red blood cell membranes and of myelin (Caspar and Kirschner, 1971; Fisher, 1976; Schroeder, 1981a; Hale and Schroeder, 1982), the inner monolayer of LM fibroblasts (Schroeder, 1981a; Hale and Schroeder, 1982), and the outer monolayer of phagosomal membranes derived from LM fibroblasts (Hale and Schroeder, 1982). Fluorescence probe methodology has greatly enhanced our ability to measure sterol transbilayer distribution conveniently (detailed in Section 3).

The asymmetric distribution of lipids in subcellular membranes other than plasma membranes (e.g., microsomes, golgi, mitochondria), viruses, and microorganism will not be dealt with herein.

1.3. Boundary Lipid

Proteins that reside within biological membranes are surrounded by lipid (Singer and Nicolson, 1972). However, the arrangement of lipids surrounding these integral membrane proteins is not random. Instead, specific lipid annuli called boundary layers (Jost et al., 1973; Hesketh et al., 1976) may surround these proteins in each of the membrane monolayers: Cytochrome c oxidase has a boundary layer of cardiolipin (Robinson et al., 1980; Cable and Powell, 1980; Kang et al., 1979); Ca^{2+}-ATPase has phosphatidylserine (Moore et al., 1978); rhodopsin may have phosphatidylserine (Watts et al., 1979; Hidalgo et al., 1982); and HMG CoA-reductase and acyl-CoA–cholesterol acyltransferase may have cholesterol (Heller et al., 1979). Fluorescence probes have been used to reveal hydrophobic pockets of membrane proteins, the interaction of myelin basic protein with phosphatidylserine, and the interaction of M13 virus coat protein with phosphatidylcholine (Mely-Goubert et al., 1979; Vadas et al., 1981; Kimelman et al., 1979).

1.4. Charge Asymmetry

The two monolayers of the cell surface membrane differ in lipid composition and exposure of charged lipid, protein, or carbohydrate constituents (Figure 1). In fact, as shown in Figure 1, divalent metal bridges between inner monolayer anionic lipids and the anionic cytoskeleton may be responsible for constraining anionic lipids to reside in the inner monolayer. It seems reasonable to expect that alterations in transbilayer charge gradients may be important in the functioning of cell membranes. Fluorescence probe molecules are excellent monitors of membrane potential (Waggoner, 1979; Whitin *et al.*, 1981; Lelkes and Miller, 1980), transmembrane pH gradients (Elema *et al.*, 1978), membrane surface charge (Okuda and Ogli, 1979; Loew *et al.*, 1979; Andley and Chakrabarti, 1981; Slavik, 1982), and permeability of membrane channels (Flagg-Newton and Loewenstein, 1980; Eidelman and Cabantchik, 1980).

2. TRANSBILAYER LIPID DISTRIBUTION

One of the more important aspects of utilizing fluorescence probe molecules as monitors of membrane properties is determination of their location or the microenvironment wherein they reside. This serves a two-fold function: (1) the transbilayer or lateral distribution of "tagged lipids" can be used to determine the distribution of lipids within the membrane (fluorescent fatty acids, fluorescent phospholipids, or fluorescent sterols); and (2) the structure and microenvironment sensed by that lipid molecule may be ascertained. The most basic way to determine the transbilayer distribution of a fluorescent probe is to design it to be a nonflipping probe with a large water-soluble and/or charged group attached to a hydrophobic end that inserts into one side of the membrane only. Examples of this type of molecule are listed in Table I and include anthroyl-, dansyl-, or pyrene rings with acidic, basic, and neutral functional groups and alkyl "spacers" of various lengths that determine the depth of insertion of the probe into the membranes (Browning and Nelson, 1979; Matayoshi, 1980; Petri *et al.*, 1981; Prendergast *et al.*, 1981; Engel and Prendergast, 1981). The location of probe molecules across membranes is ascertained by several methods. One of these is by the effect it has on erythrocyte shape: fluorescent molecules that insert in the outer monolayer of the red cell membrane are crenators ("echinocyte" or spike formers), while molecules that insert on the inner monolayer are cuppers ("stomatocyte" former) (Deuticke, 1968; Sheetz and Singer, 1974, 1976; Zwaal *et al.*, 1975; Sheetz et al., 1976). A second method is to insert a fluorescent fatty acid (Schroeder *et al.*, 1979a; Schroeder and Goh, 1979), fluorescent sterol such as cholestatrienol or fluroescent dehydroergosterol (Schroeder *et al.*, 1979a; Hale and Schroeder,

Table I
Transbilayer Distribution of Fluorescent Probe Molecules: Potential Nonflipping Probes

Name	Reference
1-Pyrenebutyrylcholinebromide	Matayoski, 1980
1-Pyrenebutyric acid	Matayashi, 1980
16-(9-Anthroyloxy)palmitoyl-glucocerebroside	Petri *et al.*, 1981
1-[4-(Trimethylamino)phenyl]-6-phenylhexa-1,3,5-triene	Prendergast *et al.*, 1981
Biphenyl Derivatives	
6-ANSA	Browning and Nelson, 1979
10-ANSA	
16-ANSA	
6-ANASAM	
10-ANASAM	
16-ANASAM	
Anthroyl Derivatives	
PA	Browning and Nelson, 1979
SA	
DMAEA	
TEAEA	
HA	
DMAHA	
TEAHA	
Pyrene Derivatives	
AMHA	Browning and Nelson, 1979
AMHS	
PBA	
PBAD	
PBAM	
PS	
AP	

1982; Patel *et al.*, 1979; Sklar *et al.*, 1980; Bergeron and Scott, 1982a,b; Yeagle *et al.*, 1982), fluorescent steryl ester such as cholestatrienyl oleate (Craig *et al.*, 1982), or fluorescent phospholipid (Molotkovsky *et al.*, 1982) into plasma lipoproteins such as very-low-density lipoprotein, low-density lipoprotein, and high-density lipoprotein. Since the lipid structure of serum lipoproteins is primarily that of a triglyceride + cholesteryl ester core surrounded by a surface monolayer of phospholipid, cholesterol, and protein (Figure 3), nonpenetrating quenching agents and trinitrophenylglycine (Schroeder *et al.*, 1979a,b; Schroeder and Goh, 1979; Hale and Schroeder, 1982) or 5(N-hexadecanoyl)aminofluorescein (Sklar *et al.*, 1980) effectively quench the fluorescence signal from fluorescent molecules located only in the surface monolayer.

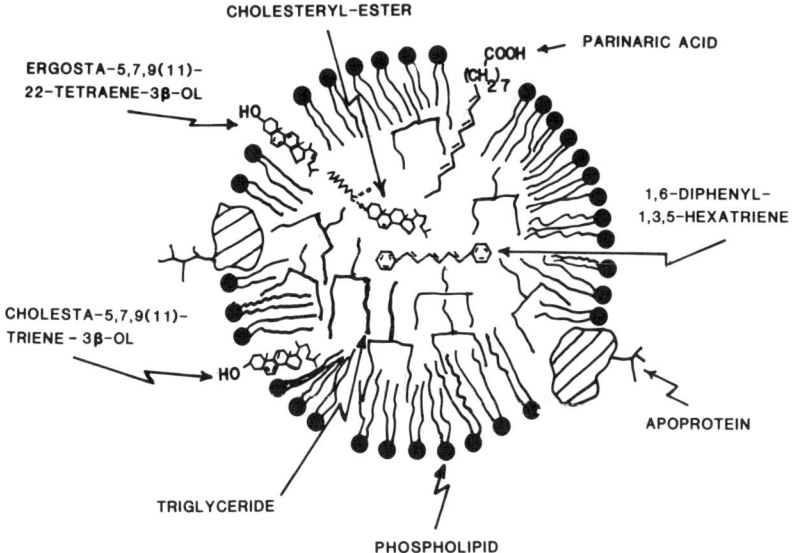

FIGURE 3. Structure of the very low density lipoprotein and location of fluorescent probes.

Fluorescence probe molecules containing a membrane-impermeant moiety (Table II) such as an oligosaccharide or glutathione (Cogan and Schachter, 1981; Flamm and Schachter, 1982; Rando et al., 1982), butyric acid (Luisetti et al., 1979), dextran (Minetti et al., 1979; Wolf et al., 1980), cytochrome b_5 (Gilmore and Glaser, 1982), or glucosamine (Schroeder, 1980a; Schroeder and

Table II
Proven Membrane Impermeant (Non-Flipping) Probe Molecules

Name	Reference
Oligosaccharide derivatives of pyrene	Cogan and Schachter, 1981
Glutathione derivatives of pyrene	Cogan and Schachter, 1981
Butyric acid derivatives	Luisetti, 1979
Fluoresceinylthiocarbamoyl-dextran	Minetti et al., 1979
Tetrasaccharide linked to pyrene butyrylhydrazide	Cogan and Schachter, 1981; Flamm and Schachter, 1982
Dansylated cytochrome b_5	Gilmore and Glaser, 1982
1-Acyl-2-[6-(7-nitro-2,1,3-benzoxadiozot-4-yl)amino]-caproyl]phosphatidylcholine	Pagano et al., 1981
NBD-cholesterol ester	Rando et al., 1982
Glucosamine-parinarate	Schroeder, 1980a; Schroeder and Kinden, 1980

Kinden, 1980) will insert in only one monolayer of erythrocyte, liposome, LM fibroblast, and phagosome membranes. Alternately, fluorescent probes that partition in both halves of the bilayer may be incorporated into membranes, and nonpenetrating quenching agent can be added on one side of the membrane (Table III). For example, hemoglobin quenches 1,6-diphenyl-1,3,5-hexatriene, 12-(9-anthroyloxy)stearate, 2-(9-anthroyloxy)stearate, and pyrene decanoic acid in the inner monolayer of red blood cells (Schachter et al., 1982); cytochrome c quenches dansylphosphatidylethanolamine in phosphatidylcholine monlayers (Teissie, 1981); paramagnetic quenching of pyrene decanoic acid and chlorophyll fluorescence in phosphatidylcholine membranes by fatty acid spin probes (Luisetti et al., 1979); N-stearoyltryptophan quenches n-(9-anthroyloxy) fatty acids in lipid bilayers (Haigh et al., 1979); trinitrophenylglycine and trinitrophenyl groups covalently attached to amino groups in one monolayer of fibroblast, erythrocyte, or phagosomal membranes effectively quench the fluorescence of parinaric acid, 1,6-diphenyl-1,3,5-hexatriene, and dehydroergosterol in that monolayer of the membrane (Schroeder, 1978, 1980a; Schroeder and Kinden, 1980; Schroeder, 1980a, 1981a; Hale and Schroeder, 1982). Recently, these techniques have been applied in a novel way

Table III
Fluorescence Probe-Impermeant Quencher Pairs

Fluorescence donor (probe)	Impermeant fluorescence acceptor (quencher)	Reference
1,6-diphenyl-1,3,5-hexatriene	Hemoglobin	Schachter et al., 1982
	Trinitrophenylglycine and trinotrophenyl-N-R	Schroeder, 1978a, 1980a; Schroeder and Kinden, 1980
12-(9-anthroyloxy)stearate	Hemoglobin	Schachter et al., 1982
2-(9-anthroyloxy)stearate	Hemoglobin	Schachter et al., 1982
Pyrene decanoic acid	Hemoglobin	Schachter et al., 1982
Dansylphosphatidylethanolamine	Cytochrome c	Teissie, 1981
Dansylphosphatidylcholine	Cytochrome c	Teissie, 1981
Pyrene decanoic acid	Spin probes	Luisetti et al., 1979
Chlorophyll	Spin probes	Luisetti et al., 1979
n-(9-anthroyloxy) fatty acids	N-Stearoyltryptophan	
Parinaric acid	Trinitrophenylglycine and trinotrophenyl-N-R	Schroeder, 1978a, 1980a; Schroeder and Kinden, 1980; Schroeder et al., 1979a,b; Schroeder and Goh, 1979
Dehydroergosterol	Trinitrophenylglycine and trinotrophenyl-N-R	Schroeder, 1981a Hale and Schroeder, 1982
Cholestatrienol	Trinitrophenylglycine and trinotrophenyl-N-R	Schroeder et al., 1979a,b Schroeder and Goh, 1979

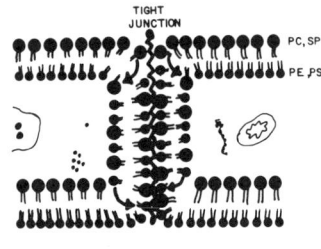

FIGURE 4. Secretion and transport in epithetial cells: the tight junction. Lipophilic molecules cannot diffuse laterally in the outer monolayer through the tight junction unless they can "flop" across the bilayer to the inner monolayer. PC, phosphatidylcholine; PE, phosphatidylethanolamine; PS, phosphatidylserine; SP, sphingomyelin.

to the problem of membrane asymmetry in epithelia (Dragsten et al., 1981, 1982). The tight junction appears to be a barrier to lateral diffusion of membrane-bound lectins and fluorescent lipid probe molecules that cannot flip-flop across the membrane bilayer (Figure 4). In fact, the barrier to lateral diffusion of lipids exists only in the outer monolayer of the membrane at the tight junction (which forms the morphological boundary between apical and basolateral surfaces) and may thereby form a barrier to lateral diffusion of membrane constituents important in secretion (Dragsten et al., 1982).

3. CHOLESTEROL TRANSBILAYER ASYMMETRY

Both the lateral and transbilayer distribution of cholesterol in membranes have been investigated by use of fluorescent sterol probe molecules. This literature has been extensively reviewed (Schroeder, 1984). The asymmetric enrichment of cholesterol in the outer monolayer of biological membranes was first demonstrated more than a decade ago in myelin and red blood cells by X-ray diffraction and freeze-fracture electron microscopy, respectively (Caspar and Kirschner, 1971; Fisher, 1976). Nitroxide-labelled sterols were used as electron spin resonance probes, but a major difficulty has been identification of suitable probe molecules which, when inserted into membranes, accurately mimic the behavior of cholesterol (Trauble and Sackman, 1972; Butler et al., 1970; Hubbel and McConnell, 1971; Mailer et al., 1974; Presti et al., 1982; Presti and Chan, 1982). Cholesterol oxidase did not react with cholesterol in intact biological membranes (Barenholz et al., 1978; Patzer et al., 1978; Gottlieb, 1977; Lange et al., 1981). Exchange methods proved useful primarily in red cells and viruses (Poznansky and Lange, 1978a, 1978b; Lenard and Rothman, 1976), but recent evidence indicates that the rapidly exchangeable and rapidly oxidizable pool of cholesterol in intenstinal brush border membranes may not reflect an inside–outside distribution, but rather an association of the membrane cholesterol with membrane proteins (Bloj and Zilversmit, 1982; Feltkamp and van der Waerden, 1982). In addition, efflux rates of cholesterol are dramatically affected by the structure and composition of the surrounding phospholipids (Bartholow and Geyer, 1982). Maximal efflux occurred with phospholipids having net zero charge [i.e., those normally found in the outer monolayer of eukaryotic plasma membranes (phosphatidylcholine and sphingomyelin)]; sterol release decreased as net charge increased [i.e., with lipids normally found in the inner monolayer (phosphatidylethanolamine, phosphatidylserine, and phosphatidylinositol; see Figure 1)]. Nuclear magnetic resonance methods have also been used (Huang *et al.*, 1974; Opella *et al.*, 1976; Morse *et al.*, 1975).

3.1. Fluorescent Sterol Distribution

In the past five years fluorescent sterols (Figure 5) such as cholestatrienol and dehydroergosterol were popularized to probe the location of sterol in lipoproteins (Schroeder *et al.*, 1979a,b; Patel *et al.*, 1979; Sklar *et al.*, 1980; Hale

NONFLUORESCENT

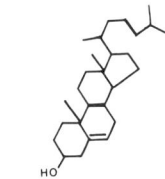
5-CHOLESTENE-3β-OL
(CHOLESTEROL)

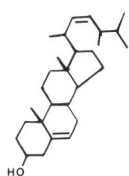
5,24-CHOLESTADIEN-3β-OL
(24-DEHYDROCHOLESTEROL)
(DESMOSTEROL)

FLUORESCENT

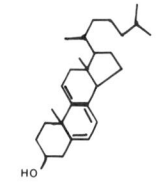
CHOLESTA-5,7,9(11)-TRIENE-3β-OL
(CHOLESTATRIENOL)

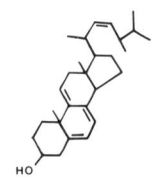
ERGOSTA-5,7,9(11)-22-TETRAENE-3β-OL
(24-DEHYDROERGOSTEROL)
(ERGOSTATRIENOL)

FIGURE 5. Structures of typical membrane nonfluorescent sterols and novel fluorescent analogues.

and Schroeder, 1982; Bergeron and Scott, 1982a,b; Craig *et al.*, 1982), sterol–apoprotein interactions (Smith and Green, 1974a; Schroeder *et al.*, 1979a,b), sterol–sterol carrier protein binding (Fischer *et al.*, 1985b; Schroeder *et al.*, 1985a); transbilayer flip-flop rates in model systems (Smith and Green, 1974b), transbilayer distribution in model membrane systems (Smith and Green, 1974b), lateral interactions between sterol and phospholipids (Schroeder and Thompson, 1985b), lateral ordering of sterols in model membranes (Rogers *et al.*, 1979; Schroeder *et al.*,1985b), and for the first time the transbilayer distribution of sterols in fibroblast and red blood cell plasma membranes (Schroeder, 1981a; Hale and Schroeder, 1982). Both cholestatrienol and dehydroergosterol have now been extensively purified by high performance liquid chromatography and their structure established by mass spectroscopy, proton NMR, and ^{13}C-NMR (Fischer *et al.*, 1985a). Despite these advances, until recently there has been little correlation of the biological properties of these fluorescent analogues of cholesterol with other native sterols. This is especially important since another fluorescent sterol, sterophenol, has a potential difficulty because of keto enol tautomerism in the A ring. In contrast, the fluorescent dehydroergosterol (ergosta-5,7,9(11),22-tetraen-3β-ol) and fluorescent cholestatrienol (cholesta-5,7,9(11)trien-3β-ol) were shown to be similar to cholesterol in lipoprotein structure as determined in circular dichroism (Yeagle *et al.*, 1982a), in altering water permeability of model membranes (Rogers *et al.*, 1979), in lipid packing (Yeagle *et al.*, 1982b), in interactions with digitonin, filipin, amphotericin B (Hale and Schroeder, 1982; Archer, 1975), and in abolishing the phase transition detected by differential scanning calorimetry in dipalmitoylphosphatidylcholine (Hale and Schroeder, 1982). In addition, these fluorescent sterols did not affect the growth of *T. pyriformis* (Rogers *et al.*, 1979), *M. mycoides* (Archer, 1975), or LM fibroblasts in culture (Schroeder, 1981a; Hale and Schroeder, 1982), did not alter membrane-bound enzyme activities of LM fibroblasts (Hale and Schroeder, 1982), and could essentially completely replace desmosterol, the native sterol of LM fibroblasts, without ill effect (Hale and Schroeder, 1982).

When dehydroergosterol was first used to probe the asymmetric distribution of cholesterol in red blood cell membranes (Schroeder, 1981a; Hale and Schroeder, 1982), the results confirmed those determined earlier by freeze-fracture electron microscopy of red blood cell membranes (Fisher, 1976). The outer monolayer of red cell plasma membranes was enriched with cholesterol. Interestingly, transformed cells such as the highly malignant LM fibroblast (Kier and Schroeder, 1981) had exactly the opposite transbilayer distribution of cholesterol as also determined by the fluorescent dehydroergosterol (Schroeder, 1981; Hale and Schroeder, 1982). Using exchange techniques, other investigators recently showed that the rate of release of [^{3}H]cholesterol from normal skin fibroblasts was greater than that from hepatoma cells (Rothblatt and Phillips, 1982). The slowly exchanging pool of cholesterol is believed to reside on the inner monolayer of the plasma membrane. Freeze fracture electron microscopy was used to demonstrate that cholesterol was enriched on the cytoplasmic face of the plasma membrane from insect K_c cells (Silberkang *et al.*, 1983). Most interesting, the same method was used to show that the transbilayer distribution of cholesterol in the lateral plane of guinea pig spermatozoa plasma membrane was not uniform: the cytoplasmic face of only the tail portion was enriched in cholesterol (Elias *et al.*, 1979). These studies point out a number of exciting areas for investigation including both lateral and transbilayer distribution of sterols in biological membranes. In addition, although the sterol content of malignant cell membranes may or may not differ from normal cells, the asymmetric, transbilayer distribution of cholesterol across the plasma membrane may represent an important aspect of cell transformation that may grossly alter the normal physiological functioning of the cell membranes.

3.2. Polyene–Sterol Interactions

In addition to the fluorescent sterols, the use of stopped-flow techniques was introduced to monitor the quenching the filipin *absorbances* when the antibiotic interacts with membrane sterols of mycoplasma (Rottem *et al.*, 1981; Clejan *et al.*, 1981) and red blood cells (Bittman, 1978). The method may, however, encounter some potential difficulties: (1) it has been demonstrated only with nonnucleated cells; (2) it is highly influenced by fatty acid chain length and charge of phospholipids (Bittman *et al.*, 1981; Blau and Bittman, 1977). As discussed in Section 1, the inner monolayer phospholipids of biological membranes appears to contain most of the charged polyunsaturated phospholipids; (3) the interaction is influenced by the structure of the sterol (Schroeder *et al.*, 1971, 1972; Clejan *et al.*, 1981); (4) inactive and active forms of filipin may coexist (Schroeder *et al.*, 1973); (5) membrane-associated proteins affect the formation of filipin–cholesterol complexes (Feltkamp and van der Waerden, 1982; McGookey *et al.*, 1983); (6) association of some polyene antibiotics may occur with phospholipids in the absence of sterol (Boland and Cheron, 1982); (7) absorbance measurements are intrinsically three or more orders of magnitude less sensitive than fluorescence measurements; and (8) it has been demonstrated that polyene antibiotics like filipin do not bind to sterol in liposomes or red blood cells except to the water-soluble cholesterol monomer (Kelly and Bieber, 1981). A 1:1 stoichiometry of high-affinity binding of cholesterol to filipin in aqueous solution was demonstrated over a decade ago (Schroeder *et al.*, 1971; 1972; 1973).

Another interesting use of filipin to probe the transbilayer distribution of cholesterol indicated that the polarity of cholesterol across the membrane bilayer of condensing vacuoles reversed on maturation to zymogen granules (Orci *et al.*, 1980).

3.3. Perturbation of Cholesterol Asymmetry

An interesting development is the availability of synthetic sterol analogues with variable hydrophilic spacer arms. This allows insertion of cholesterol molecules in one monolayer of a membrane and increasing the cholesterol content of that membrane. Such modified sterols include cholesteryl-hemisuccinate, cholesteryl-betainate, cholesteryl-bovine serum albumin, and cholesteryl-phosphoryl-choline (Shinitzky *et al.*, 1979; Lyte and Shinitzky, 1979; Pal *et al.*, 1981). If synthetic glycolipids are attached to the sterol nucleus, agglutinability of unilamellar liposomes can be investigated as well (Rando *et al.*, 1979; Rando and Bangerter, 1979; Slama and Rando, 1980).

Cholesterol oxidase has been used to determine the transbilayer distribution of cholesterol in vesicular stomatitis membranes (Pal *et al.*, 1981) under conditions of cholesterol depletion or enrichment by exchange to serum lipoproteins or PVP-albumin. In the untreated virus, 84% of the cholesterol was in the outer monolayer and 16% was in the inner monolayer. Cholesterol enrichment resulted in preferential increase in inner monolayer cholesterol. Depletion of cholesterol had little effect on the transbilayer distribution. Similarly, the transbilayer distribution of sterol in the plasma membranes of LM fibroblast genetic variants with a 50% reduction in plasma membrane sterol/phospholipid ratio was unaltered (Cowlen *et al.*, 1985).

4. TRANSBILAYER STRUCTURE

4.1. Transbilayer Coupling

As illustrated in Figure 1, the two monolayers of eukaryotic cell plasma membranes are chemically different: unsaturated fatty acids are enriched in the inner monolayer, anionic phospholipids are enriched in the inner monolayer, and cholesterol may be enriched in either monolayer, depending on the cell type. This brings up an interesting possiblity. The fluidity and other structural properties of the two monolayers may be markedly different if the two monolayers are not tightly coupled. Coupling means that lipid motion in one monolayer affects the motion of lipids in the opposite monolayers. Certainly the interior core triglycerides of the very-low-density lipoprotein may affect the surface properties of surface monolayer phospholipids (Hale and Schroeder, 1981). Sphingomyelins containing long-chain fatty acids (24:0) but not those containing shorter-chain fatty acids (18:0) illustrated coupling between monolayers when reconstituted into pure sphingomyelin bilayers (Schmidt *et al.*, 1978). However, with the exception of red blood cells (Figure 2), sphingomyelin is not the major phospholipid species in the plasma membrane and large areas of transbilayer coupling would not be expected. In fact, enrichment of the outer monolayer of red blood cell membranes with cholesterol decreases the fluidity of only the outer membrane leaflet (Flamm and Schachter, 1982) as monitored by fluorescence probe molecules. The use of nuclear magnetic resonance and impermeant paramagnetic quenching ions (Mn^{2+}, Cu^{3+}, Pr^{3+}, or Yb^{3+}) indicate that the two halves of the bilayer in phospholipid membranes are so weakly coupled that they undergo the gel to liquid-crystalline phase transition independently (Sillerud and Barnett, 1982; Bergelson and Barsukov, 1977; Bystrov *et al.*, 1971).

4.2. Membrane Depth Gradients

A variety of anthroyl-, dansyl-, or pyrene rings attached at various points along fatty acid chains has been used as fluorescence probes of membrane lipid motion, fluidity, or microviscosity with increasing depth into the center of the bilayer (Browing and Nelson, 1979; Thulborn *et al.*, 1979; Haigh *et al.*, 1979; Vincent *et al.*, 1982). These molecules indicated that as the fluorophore is moved deeper toward the center of the bilayer, it experiences greater rotational freedom because of increased in-plane rotation. These findings are perfectly in accord with results using spin labels attached at different depths along fatty acyl chains (McConnell and McFarland, 1972; Keith *et al.*, 1973; Morse *et al.*, 1975). Interestingly, both fluorescence probe molecules in blood platelets (Rotman and Heldman, 1981) and spin-label probes in sarcoplasmic reticulum (Morse *et al.*, 1975) have demonstrated that the intracellular fluid viscosity near the inner monolayer polar head groups is greater than the extracellular fluid viscosity near the outer monolayer polar head groups.

4.3. Transbilayer Fluidity Gradients

Since the two monolayer halves of biological membranes do not seem to be tightly coupled and because they have distinct lipid compositions and are asymmetric, it would seem probable that the two halves of the bilayer have different structural properties. Using differential nitroxide spin-labeled fatty acids, early investigators in this field demonstrated symmetry of probe motion across the sarcoplasmic reticulum membrane bilayer (Morse *et al.*, 1975) while others showed that the outer monolayer of red blood cell membranes was more rigid than the inner monolayers (Tanaka and Ohnishi, 1976). However, the latter investigators did not take into account the inherently different motional properties of phosphatidylcholine and phosphatidylserine spin-label probes. The greater rigidity of the red blood cell membrane outer monolayer has since been confirmed, also by ESR techniques (Seigneuret *et al.*, 1984). Differential scanning calorimetry of mixtures of inner versus outer monolayer red blood cell phospholipids indicated that the outer monolayer lipids were more rigid (Van Dijck et al., 1976). However, these investigators did not take into account the transbilayer location of cholesterol in the red blood cell membrane. Last, using enveloped viruses and phagosomal membranes to represent right-side-out and inside-out membranes, respectively, in conjunction with nonflipping spin-labeled lipid probes, Wisnieski and Iwata (1977) concluded that the outer monolayer of the LM plasma membrane was probably more fluid. These conclusions were based on the assumption that virus and phagosomal membranes represent random surface membrane areas of the parent cell. This assumption does not appear to be valid in all cases (Op den Kamp, 1979; Schroeder, 1981b, 1982a; Kier and Schroeder, 1983). Thus, the early investigations using spin-labeled probes or differential scanning calorimetry were not convincing with regard to the existence of transbilayer fluidity gradients in biological membranes.

In contrast, fluorescence probes have provided some of the first convincing data of differences in fluidity between bilayer halves of plasma cell membranes. Using the fluorescence probe *trans*-parinaric acid in conjunction with impermeant trinitrophenyl quenching agents, it was demonstrated that the outer monolayer of LM fibroblasts was more fluid than the inner monolayer (Schroeder, 1978; Schroeder and Kinden, 1979; Schroeder, 1980a). This finding was confirmed by using four different fluorescent probe molecules with and without two types of fluorescence quenching agents, trinitrophenylglycine and trinitrophenyl covalently attached to one side of the membrane only (Fontaine and Schroeder, 1979), as well as by use of nonflipping glucosamine-parinarate probe molecules in conjunction with intact cells versus phagosomal membranes. The theoretical basis of the method stems from fluorescence quenching theory and excitation energy transfer (Stryer, 1978; Fung and Stryer, 1978; Slavik, 1982). If appropriate fluorescence donor and acceptor pairs are chosen (e.g., if Ro, the distance between the donor and acceptor at which the rate constant for transfer is equal to the rate constant for spontaneous deactivation, is near 10 Å) then the efficiency of energy transfer (which decreases as a function of r^{-6}, where r is the distance between the donor and acceptor) is sufficient to quench fluorescence effectively in only one monolayer of the membrane bilayer if both donor and acceptor were present in the same monolayer. Energy transfer between donor and acceptor in opposite monolayers would be insignificant (Vanderkooi and McLaughlin, 1975; Stryer, 1978; Fung and Stryer, 1978; Schroeder, 1978; Haigh et al., 1979; Schroeder and Kinden, 1979; Schroeder, 1980a; London and Feigenson, 1981a, 1981b; Schroeder, 1981a; Hale and Schroeder, 1982). Collisional quenchers have also been used for this purpose (Wade et al., 1978). The most important features of the quenching of probe fluorescence in one monolayer is that it be nontrivial and complete. The fact that neither the emission curve nor fluorescence lifetimes shift confirms that these criteria are satisfied (Schroeder and Kinden, 1979; Schroeder, 1980a, 1981a; Hale and Schroeder, 1982). Recent results of other investigators using impermeant and permeant fluorophores in conjunction with heme quenching (Cogan and Schachter, 1981; Schachter *et al.*, 1982) also demonstrated that the outer monolayer of red blood cell membranes of some species was more fluid than the inner monolayer. In summary, impermeant fluorescence probes or probes suitably used in conjunction with fluorescence quenchers have demonstrated that a difference in motional properties of lipids (fluidity gradient) exists between the two monolayers of biological cell plasma membranes. This phenomenon has not yet been investigated in subcellular membranes except for phagosomes (Wisnieski and Iwata, 1977; Schroeder and Kinden, 1979; Schroeder, 1980a).

5. PHASE SEPARATIONS OF LIPIDS

5.1. Lateral Rearrangements

In addition to transbilayer membrane structural asymmetry, lateral lipid asymmetry or lateral phase separation of lipids occurs. This phenomenon is covered in many reviews (Slavik, 1982) and only a few newer developments will be described here. For example, calcium-induced lateral phase separations in membranes (Mayer and Nelsestuen, 1981) have recently been investigated using NBD-labeled phospholipids (Hoekstra, 1982). Fluorescence self-quenching of NBD occurred because of a Ca^{2+}-induced separation of lipid phases. Fluorescence photobleaching recovery has been used to determine the lateral diffusion of fluorescent lipid analogues and has demonstrated lateral lipid organization alterations by cholesterol (Jacobson *et al.*, 1981). Probably the most important recent development with fluorescent probes has been the quantitative formulation of methods to describe the lateral organization of fluid and solid lipid areas. This advance has come through use of *trans-* and *cis-*parinaric acid probe molecules in conjunction with fluorescence lifetime analysis. The parinaric acid probes were first introduced as natural fluorescent analogues of fatty acids (Figure 6) by Gunstone and Subbarrao (1967) and later used in model systems (Sklar *et al.*, 1975, 1976, 1977a,b,c, 1979), mammalian membranes (Schroeder *et al.*, 1976a,b; Rintoul and Simoni, 1977; Schroeder and Holland, 1978; Schroeder, 1978a, 1978b; Rintoul *et al.*, 1978; Schroeder *et al.*, 1979a, 1979b; Waring *et al.*, 1979; Sklar *et al.*, 1979a, 1979b; Schroeder, 1980a, 1982a,b; Schroeder and Kinden, 1980; Schroeder *et al.*, 1982a,b; Schroeder, 1982c, 1983; Schroeder and Soler-Argilaga, 1983), bacterial membranes (Tecoma *et al.*, 1977; Fraley *et al.*, 1978) and serum lipoproteins (Schroeder and Goh, 1979; Schroeder *et al.*, 1979a,b; Sklar *et al.*, 1980; Hale and Schroeder, 1981; Schroeder *et al.*, 1982c).

cis-PARINARIC ACID

trans-PARINARIC ACID   FIGURE 6. Structure of the fluorescent parinaroyl fatty acids.

Although the presence of breakpoints in Arrhenius plots of biophysical probe parameters (fluorescence or electron spin resonance) may be indicative of phase alterations, the existence of fluid and solid domains in many plasma membranes has not been adequately demonstrated. Breakpoints in Arrhenius plots may also be due to the presence of microdomains, clusters, or boundary lipid (Marsh et al., 1976; Watts et al., 1979) as well as lateral phase alterations. Fluorescence methods using parinaric acids were derived to demonstrate the coexistence of both fluid and solid phases and to quantitate the extent of these domains in model membranes (Sklar et al., 1977c, 1979a), retinal rod outer segment membranes (Sklar et al., 1979b, 1979c), and rat liver plasma membranes (Schroeder, 1983; Schroeder and Soler-Argilaga, 1983). The partitioning of fluorescent parinaric acid probe molecules between coexisting solid and fluid phases can be expressed by:

$$K_P^{s/f} = K_P^{2/1} = (X_2^P/X_2)/(X_1^P/X_1) \tag{1}$$
$$K_P^{2/1} = (X_2^P/X_1^P) \cdot (X_1/X_2) \tag{2}$$

where $X_2^P + X_1^P = 1$ represents all of the bound probe. X_2 and X_1 represent the fractions of solid and fluid phase. X_2^P and X_1^P are determined from multiple fluorescence lifetime analysis. The longer lifetime component, τ_2, has been correlated to the probe molecule in solid phase, while the shorter lifetime component, τ_1, was correlated to the probe molecule in fluid phase (Sklar et al., 1977a). The parameters X_2 and X_1 may be obtained from a phase diagram in simple model systems (Sklar et al., 1977a,b), but in biological membranes this was not possible. However, other investigators (Sklar et al., 1977b) have shown that:

$$K_p^{s/f} = K_p^s/K_p^f \tag{3}$$

Biological membranes such as liver plasma membranes, for example, undergo a phase alteration with beginning and end points near 18° and 31°C (Watts et al., 1979; Schroeder, 1983). Thus, at temperatures considerably above 31°

(i.e., 37°) and below 19° (i.e., 10°) the liver plasma membranes may be considered to be mostly in the fluid and solid state, respectively. It was demonstrated that even 10° above the phase transition *trans*-parinarate is sensitive to the presence of minute amounts (<1%) of solid lipid clusters (Sklar *et al.*, 1979a). For the present calculations this small percentage of solid will not be considered. Determination of K_P at 10° and 37° by a direct binding assay for *trans*-parinaric and for *cis*-parinaric acid would approximate K_p^s and K_p^f, respectively, for each probe. The $K_p^{s/f}$ calculated from *trans*- and *cis*-parinaric acid by Eq. (3) was 3.30 for *trans*-parinaric and 0.92 for *cis*-parinaric acid in liver plasma membranes. These values were similar to those found in simple phospholipid model systems (Sklar *et al.*, 1977); the K_p was 2.94 for *trans*-parinarate and 0.59 for *cis*-parinarate. Thus, *trans*-parinarate partitioned preferentially into solid areas of liver plasma membranes while *cis*-parinarate partitioned nearly equally between solid and fluid lipids. Therefore, these ratios may be inserted into Eq. (4) to determine X_2/X_1:

$$\left(\frac{X_2}{X_1}\right) = \left(\frac{X_2^p}{X_1^p}\right) \times \left(\frac{1}{K_p^{s/f}}\right) \qquad (4)$$

Then the mole fraction of each phase, X_1 and X_2, can be simply calculated using $(X_1 + X_2 = 1)$.

The above data treatment requires that parinaric acid isomers decay monoexponentially in pure solvents, a requisite verified by a pulsed lifetime study using synchrotron radiation (Wolber & Hudson, 1981). However, the latter finding was most recently contradicted by a multifrequency phase flourometry investigation (Parasossi *et al.*, 1984).

5.2. Independent Monolayers

Lateral phase separations of lipid can occur in the outer, the inner, or both monolayers of biological as well as model membranes since the two monolayers are poorly coupled. Considerable evidence for independent modulation of phase transitions in the two membrane halves has been acquired through use of fluorescence probes. The fluorescence probe NBD-phosphatidylethanolamine when inserted into the outer monolayer of red blood cells exhibited linear Arrhenius plots of the lateral diffusion coefficient determined by fluorescence photobleaching recovery (Henis *et al.*, 1982). In contrast, labeling on both sides, however, yielded a breakpoint near 30°C. Other investigators using dansylphosphatidylethanolamine revealed discontinuities in Arrhenius plots of dansyl fluorescence in red blood cell membranes for the inner monolayer at 35, 24, and 10°C, but none for the outer monolayer (Rimon *et al.*, 1980). In rod outer segment disc membranes the primary thermal reorganization extends below 5°C and is attributed to phosphatidylethanolamine and phosphatidylserine, which are thought to be largely in the outer (cytoplasmic) monolayer (Sklar *et al.,* 1979b,c); the inner monolayer did not display a thermal transition, possibly because of enrichment of cholesterol in this monolayer. The fluorescence probe molecules, *trans*-parinaric acid and glucosamine-*trans*-parinarate, displayed five breakpoints in Arrhenius plots near 17, 24, 30, 35, and 41°C for each of the two monolayers of the LM fibroblast plasma membrane (Schroeder, 1980). Using electron spin resonance probes, enveloped viruses, and phagosomal membranes from LM fibroblasts, Wisnieski and Iwata (1977) concluded that two breakpoints in Arrhenius plots near 14 and 33° were in the outer monolayer while two others near 23 and 38°C were in the inner monolayer of LM plasma membranes.

In addition to the two monolayers of cell membrane having independent lateral phase separation, it is possible that different patches or areas of membrane in a single cell type may not have the same lateral phase separations. For example, the rat intestinal enterocyte microvillar membrane has a reversible lateral phase separation between 23 and 31° as detected by fluorescence probes (2-anthroylstearate, 12-anthroylstearate, dansylphosphatidylethanolamine, diphenylhexatriene, and retinol), differential scanning calorimetry, or breakpoints in Arrhenius plots of membrane-bound enzyme activities (Brasitus *et al.,* 1980; Brasitus and Schachter, 1980). In contrast, the basolateral membrane area had a lateral phase transition at a slightly higher temperature range, 27–40°C. In addition, the microvillus membrane was less fluid than the basolateral membrane.

6. PHYSIOLOGICAL FUNCTION OF ASYMMETRIC STRUCTURE OF MEMBRANES

Although we know that the phenomenon of membrane lipid asymmetry exists in biological membranes, its origin, regulation, and function are poorly understood. Recent advances in our understanding of bilayer transmembrane structure have allowed correlations of membrane lipid asymmetry to be made with certain physiological membrane functions.

6.1. Receptor Modulation

At least three types of receptor–membrane lipid interactions are possible: (1) the affinity of the receptor for its ligand may be modulated by alterations in membrane lipid. This is best illustrated by modulation of the insulin receptor by cholesterol (Figure 7). Decreased insulin receptor number and affinity for insulin in the liver plasma membrane may be due to increased membrane rigidity caused by elevated membrane cholesterol or decreased unsaturated fatty acid levels (Luly and Schinitzky, 1979; McCaleb and Donner, 1981; Ginsburg

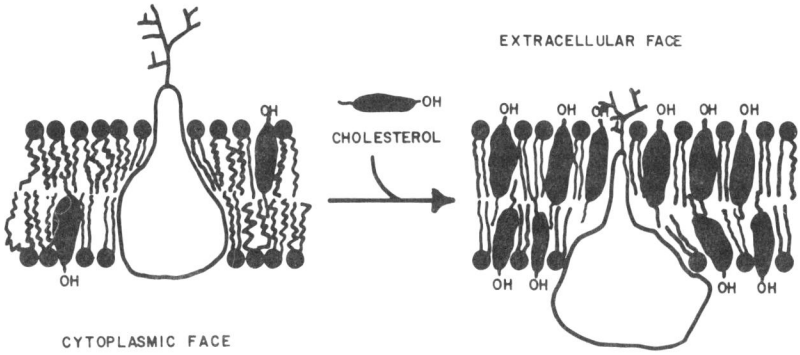

FIGURE 7. Potential modulation of the insulin receptor number and/or binding affinity by cholesterol.

et al., 1981; Gould *et al.*, 1982). This effect may be rationalized by changes in the vertical displacement of membrane proteins induced by changes in lipid fluidity. Such alterations in fluidity could be elicited by changes in lipid composition and/or transbilayer distribution. The latter mechanism would not involve lipid compositional alterations but would simply involve an alteration in transbilayer lipid distribution. The passive modulation of receptor protein exposure by altered lipid composition and/or distribution (laterally or vertically) has also been invoked for several other receptors, including blood group antigens (Shinitzky and Souroujon, 1979), platelet phospholipase (Kramer *et al.*, 1982), (Ca^{2+} + Mg^{2+})ATPase activity (Golo *et al.*, 1981), generation of platelet agglutination by transbilayer redistribution of phosphatidylserine (see Figure 8; Bevers *et al.*, 1982), and enhancement of tumor cells to antibody-C-mediated killing (Schlager, 1982).

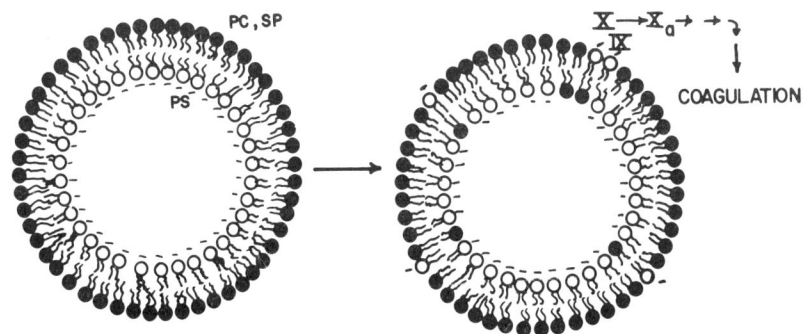

FIGURE 8. Potential role of platelet phosphatidylserine asymmetry in platelet coaggulation.

(2) A second type of modulation is one in which the receptor binding ligand elicits a direct structural change in the membrane (Kury *et al.,* 1974; Chen and Hoch, 1977; David *et al.,* 1978; Dipple and Houslay, 1978; Lee *et al.,* 1978; Luly and Shinitzky, 1979; Schroeder, 1982b). Such direct effects may be due to intracellular redistribution of Ca^{2+} (Shlatz and Marinetti, 1972; Andia-Waltenbaugh *et al.,* 1978; Kiss, 1979; Luly and Shinitzky, 1979; Rahwan *et al.,* 1979; Livingstone and Schachter, 1980; Schroeder, 1982b; Schroeder and Soler-Argilaga, 1983). Insulin (10^{-8}–10^{-6} M) decreased Ca^{2+} binding to liver plasma membranes by 10–30%, while glucagon increased Ca^{2+} binding 80–100% (Shlatz and Marinetti, 1972). Epinephrine and hydrocortisone also increased Ca^{2+} binding (Shlatz and Marinetti, 1972). As shown elsewhere, the uptake of fatty acids or fluorescent fatty acid analogues by hepatocytes, liver plasma membranes, or KB cells is modulated by Ca^{2+} (Moskal *et al.,* 1977). Insulin (Nelson, 1980) and thyroid hormone (Hulbert, 1978) may also have indirect effects on cell membrane fluidity. These effects appear to be mediated via altered fatty acid composition of membrane phosphoglycerides (Peifer, 1968; Platner *et al.,* 1972; Chen and Hoch, 1977; Faas and Carter, 1981) and can be accounted for at least in part by altered fatty acid desaturase enzymes (Faas and Carter, 1981) or altered secretion of very-low-density lipoproteins (Keyes and Heimberg, 1979; Schroeder *et al.,* 1981).

(3) A third possibility is that a receptor binding agonist alters the structure of the membrane and thereby alters the exposure of a secondary protein or receptor. Such a mechanism has been demonstrated for enhanced antigen binding in T cell membranes induced by lymphocyte activating factor (Puri *et al.,* 1980) and for bradykinin modification of prolactin binding in hepatic membranes (Davis *et al.,* 1981).

6.2. Receptor–Adenylate Cyclase Coupling

It has been recognized for some time now that, based on fluorescence photobleaching recovery measurements, certain receptor proteins are free to diffuse laterally in the plane of the plasma membrane. The mobile receptor model developed separately by Cuatrecasas (Bennet *et al.,* 1975; Cuatrecasas, 1974; Cuatrecasas and Hollenberg, 1975, 1976; Jacobs and Cuatrecasas, 1976; 1977), Dumont (Boeynaems and Dumont, 1977), and DeHaen (1976) hypothesizes that a receptor–hormone complex in one monolayer may interact with a number of effectors distributed in the opposite monolayer or in the same monolayer of the membrane. This accounts for the ability of a variety of specific receptor binding agonists in the outer membrane monolayer to independently stimulate a unique adenylate cyclase enzyme rather than having each receptor clustered at the cyclase or physically linked to the cyclase (Hollenberg, 1979).

One of the most exciting findings in this field has been the recent demonstration that the transbilayer fluidity gradient (determined by fluorescence

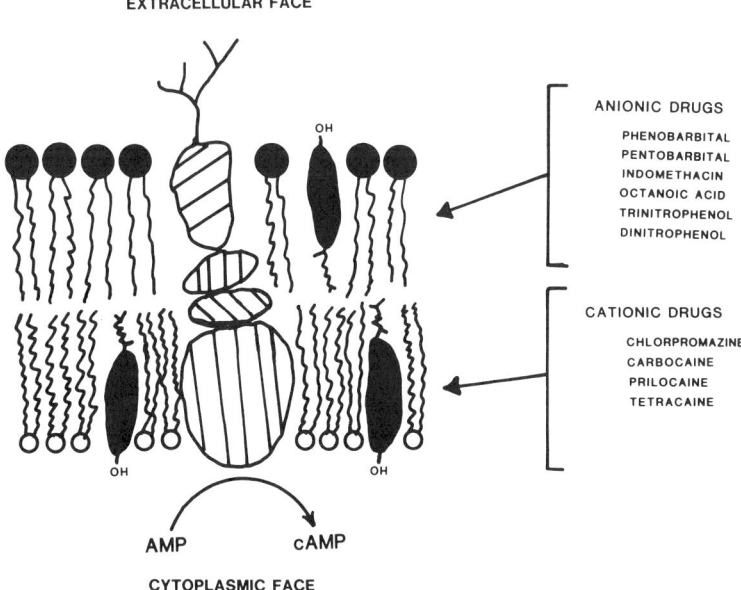

FIGURE 9. Effect of membrane-perturbing agents on receptors or adenylate cyclase. The branched material on the extracellular face is polysaccharide.

(see Figure 9; Salesse and Garnier, 1979; Rimon et al., 1980; Houslay et al., 1980, 1981; Salesse et al., 1982a, 1982b; Gwynne and and Toft, 1982). In rat liver plasma membranes, for example, a lipid phase separation occurs with an onset temperature near 28°C and is believed to be due to "clustering of lipid" (Gordon et al., 1978, 1980) in the outer monolayer of the lipid bilayer (Houslay, 1979). This phase transition appears to occur in cholesterol-poor regions of the outer monolayer (Houslay and Palmer, 1978; Dipple and Houslay, 1979; Kremer et al., 1973). Cholesterol appears to be enriched in the inner monolayer of fibroblast membranes (Schroeder, 1981; Hale and Schroeder, 1982). Phenobarbital, an anionic drug that selectively interacts with the outer monolayer of the plasma membrane (fluidizes it), alters the glucagon-ligand interaction, while prilocaine, which interacts with the inner monolayer, causes an inner monolayer phase separation at 11°C that affects the adenylate cyclase component. In general, cationic drugs (chlorpromazine, tetracaine, *n*-octylamine, carbocaine, and prilocaine) did not abolish or alter ligand–receptor binding by interaction with the inner monolayer of the membrane but altered adenylate cyclase activity. In contrast, anionic drugs (di- and trinitrophenols, indomethacin, octanoic acid, phenobarbital, and pentobarbital) affected ligand–receptor interactions in the outer monolayer but did not modify the adenylate cyclase activity stimulated by the hormone or ligand. Such a control mechanism is important not only for the receptor–adenylate cyclase system but also for the activity of 5′-nucleotidase, which resides primarily in the outer monolayer (Dipple et al., 1982). These observations display the important role of an asymmetric control of the adenylate cyclase system in the membrane by the two halves of the bilayer.

6.3. Phospholipid Methylation and Receptor Functions

As indicated in Section 1, phosphatidylethanolamine is enriched in the inner monolayer of cell surface plasma membranes. This phosphatidylethanolamine can be converted to phosphatidylcholine by two membrane-bound methyltransferase enzymes that concomitantly translocate the phospholipid molecule across the membrane bilayer (Figure 10; Hirata et al., 1978). The first enzyme, methyltransferase I, in the pathway is rate limiting and converts phosphatidylethanolamine to phosphatidylmonomethylethanolamine. The second enzyme, methyltransferase II, sequentially adds two more methyl groups to form phosphatidyl(N, N')dimethylethanolamine, and phosphatidyl-N,N',N''-trimethylethanolamine (or phosphatidylcholine). S-adenosyl-methionine is the methyl donor for this reaction. This enzymatic methylation produces phospholipids that are more fluid (Hirata and Axelrod, 1978; Vaughan and Keough, 1974; Blume and Ackerman, 1974). However, the total methylation usually only accounts for less than 0.1% of the membrane phospholipid. In contrast, other investigators have demonstrated that as much as 40–50% of membrane phospholipids may be replaced by phosphatidyl-monomethylethanolamine or

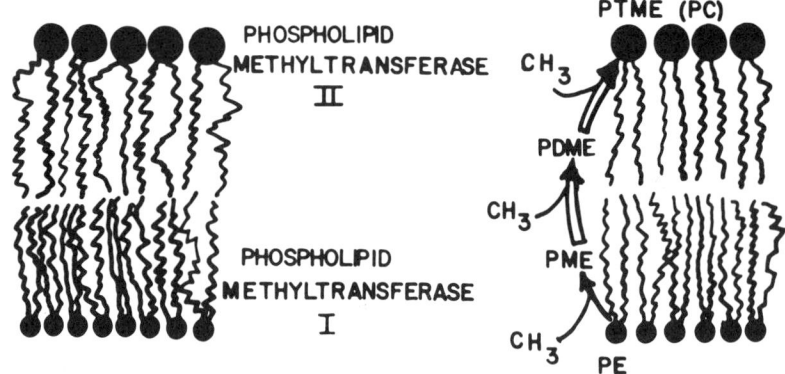

FIGURE 10. Phospholipid methylation in the plasma membrane. PTME, phosphatidyl-N,N',N''-trimethylethanolamine or phosphatidylcholine; PDME, phosphatidyl-N,N'-dimethylethanolamine; PME, phosphatidylmonomethylethanolamine; PE, phosphatidylethanolamine.

phosphatidyl-N,N'-dimethylethanolamine without altering membrane structure (Schroeder *et al.*, 1976a,b; Schroeder, 1978b, 1982c). Membrane fluidity appears to be maintained by isothermal regulation mechanisms (Schroeder, 1978b). The methylation of phosphatidylethanolamine has been correlated to β-receptor stimulation of adenylate cyclase (see Strittmatter *et al.*, 1981, for a review). Yet other results indicate that such a relationship is absent (Colar and Breton, 1981). There is evidence that the peripheral-type benzodiazepine binding site stimulates phospholipid methylation of C_6 astrocytoma cells (Strittmatter *et al.*, 1979). In another tumor cell line LM fibroblasts the existence of the peripheral benzodiazepine receptor and the methylation pathway has been demonstrated (Feller *et al.*, 1983; Maeda *et al.*, 1981), but the two were not coupled (Feller *et al.*, 1983). Enhanced phospholipid methylation has also been related to histamine release (Crews *et al.*, 1981), induction of liver microsomal cytochrome P-450 by phenobarbital and 3-methylcholanthrene (Sastry *et al.*, 1981); viral transformation of cultured cells (Maziere *et al.*, 1981); and concanavalin A stimulated histamine release in mast cells (Hirata *et al.*, 1979). In contrast, decreased methylation of phospholipids has been associated with macrophage chemotaxis (Pike *et al.*, 1979). A number of problems associated with these studies have recently been pointed out (Audubert and Vance, 1983).

6.4. Role of Cholesterol Asymmetry in Secretion, Endocytosis, Transport, and Aging

Alterations in sterol asymmetry may be important in a number of processes, including

(1) *Secretion.* Unequal partitioning of cholesterol between the bilayer leaflets occurs during the transformation of the Golgi condensing vacuole into the mature zymogen granule (Orci *et al.*, 1980a). By analogy, such redistribution of cholesterol may occur in endocytic membranes (Schroeder, 1982a) and in other secretory vesicles such as synaptic vesicles, important to secretion of neurotransmitters (Moroni *et al.*, 1976; Sun, 1976; Perrelet *et al.*, 1982).

(2) *Receptor function.* Vertical displacement of membrane proteins such as the insulin receptor, concanavalin A receptor, and the glucose transport system is affected by the cholesterol content of the membrane (Yuli *et al.*, 1981; Borochov and Shinitzky, 1976; Borochov *et al.*, 1979; Shinitzky and Rivnay, 1977). An important role for an unequal distribution of cholesterol among membranes has also been determined for the acetylcholine receptor (Perrelet *et al.*, 1982).

(3) *Endocytosis.* Alterations in sterol asymmetry may be crucial in endocytosis (Figure 11). In fact, specialized areas of fibroblast cell membranes are involved in phagocytosis of latex beads (Schroeder 1981b, 1982a; Schroeder and Kinden, 1983; Montessano *et al.*, 1974). These membrane areas may be enriched in sterol (Schroeder, 1982a; Seed and Kreier, 1972; Beach *et al.*,

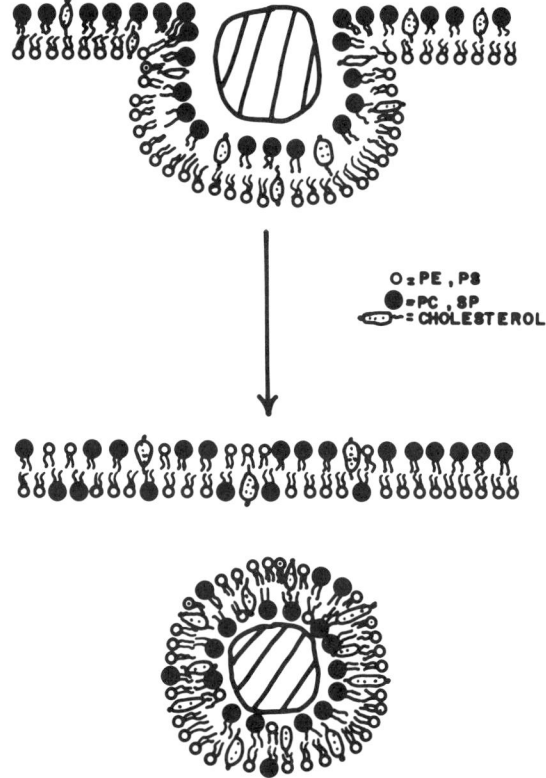

FIGURE 11. Endocytosis occurs at specialized membrane areas and/or rearrangement of lipids occurs.

1977), and the extent of endocytosis is also closely regulated by cholesterol, phospholipid, and fatty acid composition (Schroeder, 1981b; Kier and Schroeder, 1983a; Heininger and Marshall, 1979; Roberts and Quastel, 1963; Mahoney *et al.*, 1977; Schroit and Gallily, 1979).

(4) *Aging.* The potential role of lipid asymmetry and lipid peroxidative alterations in aging has recently been reviewed (Schroeder, 1985). It is well known that the sterol content of plasma and of cell membranes is highly dependent on age (Rivnay *et al.*, 1980). An enrichment of cholesterol occurs in the outer monolayer of aging *M. Gallisepticum* (Bittman *et al.*, 1981). Membrane cholesterol and plasma cholesterol interrelationships of human red blood cells are also altered as a function of age. Older red blood cells had consistently less uptake of [^{14}C]cholesterol from plasma in comparison with young erythrocytes (Jain and Shohet, 1982). Decrease in sterol–phospholipid ratio of red blood

cells and of *M. mycoides* with age drastically altered membrane architecture such that phosphatidylethanolamine (Gupta *et al.*, 1981), phosphatidylglycerol, and cardiolipin (Rigaud and LeBlanc, 1980), which are normally inaccessible to hydrolysis by phospholipase A_2, become hydrolyzed by the enzyme.

In summary, the asymmetric distribution of cholesterol in cell membranes can be determined by use of fluorescent sterol analogues. This asymmetric arrangement of cholesterol may be important in regulation of a variety of transmembrane processes. Facile determination of sterol location in membranes may provide excellent potential for investigation of membrane disorders in which regulation of the asymmetric transbilayer distribution of cholesterol in membranes is abnormal.

7. POTENTIAL ARTIFACTS

A number of potential artifacts may complicate use of fluorescent probe molecules in determining transbilayer membrane structure.

7.1. Membrane Preparation

The use of whole cells will not generally allow determination of membrane transbilayer asymmetry since the inner monolayer is inaccessible. The use of viral membranes and phagosomal membranes to represent right-side-out and inside-out plasma membranes (Wisnieski and Iwata, 1977; Sandra and Pagano, 1978; Kramer and Branton, 1979; Op den Kamp, 1979; Schroeder, 1980a, 1981a; Hale and Schroeder, 1982) may not be appropriate (Schroeder, 1982a) since these membranes do not represent random areas of the cell surface membrane. Inside-out membranes prepared from red blood cell ghosts may have an altered structure as compared to the intact cell (Steck *et al.*, 1971; Wise, 1982; Bloj and Zilversmit, 1976). Affinity chromatography of plasma membrane vesicles may not separate right-side-out from inside-out fractions but instead yield different right-side-out membrane subpopulations (Schroeder *et al.*, 1982).

7.2. Probe Location

Many probe molecules (e.g., perylene, diphenylhexatriene, parinaric acid) will partition throughout the intact cell such that measurements of fluorescence reveal only average values for all cell membranes (plasma membranes + microsomes + mitochondria + lipid droplets + etc.) and will not reveal anything about transbilayer structure (Pagano *et al.*, 1977; Esko *et al.*, 1977; Moskal *et al.*, 1978; Radda and Vanderkooi, 1972; Kaye and West, 1967). Therefore, it is vital either that cell fractionation be performed to isolate plasma membrane subfractions or that nonflipping fluorescence probes such as glucosamine parinarate or others be used (Schroeder, 1981a).

7.3. Quenching and Other Artifacts of Fluorescence Determinations

Several potential artifacts are possible because of fluorescence quenching. First, local anesthetics used to fluidize the inner or the outer monolayer of a membrane may quench intramembranous fluorescence probes (Surewicz and Leyko, 1982). Similar results have been obtained with pesticides (Lakowicz *et al.*, 1977; Lakowicz and Hogan, 1977) and antidepressive drugs (Romer and Bickel, 1979). Second, fluorescence quenching can lower the fluorescence lifetime of a probe molecule, thereby increasing polarization values, without change in rotational correlation times (Schroeder, 1980a). This is an especially important variable to consider when determining fluidity gradients (Schroeder, 1978a; Schroeder and Kinden, 1980; Schroeder, 1980a). The fluorescence lifetime of probe molecules will be substantially reduced if only partial quenching of fluorescence by trinitrophenyl groups covalently linked to the plasma membranes occurred. Under the conditions used by Schroeder (1980a), the outer monolayer fluorescence probe molecules are maximally quenched because of a large excess of trinitrophenyl groups covalently attached to the outer monolayer (Figure 12). If as little as 10% of the inner monolayer fluorescence probe molecules were quenched by trinitrobenzenesulfonic acids that had penetrated

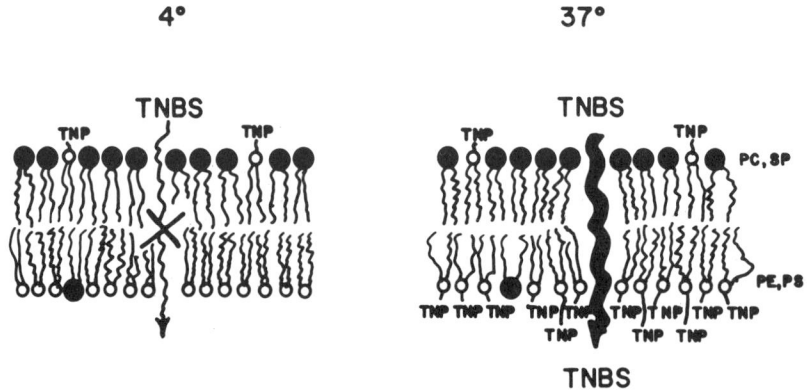

FIGURE 12. Labeling of plasma membrane amino groups with trinitrobenzene–sulfonic acid under penetrating (37°C) and nonpenetrating (4°C) conditions. TNBS, trinitrobenzenesulfonic acid; TNP, trinitrophenyl groups covalently attached to membrane aminolipids and other primary amino containing molecules.

to the inner monolayer, the fluorescence lifetime would be significantly reduced. This was not the case with mouse fibroblasts (Schroeder, 1978a, 1980a, 1981a; Hale and Schroeder, 1981; Schroeder and Kinden, 1980).

Light scattering is an important artifact in fluorescence measurement of probes incorporated into cell membrane vesicles. This problem can become acute in fluorescence lifetime measurements. Therefore, all fluorescence lifetimes should be determined using appropriate cutoff filters, a 55° polarizer in the emission beam, and a reference solution. The fluorescence lifetimes should be measured simultaneously relative to a reference solution such as dimethyl-p-bis[2-(5-phenyloxazolyl)] benzene (i.e., dimethylPOPOP) in absolute ethanol as opposed to the usual glycogen scattering solution (Lakowicz et al., 1981). This solution minimizes wavelength and geometry-dependent time response of the photomultiplier tubes. The fluorescence lifetime of dimethyl POPOP in ethanol at 24° is 1.45 nsec, independent of emission wavelength or temperature (Lakowicz et al., 1981). Because of this lifetime (τ_R) the phase angle of dimethyl POPOP (θ_R) lags behind the exciting light by 3.13° at 6 MHz, 9.31° at 18 MHz, and 15.29° at 30 MHz. Using $\tan \theta = w\tau_P$ phase angles can be corrected for these shifts and are, therefore, absolute phase angles relative to the phase of the modulated excitation.

The use of steady-state fluorescence polarization values to calculate rotational correlation times is subject to certain limitations. Steady-state instruments will not provide the appropriate values necessary to discriminate between rate and range of wobbling of a fluorescence molecule. The rationale used is as follows: Alterations in polarization can be converted to rotational correlation times, $\overline{P}$, which can be calculated from the Perrin equation (Weber, 1953; Shinitzky and Inbar, 1974):

$$\left(\frac{1}{p} - \frac{1}{3}\right) = \left(\frac{1}{p_0} - \frac{1}{3}\right)\left(1 + \frac{3\tau}{\overline{P}}\right) \qquad (1)$$

where p is the polarization, τ is the lifetime of the excited state, and p_0 is the polarization in the absence of rotational motion: $p_0 = 0.492$ for 1,6-diphenyl-1,3,5-hexatriene (Lakowicz et al., 1979); $p_0 = 0.490$ and 0.479 for trans- and cis-parinaric acid, respectively (Wolber and Hudson, 1981). This equation applies to nonspherical molecules when the absorption and emission oscillators are parallel (Kinosita et al., 1981). The anisotropy r is equivalent to $2p/(3 - p)$, where p is the polarization. However, it should be noted that the steady-state anisotropy, which is calculated according to the preceding equation, actually depends on both rate and range of wobbling motion of the probe molecule, as well as on the fluorescence lifetime (Wolber and Hudson, 1981; Kinosita et al., 1981). The steady-state anisotropy r^s can be expressed as the sum of a kinetic term $r^s(t)$ and a static term, r_∞. Differential polarized phase fluorometry can be used to obtain the rotational rate and limiting anisotropy of fluorescence probes in membranes (Weber, 1978; Lakowicz et al., 1979). Determination of rotational relaxation times or microviscosity according to steady-state techniques (Weber, 1953; Shinitzky and Inbar, 1974) is limited, since the steady-state anisotropy used in these calculations is comprised of two components: r_∞, the limiting anisotropy, which is determined primarily by the degree to which the rotations are hindered, and R, the rotational rate (Lakowicz et al., 1979; Kinosita et al., 1981; Kawato et al., 1977). The method of Weber (1978) as extended by Lakowicz et al. (1979) can be used to obtain the limiting anisotropy and rotational rate directly as follows:

$$r_\infty = r + (r - r_0)/6\, R\tau \qquad (2)$$

where r_∞ is the limiting anisotropy, r is the steady-state anisotropy, r_0 is the anisotropy in the absence of rotational motion, τ is the lifetime, and R is the rotational rate in rad/sec. The relaxation time expressed in nanoseconds is equivalent to $(6R)^{-1}$. The r_0 for diphenylhexatriene and *trans*-parinaric acid is 0.392 and 0.390, respectively (Lakowicz, 1979; Wolber and Hudson, 1981). The value for R can be obtained from differential polarized phase measurements as follows (Lakowicz et al., 1979b):

$$(m \tan \Delta)(2R\tau)^2 + (C \tan \Delta - A)(2R\, \tau) + (D \tan \Delta - B) = 0 \qquad (3)$$

where

$$A = 3B = w\tau(r_0 - r)$$
$$C = (\tfrac{1}{3})(2r - 4r^2 + 2)$$
$$D = (\tfrac{1}{9})(m + m_0 w^2 \tau^2)$$
$$m = (1 + 2r)(1 - r)$$

in which Δ is the phase shift angle, w is the circular modulation frequency, and the steady-state anisotropy r is equivalent to $[2p/(3 - p)]$, where p is the polarization. In biological membranes, the r is dependent primarily on the range of motion rather than on the rate, and alterations in membrane components such as cholesterol, proteins, and so on, largely reflect alterations in range of motion rather than rate (Kinosita et al., 1981; Jahnig, 1979; Jahnig et al., 1982). Such calculations can also be performed with probe molecules that have multiple lifetimes (i.e., *trans*-parinaric acid) by using Weber's law for the additivity of polarization anisotropy (Weber, 1953). Weber's law of anisotropy requires intensity weighting for the addition of fractional anisotropies. Fractional fluorescence intensities (F_1 and F_2) are obtained from the multiple lifetime analysis. Multiple lifetimes obtained by phase and modulation techniques can be resolved by Weber's closed form heterogeneity analysis (Weber, 1981). This method also provides fractional fluorescence values. The latter can be converted to mole fractions according to:

$$X_i = (F_i/\tau_i) / \left(\sum_j F_j/\tau_j \right) \quad (4)$$

where X_i, F_i, and τ_i are the mole fraction, fractional fluorescence, and lifetime of component i. Present commercially available phase and modulation instruments have only three frequencies. This restriction allows resolution of only two fluorescence lifetimes. Recently, Gratton and Limkeman (1983) constructed a continuously variable frequency phase and modulation instrument that circumvents this problem. Alternately, photon decay-type instrument can be used to resolve multiple lifetimes (Kinosita et al., 1981; Kawato et al., 1977).

The contribution of probe molecule in several lipid domains with different lifetimes to an average or τ_{mix} can be obtained from $\tau_1 F_1$ plus $\tau_2 F_2$. Thus, an r_{mix} obtained from different polarized phase fluorescence measurements and a τ_{mix} may be inserted into Eq. (2) to provide an R_{mix} or average rotational relaxation time. The latter is only an approximation and subject to the assumption that r_∞ is the same in both lipid environments. Another limitation is the error involved in determination of lifetimes and fractional fluorescence intensities with the phase method. Recently, many of these problems have been circumvented by development of multifrequency phase and modulation fluorometers with continuously variable frequency from one to several hundred megaherz (reviewed by Gratton et al., 1983; Gratton and Limkeman, 1983).

It should be noted that when the data are analyzed by the Perrin equation 1 (Weber, 1953; Shinitzky and Inbar, 1974) using only the steady-state anisotropy, differences may be obtained between control and experimental membranes with probe molecules. This is not surprising since the steady-state anisotropy is comprised of two terms, r_∞ the limiting anisotropy and R the rotational rate. Since the rotational rate of the probes usually does not change and is not significantly affected by membrane additives such as cholesterol or proteins (Kinosita et al., 1981; Jahnig et al., 1982, Jahnig, 1979), changes in r_∞ should be directly correlated in changes in r. One last point in this regard relates to the absolute values for rotational rates. The methods of Lakowicz et al. (1979) and Kinosita et al. (1981) provide one value; the Perrin equation provides another value (Weber, 1953; Shinitzky and Inbar, 1974); the method of oxygen quenching provides yet a third value (Lakowicz et al., 1979b). Each method provides a different value, with the range in values being nearly three orders of magnitude (Lakowicz et al., 1979b). It would be naive at this point to assert that any of these methods provides the "true" microviscosity or rotational relaxation time of a membrane. Instead, each method when appropriately used may demonstrate "differences" in relaxation rate between membranes, and all comparisons must at this time be regarded strictly as relative differences.

Microviscosity determination from steady-state anisotropy values are subject to certain basic mathematical assumptions that are not valid for many probe molecules in biological membranes. Molecular rotations of probes such as perylene and diphenylhexatriene are not the same in lipid bilayers as they are in solvents such as paraffin oil (Lakowicz and Knutson, 1980; Lackowicz and Prendergast, 1978; Lakowicz et al., 1979a,b). Thus, microviscosity $\bar{\eta}$ determinations from steady-state polarization may be erroneous. In evaluating the properties of membrane lipids it is important to keep concepts of orientational order and dynamics of these molecules apart (Heyn, 1979; Jahnig, 1979). Steady-state fluorescence anisotropy contains information on not only dynamics (viscosity) but also statics (order) of the label, and such simple interpretations of microviscosity have to be revised when applied to a biological system. The order parameters S may be evaluated from the $(r_\infty/r_0)^{1/2}$ (Jahnig, 1979). The value of (r_∞/r_0) is directly available from time dependent fluorescence depolarization experiments. Because of the preceding problems with the use of $\bar{\eta}$, only rotational relaxational time, $\bar{p}$, or r_∞ and R should be calculated instead.

Corrections of fluorescence excitation for the instrumental variables, inner filter effect, and secondary absorbance artifact should be performed for all fluorescence intensity determinations (Christman et al., 1980, 1981; Holland et al., 1973, 1977). Serial dilutions should be used to correct polarization measurements for membrane absorbance such that a limiting polarization value is obtained (Lentz et al., 1979; Chong and Colbow, 1976).

7.4. Use of Filipin to Detect Cholesterol Location

Although filipin and several other polyenes are fluorescent, recently advantage was taken of the absorbance spectral changes in filipin–sterol interactions to probe sterol location in membranes (Rottem et al., 1981; Clejan et al., 1981; Bittman, 1978; Bittman, 1981). In addition, freeze fracture electron microscopy has been used to determine the distribution of cholesterol in the two monolayer halves of membranes from secretory granules (Orci et al., 1980), tissue cultured cells (Friend and Bearer, 1981; Robinson and Karnovsky, 1980; Montesano et al., 1979; Hennache et al., 1982; Kruth and Vaughan, 1980), and intercellular junctions (Robenek et al., 1982). One of the primary interpretations of these results is that filipin or amphotericin form complexes with the sterol in the membrane being studied and that the quantity of complex is proportional to the amount of sterol in that membrane monolayer. These data must be viewed with caution since polyene–sterol complexes also form in water (Schroeder *et al.*, 1971, 1972, 1973). The antibiotic filipin apparently does not cross membranes (Schroeder and Bieber, 1975). Thus, it seems possible that filipin–sterol complexes may be formed in water and not in membranes. Indeed, radiolabeled filipin was not associated with model membranes containing cholesterol or with red blood cell membranes (Kelly and Bieber, 1981). Instead, the filipin removed radiolabeled cholesterol from these membranes to the aqueous phase. In another study using amphotericin B and spin labels, the interaction of the polyene with sterol in model membranes also did not occur in the membrane (Aracava *et al.*, 1981). Amphotericin B removed cholesterol from red blood cells (d'Hollander, 1972). Last, the fluidity of the membrane dramatically affects the formation of polyene–sterol complexes (Bolard and Cheron, 1982; Sekiya *et al.*, 1979). When these data are taken together with the observation that sterol exchanges into the aqueous phase (Jain and Shohet, 1982) by desorbing from membranes as a monomer (McLean and Phillips, 1981), the following possibility seems much more probable: Filipin and other polyenes form aqueous complexes with sterols as the sterols desorb as monomers from the membrane. This then may cause a lateral and/or transbilayer redistribution of sterol that could be responsible for pits or protuberances in freeze fracture electron micrographs. Since perturbation of sterol equilibria across membranes can dramatically enhance transbilayer flip-flop (Poznansky and Lange, 1978), it seems questionable that an accurate determination of transbilayer sterol distribution can be made by this method.

8. FUTURE DIRECTIONS

One of the primary aims of membraneologists is to understand the origin, regulation, and function of the transbilayer distribution of lipids in biological membranes. The answers to these questions are not known at this time. The data reviewed here indicate that the necessary fluorescence technology has now been developed to investigate these newer aspects of lipid dynamics, not just structure, of membranes.

ACKNOWLEDGMENTS

This work was supported in part by grants from the U.S. Public Health Service (GM 31651 and CA 24339). The helpful technical assistance of Eugene Hubert, John Hale, Daniel Feller, Jack Gardiner, Diana Cartwright, H.-M. Wiedemeyer, L. Whitmer, E. Moezpoor, L. Schuster, and B. Fischer was much appreciated. The collaborations of several colleagues, Drs. A. B. Kier, T. E. Thompson, F. Stephenson, R. N. Fontaine, R. A. Harris, W. D. Sweet, M. Heimberg, H. G. Wilcox, and C. Soler-Argilaga were also very helpful, and the secretarial assistance of Genie Eckenfels was also appreciated.

9. REFERENCES

Audubert, F. and D. E. Vance, 1983, Pitfalls and problems in studies on the methylation of phosphatidylethanolamine. *J. Biol. Chem* **258**: 10695–10701.
Andia-Waltenbaugh, A. M., Kimura, S., Wood, J., Divakaran, P., and Friedmann, N., 1978, Effects of glucagon, insulin and cAMP on mitochondrial calcium uptake in the liver, *Life Sci.* **23**:2437–2443.
Andley, U. P., and Chakrabarti, B., 1981, Interaction of ANS with rod outer segment membrane, *Biochemistry* **20**:1687–1693.
Aracava, Y., Smith, I. C., and Schreier, S., 1981, Effect of amphotericin B on membranes: A spin probe study, *Biochemistry* **20**:5702–5707.
Archer, D. B., 1975, The use of a fluorescent sterol to investigate the mode of action of amphotericin methyl ester, a polyene antibiotic. *Biochem. Biophys. Res. Commun.* **66**:195–201.
Axelrod, D., Koppel, D. E., Schlessinger, J., Elson, E. L., and Webb, W. W., 1976, Mobility measurement by analysis of fluorescence photobleaching recovery kinetics, *Biophys. J.* **16**:1055–1069.
Badley, R. A., 1976, Fluorescent probing of dynamic and molecular organization of biological membranes, in *Modern Fluorescence Spectroscopy*, Vol. 2 (E. L. Wehry, ed.), pp. 91–168, Plenum Press, New York.
Barenholz, Y., Patzer, E. J., Moore, N. F., and Wagner, R. R., 1978, Cholesterol oxidase as a probe for studying membrane composition and organization, *Adv. Exp. Med. Biol* **101**: 45–56.
Beach, D. H., Sherman, I. W., and Holz, G. G., 1977, Lipids of plasmodium-Lophurae and of erythrocytes and plasmas of normal and plasmodium Lophurae infected Pekin ducklings, *J. Parasitol.* **63**:62–75.
Bennett, V., O'Keefe, E., and Cuatrecasas, P., 1975, Mechanism of action of cholera toxin and the mobile theory of hormone receptor-adenylate cyclase interactions, *Proc. Natl. Acad. Sci. USA* **72**:33–37.
Bergelson, L. D., and Barsukov, L. I., 1977, Topological asymmetry of phospholipids in membranes, *Science* **197**:224–229.
Bergeron, R., and Scott, J., 1982, Fluorescent lipoprotein probe, *Anal. Biochem.* **119**:128–134.
Bergeron, R. J., and Scott, J., 1982b, Cholestatriene and ergostatetraene as *in vivo* and *in vitro* membrane and lipoprotein probes, *J. Lipid Res.* **23**:391–404.
Bevers, E. M., Op den Kamp, J. A., and van Deenen, L. L., 1978, The distribution of molecular classes of phosphatidylglycerol in the membrane of *Acholeplasma Laidlawii, Biochim. Biophys. Acta* **511**:509–512.
Bevers, E. M., Comfurius, P., van Rijn, J. L., Hemker, H. C., and Zwaal, R. F., 1982, Generation of prothrombin-converting activity and the exposure of phosphatidylserine at the outer surface of platelets, *Eur. J. Biochem.* **122**:429–436.
Bittman, R., 1978, Sterol-polyene antibiotic complexation: Probe of membrane structure, *Lipids* **13**:686–691.
Bittman, R., Blau, L., Clejan, S., and Rottem, S., 1981, Determination of cholesterol asymmetry

by rapid kinetics of filipin-cholesterol association: Effect of modification in lipids and proteins, *Biochemistry* **20**:2425–2432.
Blau, L., and Bittman, R., 1977, Interaction of filipin with cholesterol in vesicles of saturated phospholipids, *Biochemistry* **16**:4139–4144.
Bloj, B., and Zilversmit, D. B., 1976, Asymmetry and transposition rates of phosphatidylcholine in rat erythrocyte ghosts, *Biochemistry* **15**:1277–1283.
Bloj, B., and Zilversmit, D. B., 1982, Heterogeneity of rabbit intestine brush border plasma membrane cholesterol, *J. Biol. Chem.* **257**:7608–7614.
Blume, A., and Ackermann, T., 1974, A calorimetric study of the lipid phase transitions in aqueous dispersions of phosphorylcholine-phosphorylethanolamine mixtures, *FEBS Lett.* **43**:71–74.
Boeynaems, J. M., and Dumont, J. E., 1977, The two-step model of ligand-receptor interaction, *Mol. Cell Endocrinol.* **7**:33–47.
Bolard, J., and Cheron, M., 1982, Association of the polyene antibiotic amphotericin B with phospholipid vesicles: Perturbation by temperature changes, *Can. J. Biochem.* **60**:782–789.
Borochov, H., and Shinitzky, M., 1976, Vertical displacement of membrane proteins mediated by changes in microviscosity, *Proc. Natl. Acad. Sci. USA* **73**:4526–4530.
Borochov, H., Abbott, R. E., Schachter, D., and Shinitzky, M., 1979, Modulation of erythrocyte membrane proteins by membrane cholesterol and lipid fluidity, *Biochemistry* **18**:251–255.
Browning, J. L., and Nelson, D. L., 1979, Fluorescent probes for asymmetric lipid bilayers: Synthesis and properties in phosphatidyl choline liposomes and erythrocyte membranes, *J. Membrane Biol.* **49**:75–103.
Butler, K. W., Dugas, H., Smith, I. C. P., and Schneider, H., 1970, Cation-induced organization changes in a lipid bilayer model membrane, *Biochem. Biophys. Res. Commun.* **40**:770–776.
Bystrov, V. F., Dubrovina, N. I., Barsukov, L. I., and Bergelson, L. D., 1971, NMR differentiation of the internal and external phospholipid membrane surfaces using paramagnetic Mn^{+2} and Eu^{+3} ions, *Chem. Phys. Lip.* **6**:343–350.
Cable, M. B., and Powell, G. L., 1980, Spin-labeled cardiolipin: Preferential segregation in the boundary layer of cytochrome c oxidase, *Biochemistry* **19**:5679–5686.
Caspar, D. L. D., and Kirschner, D. A., 1971, Myelin membrane structure at 10 Å resolution, *Nature* **231**:46–52.
Chen, Y.-D. I., and Hoch, F. L., 1977, Thyroid control over biomembranes: Rat liver mitochondrial inner membranes, *Arch. Biochem. Biophys.* **181**:470–483.
Cherry, R. J., 1979. Rotational and lateral diffusion of membrane proteins, *Biochim. Biophys. Acta* **559**:289–327.
Cherry, R. J., Nigg, E. A., and Beddard, G. S., 1980, Oligosaccharide motion in erythrocyte membranes investigated by picosecond fluorescence polarization and microsecond dichroism of an optical probe, *Proc. Natl. Acad. Sci. USA* **77**:5899–5903.
Chong, C. S., and Colbow, K., 1976, Light scattering and turbidity measurements on lipid vesicles, *Biochim. Biophys. Acta* **436**:260–282.
Christmann, D. R., Crouch, S. R., Holland, J. F., and Timnick, A., 1980, Correction of right-angle molecular fluorescence measurements for absorption of fluorescence radiation, *Anal. Chem.* **52**:291–295.
Christmann, D. R., Crouch, S. R., and Timnick, A., 1981, Precision and accuracy of absorption-corrected molecular fluorescence measurements by the cell shift method, *Anal. Chem.* **53**:2040–2044.
Cogan, U., and Schachter, D., 1981, Asymmetry of lipid dynamics in human erythrocyte membranes studied with impermeant fluorophores, *Biochemistry* **20**:6396–6403.
Colard, O., and Breton, M., 1981, Rat liver plasma membrane phospholipids methylation; Its

absence of direct relationship to adenylate cyclase activities, *Biochem. Biophys. Res. Commun.* **101**:727–733.

Cowlen, M. S., Kier, A, B., and Schroeder, F., 1985, Stability of transbilayer sterol asymmetry in LM fibroblast plasma membranes. *Fed. Proc.* **44**:1350.

Craig, I. F., Via, D. P., Sherrill, B. C., Sklar, L. A., Mantulin, W. W., Gotto, A. M. Jr., and Smith, L. C., 1982, Incorporation of defined cholesteryl-esters into lipoproteins using cholesteryl-ester-rich microemulsions, *J. Biol. Chem.* **257**:330–335.

Crews, F. T., Morita, Y., McGivney, A., Hirata, F., Siraganian, R. P., and Axelrod, D., 1981, IgE-mediated histamine release in rat basophilic leukemia cells: Receptor activation, phospholipid methylation, Ca^{2+} flux, and release of arachidonic acid, *Arch. Biochem. Biophys.* **212**:561–571.

Cuatrecasas, P., 1974, Membrane receptors, *Annu. Rev. Biochem.* **43**:169–214.

Cuatrecasas, P., and Hollenberg, M. D., 1976, Membrane receptors and hormone action, *Advan. Protein. Chem.* **30**:251–451.

Cullis, P. R., and DeKruijff, B., 1979, Lipid polymorphism and the functional roles of lipids in biological membranes, *Biochim. Biophys. Acta* **559**:339–420.

Dave, J. R., Knazek, R. A., and Liu, S. C., 1981, Arachidonic acid, bradykinin and phospholipase A_2 modify both prolactin binding capacity and fluidity of mouse hepatic membranes, *Biochem. Biophys. Res. Comm.* **103**:727–738.

Davis, R. A., Kern, F., Jr., Showalter, R., Sutherland, E., Sinensky, M., and Simon, F. R., 1978, Alterations of hepatic Na^+, K^+-ATPase and bile flow by estrogen: Effects on liver surface membrane lipid structure and function, *Proc. Natl. Acad. Sci. USA* **75**:4130–4134.

Davson, H., 1967, in *Physiology of the Cerebrospinal Fluid* (H. Davson, ed.), p. 282, J. A. Churchill Ltd., London.

DeHaen, C., 1976, The non-stoichiometric floating receptor model for hormone-sensitive adenylyl cyclase, *J. Theor. Biol.* **58**:383–400.

DeMendoza, D., Moreno, H., Massa, E. M., Morero, R. E., and Farias, R. N., 1977, Thyroid hormone actions and membrane fluidity: Blocking action thyroxine on triiodothyronine effect, *FEBS Lett.* **84**:199–203.

Deuticke, B., 1968, Transformation and restoration of biconcave shape of human erythrocytes induced by amphiphilic agents and changes of ionic environment, *Biochim. Biophys. Acta* **163**:494–500.

Dipple, I., and Houslay, M. D., 1978, The activity of glucagon-stimulated adenylate cyclase from rat liver plasma membranes is modulated by the fluidity of its lipid environment, *Biochem. J.* **174**:179–190.

Dipple, I., and Houslay, M. D., 1979, Amphoptericin B has very different effects on the glucagon and fluoride-stimulated adenylate cyclase activities of rat liver plasma membranes, *FEBS Lett.* **106**:21–24.

Dragsten, P. R., Blumenthal, R., and Handler, J. S., 1981, Membrane asymmetry in epithelia: Is the tight junction a barrier to diffusion in the plasma membrane? *Nature* **294**:718–722.

Dragsten, P. R., Handler, J. S., and Blumenthal, R., 1982, Fluorescent membrane probes and the mechanism of maintenance of cellular asymmetry in epithelia, *Fed. Proc.* **41**:48–53.

Edidin, M., and Fambrough, D., 1973, Fluidity of the surface of cultured muscle fibers. Rapid lateral diffusion of marked surface antigens, *J. Cell Biol.* **57**:27–37.

Eidelman, O., and Cabantchik, Z. I., 1980, A method for measuring anion transfer across red cell membranes by continuous monitoring of fluorescence, *Anal. Biochem.* **106**:335–341.

Elema, R. P., Michels, P. A., and Konings, W. N., 1978, Response of 9-aminoacridine fluorescence to transmembrane pH-gradients in chromatophores from *R. sphaeroides*, *Eur. J. Biochem.* **92**:381–387.

Elias, P. M., Friend, D. S., and Goerke, J., 1979, Membrane sterol heterogeneity. Freeze-fracture detection with saponins and filipin. *J. Histochem. Cytochem.* **27**:1247–1260.

Emmelot, P., and Van Hoeven, R. P., 1975, Phospholipid unsaturation and plasma membrane organization, *Chem. Phys. Lip.* **14**:236–246.

Engel, L. W., and Prendergast, F. G., 1981, Values for and significance of order parameters and "cone-angles" of fluorophore rotation in lipid bilayers, *Biochemistry* **20**:7338–7345.

Esko, J. D., Gilmore, J. R., and Glaser, M., 1977, Use of a fluorescent probe to determine the viscosity of LM cell membranes with altered phospholipid compositions, *Biochemistry* **16**:1881–1890.

Etemadi, A.-H, 1980, Membrane asymmetry. A survey and critical appraisal of the methodology I. Methods for assessing the asymmetric orientation and distribution of proteins, *Biochim. Biophys. Acta* **604**:347–422.

Etemadi, A.-H., 1980, Membrane asymmetry. A survey and critical appraisal of the methodology II. Methods for assessing the unequal distribution of lipids, *Biochim. Biophys. Acta* **604**:423–475.

Faas, F. H., and Carter, W. J., 1981, Fatty acid desaturation and microsomal lipid fatty-acid composition in experimental hyperthyroidism, *Biochem. J.* **193**:845–852.

Feller, D. J., Schroeder, F., and Bylund, D., 1983, Binding of [^{3}H]flunitrazepam to the LM cell, a transformed murine fibroblast, *Biochem. Pharmacol.* **32**:2217–2223.

Feltkamp, C. A., and van der Waerden, A. W., 1982, Membrane-associated proteins affect the formation of filipin-cholesterol complexes in viral membranes, *Exp. Cell Res.* **140**:289–297.

Fisher, K. A., 1976, Analysis of membrane halves: Cholesterol, *Proc. Natl. Acad. Sci. USA* **73**:173–177.

Fischer, R. T. Stephenson, F. A., Shafiee, A., and Schroeder, F., 1984, $\Delta^{5,7,9(11)}$-Cholestatrien-3β-ol: A fluorescent cholesterol analogue, *Chem. Phys. Lip.* **36**:1–14..

Fischer, R. T., Stephenson, F. A., Shafiee, A., and Schroeder, F., 1985a, Structure and dynamic properties of dehydroergosterol, $\Delta^{5,7,9(11),22}$-ergostatetraen-3β-ol, *J. Biol. Phys.*, in press.

Fischer, R. T., Cowlen, M. S., Dempsey, M. E., and Schroeder, F., 1985b, Fluorescence of $\Delta^{5,7,9(11)22}$-ergostatetraen-3β-ol in micelles, sterol carrier protein complexes and plasma membranes, *Biochemistry,* in press.

Flagg-Newton, J. L., and Loewenstein, W. R., 1980, Asymmetrically permeable membrane channels in cell junction, *Science* **207**:771–773.

Flamm, M., and Schachter, D., 1982, Acanthocytosis and cholesterol enrichment decrease lipid fluidity of only the outer human erythrocyte membrane leaflet, *Nature* **298**:290–292.

Fong, B. S., and Brown, J. C., 1978, Asymmetric distribution of phosphatidylethanolamine fatty acyl chains in the membrane of vesicular stomatitis virus, *Biochim. Biophys. Acta* **510**:230–241.

Fontaine, R. N., Harris, R. A., and Schroeder, F., 1979, Neuronal membrane lipid asymmetry, *Life Sci.* **24**:395–399.

Fontaine, R. N., Harris, R. A., and Schroeder, F., 1980, Aminophospholipid asymmetry in murine synaptosomal plasma membrane, *J. Neurochem.* **34**:269–277.

Fontaine, R. N., and Schroeder, F., 1979, Plasma membrane aminophospholipid distribution in transformed murine fibroblasts, *Biochim. Biophys. Acta* **558**:1–12.

Fraley, R. T., Jameson, D. M., and Kaplan, S., 1978, The use of the fluorescent probe α-parinaric acid to determine the physical state of the intercytoplasmic membranes of the photosynthetic bacterium *R. sphaeroides*, *Biochim. Biophys. Acta* **511**:52–69.

Friend, D. S., and Bearer, E. L., 1981, β-hydroxysterol distribution as determined by freeze fracture cytochemistry, *Histochem. J.* **13**:535–546.

Fung, B. K., and Stryer, L., 1978, Surface density determination in membranes by fluorescence energy transfer, *Biochemistry* **17**:5241–5248.

Galo, M. G., Unates, L. E., and Farias, R. N., 1981, Effect of membrane fatty acid composition on the action of thyroid hormones on (Ca^{++} + Mg^{++})-ATPase from rat erythrocyte, *J. Biol. Chem.* **256**:7113–7114.

Gilmore, R., and Glaser, M., 1982, Preparation of a fluorescent derivative of cytochrome b_5 and its interaction with phospholipids, *Biochemistry* **21**:1673–1680.

Ginsberg, B. H., Brown, T. J., Simon, I., and Spector, A. A., 1981, Effect of the membrane lipid environment on the properties of insulin receptors, *Diabetes* **30**:773–780.

Golan, D. E., and Veatch, W., 1980, Lateral mobility of band 3 in the human erythrocyte membrane studied by fluorescence photobleaching recovery: Evidence for control by cytoskeletal interactions, *Proc. Natl. Acad. Sci. USA* **77**:2537–2541.

Gordon, L. M., Sauerheber, R. D., and Esgate, J. A., 1978, Spin label studies on rat liver and heart plasma membranes: Effects of temperature, calcium and lanthanum on membrane fluidity, *J. Supramol. Struct.* **9**:299–326.

Gordon, L. M., Sauerheber, R. D., Esgate, J. A., Dipple, I., Marchmont, R. J., and Houslay, M. D., 1980, The increase in bilayer fluidity of rat liver plasma membranes achieved by the local anesthetic benzyl alcohol affects the activity of intrinsic membrane enzymes, *J. Biol. Chem.* **255**:4519–4527.

Gottlieb, M. H., 1977, The reactivity of human erythrocyte membrane cholesterol with a cholesterol oxidase, *Biochim. Biophys. Acta* **466**:422–428.

Gould, R. J., Ginsberg, B. H., and Spector, A. A., 1982, Lipid effects on the binding properties of a reconstituted insulin receptor, *J. Biol. Chem.* **257**:477–484.

Gratton, E., and Limkeman, M., 1983, A continuously variable frequency cross-correlation phase fluorometer with picosecond resolution. *Biophys. J.* **44**:315–324.

Gratton, E., Jameson, D. M., and Hall, R. D., 1984, Multifrequency phase and modulation fluorometry, *Ann. Rev. Biophys. Bioeng.* **13**:105–124.

Grynne, B. H., and Toft, I.-L., 1982, The action of six lipid perturbing substances on hormone receptor adenylate cyclase complex in turkey erythrocytes and intestinal mucosa cells, *Acta Pharmacol. Toxicol.* **50**:283–293.

Gunstone, F. D., and Subbarao, R., 1967, New tropical seed oils. Part I. Conjugated trienoic and tetraenoic acids and their oxo derivatives in the seed oils of *C. icaco* and *P. laurinum*, *Chem. Phys. Lip.* **1**:349–359.

Gupta, C. M., and Mishra, G. C., 1981, Transbilayer phospholipid asymmetry in *P. knowlesi*-infected host cell membrane, *Science* **212**:1047–1049.

Haigh, E. A., Thulborn, K. R., and Sawyer, W. H., 1979, Comparison of fluorescence energy transfer and quenching methods to establish the position and orientation of components within the transverse plane of the lipid bilayer. Application to the Gramicidin A-bilayer interaction, *Biochemistry* **18**:3525–3532.

Hale, J. E., and Schroeder, F., 1981, Differential scanning calorimetry and fluorescence probe investigations of very low density lipoprotein from the isolated perfused rat liver, *J. Lipid Res.* **22**:838–851.

Hale, J. E., and Schroeder, F., 1982, Asymmetric transbilayer distribution of sterol across plasma membranes determined by fluorescence quenching of dehydroeregosterol, *Eur. J. Biochem.* **122**:649–661.

Hampton, R. Y., Holz, R. W., and Goldstein, I. J., 1980, Phospholipid, glycolipid and ion dependencies of concanavalin A- and *Ricinus commonis* agglutin I-induced agglutination of lipid vesicles, *J. Biol. Chem.* **255**:6766–6771.

Heininger, H.-J., and Marshall, J. D., 1979, Pinocytosis in L cells: Its dependence on membrane sterol and the cytoskeleton, *Cell Biol. Intern. Rept.* **3**:409–420.

Heller, R. A., Klotzbucher, R., and Stoffel, W., 1979, Interactions of a photosensitive analog of cholesterol with hydroxymethyglutaryl CoA reductase (NADPH) and acyl-CoA: Cholesterol acyltransferase, *Proc. Natl. Acad. Sci. USA* **76**:1721–1725.

Henis, Y. I., Rimon, G., and Felder, S., 1982, Lateral mobility of phospholipids in turkey erythrocytes: implications for adenylate cyclase activation, *J. Biol. Chem.* **257**:1407–1411.

Hennache, B., Torpier, G., and Boulanger, P., 1982, Adenovirus adsorption and sterol redistribution in KB cell plasma membrane, *Exp. Cell Res.* **137**:459–463.

Hesketh, T. R., Smith, G. A., Houslay, M. D., McGill, K. A., Birdsall, N. J. M., Metcalfe, J. C., and Warren, G. B., 1976, Annular lipids determine the ATPase activity of a calcium transport protein complexed with dipalmitoyllecithin, *Biochemistry* **15**:4145–4151.

Heyn, M. P., 1979, Determination of lipid order parameters and rotational correlation times from fluorescence depolarization experiments, *FEBS Lett.* **108**:359–364.

Hidalgo, C., Petrucci, D. A., and Vergara, C., 1982, Uncoupling of Ca^{++} transport in sarcoplasmic reticulum as a result of labelling lipid amino groups and inhibition of Ca^{++}-ATPase activity by modification of lysine residues of the Ca^{++}-ATPase polypeptide, *J. Biol. Chem.* **257**:208–216.

Hirata, F., and Axelrod, J., 1978, Enzymatic methylation of phosphatidylethanolamine increases erythrocyte membrane fluidity, *Nature* **275**:219–220.

Hirata, F., Axelrod, J., and Crews, F. T., 1979, Concanavalin A stimulates phospholipid methylation and phosphotidylserine decarboxylation in rat mast cells, *Proc. Natl. Acad. Sci. USA* **76**:4813–4816.

Hirata, F., Viveros, O. H., Diliberto, E. J., and Axelrod, J., 1978, Identification and properties of two methyltransferases in conversion of phosphatidylethanolamine to phosphatidylcholine, *Proc. Natl. Acad. Sci. USA* **75**:1718–1721.

Hoekstra, D., 1982, Fluorescence method for measuring the kinetics of Ca^{++}-induced phase separations in phosphatidylserine-containing lipid vesicles, *Biochemistry* **21**:1055–1061.

Holland, J. F., Teets, R. E., and Timnick, A., 1973, A unique computer centered instrument for simultaneous absorbance and fluorescence measurements, *Anal. Chem.* **45**:145–153.

Holland, J. F., Teets, R. E., Kelly, P. M., and Timnick, A., 1977, Correction of right-angle fluorescence measurements for the absorption of excitation radiation, *Anal. Chem.* **49**:706–710.

d'Hollander, F., and Chevallier, F., 1972, Movement of cholesterol *in vitro* in rat blood and quantitation of exchange of free cholesterol between plasma and erythrocytes, *J. Lipid Res.* **13**:733–744.

Hollenberg, M. D., 1979, Hormone receptor interactions at the cell membrane, *Pharmacol. Rev.* **30**:393–410.

Horwitz, A. F., Hatten, M. E., and Burger, M. M., 1974, Membrane fatty acid replacements and their effect on growth and lectin-induced agglutinability, *Proc. Natl. Acad. Sci. USA* **71**:3115–3119.

Houslay, M. D., 1979, Coupling of the glucagon receptor to adenylate cyclase, *Biochem. Soc. Trans.* **7**:843–846.

Houslay, M. D., and Palmer, R. W., 1978, Changes in the form of Arrhenius plots of the activity of glucagon-stimulated adenylate cyclase and other hamster liver plasma-membrane enzymes occurring on hibernation, *Biochem. J.* **174**:909–919.

Houslay, M. D., Dipple, I., Rawal, S., Sauerheber, R. D., Esgate, J. A., and Gordon, L. M., 1980, Glucagon-stimulated adenylate cyclase detects a selective perturbation of the inner half of the liver plasma-membrane bilayer achieved by the local anaesthetic prilocaine, *Biochem. J.* **190**:131–137.

Houslay, M. D., Dipple, I., and Gordon, L. M., 1981, Phenobarbital selectively modulates the glucagon-stimulated activity of adenylate cyclase by depressing the lipid phase separation

occurring in the outer half of the bilayer of liver plasma membranes, *Biochem. J.* **197**:675–681.
Huang, C.-H., Sipe, J. P., Chow, S. T., and Martin, R. B., 1974, Differential interaction of cholesterol with phosphatidylcholine on the inner and outer surfaces of lipid bilayer vesicles, *Proc. Natl. Acad. Sci. USA* **71**:359–362.
Hubbell, W. L., and McConnell, H. M., 1981, Molecular motion in spin-labeled phospholipids and membranes, *J. Am. Chem. Soc.* **93**:314–326.
Hulbert, A. J., Augee, M. L., and Raison, J. D., 1976, The influence of thyroid hormones on the structure and function of mitochondrial membranes, *Biochim. Biophys. Acta* **455**:597–601.
Hulbert, A. J., 1978, The thyroid hormones: A thesis concerning their action, *J. Theor. Biol.* **73**:81–100.
Isenberg, I., 1975, Time decay fluorometry by photon counting, in: *Biochemical Fluorescence*, Vol. 1 (R. F. Chen and H. Edelhock, eds.), pp. 43–77, Marcel Dekker, New York.
Jacobs, S., and Cuatrecasas, P., 1976, The mobile receptor hypothesis and "cooperativity" of hormone binding. Application to insulin, *Biochim. Biophys. Acta* **433**:482–495.
Jacobs, S., and Cuatrecasas, P., 1977, The mobile receptor hypothesis for cell membrane receptor action, *Trends Biochem. Sci.* **2**:280–282.
Jacobson, K., Elson, E., Koppel, D., and Webb, W., 1982, Fluorescence photobleaching in cell biology, *Nature* **295**:283–284.
Jacobson, K., Hou, Y., Derzko, Z., Wojcieszyn, J., and Organisciak, D., 1981, Lipid lateral diffusion in the surface membrane of cells and in multibilayers formed from plasma membrane lipids, *Biochemistry* **20**:5268–5275.
Jahnig, F., 1979, Structural order of lipids and proteins in membranes: Evaluation of fluorescence anisotropy data, *Proc. Natl. Acad. Sci. USA* **76**:6361–6365.
Jahnig, F., Vogel, H., and Best, L., 1982, Unifying description of the effect of membrane proteins on lipid order. Verification for the melittin–dimyristoylphosphatidylcholine system, *Biochemistry* **21**:6790–6798.
Jain, S. K., and Shohet, S. B., 1982, Red blood cell [^{14}C]cholesterol exchange and plasma cholesterol esterifying activity of normal and sickle cell blood, *Biochim. Biophys. Acta* **688**:11–15.
Jokinen, M., and Gahmberg, C. G., 1979, Phospholipid composition and external labeling of aminophospholipids of human En (a$^-$) erythrocyte membranes which lack the major sialoglycoprotein (glycophorin A), *Biochim. Biophys. Acta* **554**:114–124.
Jost, P. C., Griffith, O. H., Capaldi, R. A., and Vanderkooi, G., 1973, Evidence for boundary lipid in membranes, *Proc. Natl.Acad. Sci. USA* **70**:480–484.
Kang, S. Y., Gutawsky, H. S., Hsung, J. C., Jacobs, R., King, T. E., Rice, D., and Oldfield, E., 1979, NMR investigation of the cytochrome oxidase–phospholipid interaction: A new model for boundary lipid, *Biochemistry* **18**:3257–3267.
Kao, Y. J., Soutar, A. K., Hong, K.-Y., Pownall, H. J., and Smith, L. C., 1978, N-(2-Naphthyl)-23,24-dinor-5-cholen-22-amin-3β-ol, a fluorescent cholesterol analogue, *Biochemistry* **17**:2689–2696.
Kavet, R. I., and Brain, J. D., 1980, Methods to quantify endocytosis: A review, *J. Reticuloendothelial Soc.* **27**:201–221.
Kawato, S., Kinosita, K., and Ikegami, A., 1977, Dynamic structure of lipid bilayer studied by nanosecond fluorescence techniques, *Biochemistry* **16**:2319–2324.
Kaye, W., and West, D., 1967, Fluorescence polarization by modulation techniques. in: *Fluorescence: Theory, Instrumentation and Practice* (G. G. Guilbault, ed.), pp. 255–273, Marcel Dekker, New York.
Keith, A. D., Melhorn, R. J., Freeman, N. K., and Nichols, A. V., 1973, Spin labeled lipid probes in serum lipoproteins, *Chem. Phys. Lip.* **10**:223–236.

Kelly, P. M., Holland, J. F., and Bieber, L. L., 1979, Interaction of isotopically labeled and unlabeled filipin with egg lecithin vesicles and rat erythrocytes, *Biochemistry* **18**:4769–4775.

Keyes, W. D., and Heimberg, M., 1979, Influence of thyroid status on lipid metabolism in the perfused rat liver, *J. Clin Invest.* **63**:182–190.

Kier, A. B., and Schroeder, F., 1982, Development of metastatic tumors in athymic (nude) mice from LM cells grown *in vitro, Transplantation* **33**:274–279.

Kier, A. B., and Schroeder, F., 1983a, Lipid composition alters phagocytosis of fluorescent latex beads. *J. Immunol. Meth.* **57**:363–372.

Kier, A. B., and Schroeder, F., 1983b, Membrane properties of primary tumors and metastatic tumors derived from LM fibroblasts in C_3H mice, *Cancer Res.*, submitted.

Kier, A. B., and Schroeder, F., 1985, Membrane properties of primary tumors and metastatic tumors derived from LM fibroblasts in C_3H mice, *Cancer Res.*, submitted.

Kimelman, D., Tecoma, E. S., Wolber, P. K., Hudson, B. S., Wickner, W. T., and Simoni, R. D., 1979, Protein-lipid interactions. Studies of the M13 coat protein in dimyristoylphosphatidylcholine vesicles using parinaric acid, *Biochemistry* **18**:5874–5880.

Kinosita, K., Jr., Kataoka, R., Kimura, Y., Gotoh, O., and Ikegami, A., 1981, Dynamic structure of biological membranes as probed by 1,6,-diphenyl-1,3,5-hexatriene: A nanosecond fluorescence depolarization study, *Biochemistry* **20**:4270–4277.

Kiss, Z., 1979, Involvement of calcium in the inhibition by insulin of the glucagon-stimulated adenylate cyclase activity, *Eur. J. Biochem.* **95**:607–611.

Kramer, R. M., and Branton, D., 1979, Retention of lipid asymmetry in membranes on polyserine-coated polyacrylamide beads, *Biochim. Biophys. Acta* **556**:219–232.

Kramer, R. M., Jakubowski, J. A., Vaillancourt, R., and Deykin, D., 1982, Effect of membrane cholesterol on phospholipid metabolism in thrombin-stimulated platelets, *J. Biol. Chem.* **257**:6844–6849.

Kreiner, P. W., Keirns, J. J., and Bitensky, M. W., 1973, A temperature-sensitive change in the energy of activation of hormone-stimulated hepatic adenylyl cyclase, *Proc. Natl. Acad. Sci. USA* **70**:1785–1789.

Kruth, H. S., and Vaughan, M., 1980, Quantification of low density lipoprotein binding and cholesterol accumulation by single human fibroblasts using fluorescence microscopy *J. Lipid Res.* **21**:123–130.

Kury, P. G., Ramwell, P. W., and McConnell, H. M., 1974, The effect of prostaglandins E_1 and E_2 on the human erythrocyte as monitored by spin labels, *Biochem. Biophys. Res. Commun.* **56**:478–483.

Lakowicz, J. R., Cherek, H., and Balter, A., 1981, Correction of timing errors in photomultiplier tubes used in phase-modulation fluorometry, *J. Biochem. Biophys. Meth.* **5**:131–146.

Lakowicz, J. R., and Hogan, D., 1977, in: *Membrane Toxicity* (W. M. Miller and A. E. Shamoo, eds.), pp. 509–544, Plenum Press, New York.

Lakowicz, J. R., Hogen, D., and Omann, G., 1977, Diffusion and partitioning of a pesticide, lindane, into phosphatidylcholine bilayers, *Biochim. Biophys. Acta* **471**:401–411.

Lakowicz, J. R., and Knutson, J. R., 1980, Hindered depolarizing rotations of perylene in lipid bilayers. Detection of lifetime-resolved fluorescence anisotropy measurements, *Biochemistry* **19**:905–911.

Lakowicz, J. R., and Prendergast, F. G., 1978, Quantitation of hindered rotations of diphenylhexatriene in lipid bilayers by differential polarized phase fluorometry, *Science* **200**:1399–1401.

Lakowicz, J. R., Prendergast, F. G., and Hogan, D., 1979a, Differential polarized phase fluorometric investigations of diphenylhexatriene in lipid bilayers. Quantitation of hindered depolarizing rotations, *Biochemistry* **18**:508–519.

Lakowicz, J. R., Prendergast, F. G., and Hogan, D., 1979b, Fluorescence anisotropy measurements under oxygen quenching conditions as a method to quantify the depolarizing rotations of fluorophores. Application to diphenylhexatriene in isotropic solvents and in lipid bilayers, *Biochemistry* **18**:520–527.

Lange, Y., Dolde, J., and Steck, T. L., 1981, The rate of transmembrane movement of cholesterol in the human erythrocyte, *J. Biol. Chem.* **256**:5321–5323.

Lee, E., Baba, A., Ohta, A., and Iwata, H., 1982, Solubilization of adenylate cyclase of brain membranes by lipid peroxidation, *Biochim. Biophys. Acta* **689**:370–374.

Lee, G., Consiglio, E., Habig, W., Dyer, S., Hardegree, C., and Kohn, L. D., 1978, Structure: Function studies of receptors for thyrotropin and tetanus toxin: Lipid modulation of effector binding to the glycoprotein receptor component, *Biochem. Biophys. Res. Commun.* **83**:313–320.

Lelkes, P. I., and Miller, I. R., 1980, Perturbations of membrane structure of optical probes I. Location and structural sensitivity of merocyanine 540 bound to phospholipid membranes, *J. Memb. Biol.* **52**:1–15.

Lenard, J., and Rothman, J. E., 1976, Transbilayer distribution and movement of cholesterol and phospholipid in the membrane of influenza virus, *Proc. Natl. Acad. Sci. USA* **73**:391–395.

Lentz, B. R., Moore, B. M., and Barrow, D. A., 1979, Light-scattering effects in the measurement of membrane microviscosity with diphenylhexatriene, *Biophys. J.* **25**:489–494.

Livingstone, C. J., and Schachter, D., 1980, Calcium modulates the lipid dynamics of rat hepatocyte plasma membranes by direct and indirect mechanisms, *Biochemistry* **19**:4823–4827.

Loew, L. M., Scully, S., Simpson, L., and Waggoner, A. S., 1979, Evidence for a charge-shift electrochromic mechanism in a probe of membrane potential, *Nature* **281**:497–499.

London, E., and Feigenson, G. W., 1981b, Fluorescence quenching in model membranes. 2. Determination of local lipid environment of the Ca^{++}-ATPase from sarcoplasmic reticulum, *Biochemistry* **20**:1939–1948.

London, E., and Feigenson, G. W., 1981, Fluorescence quenching in model membranes. 1. Characterization of quenching caused by spin labeled phospholipid, *Biochemistry* **20**:1932–1938.

Low, P. S., Cramer, W. A., Abraham, G., Bone, R., and Ferguson-Segall, M., 1982, Evidence for restricted oligosaccharide mobility at the erythrocyte membrane surface; A fluorescence study, *Arch. Biochem. Biophys.* **214**:675–680.

Luisetti, J., Mohwald, H., and Galla, H.-J., 1979, Monitoring the location profile of fluorophores in phosphatidylcholine bilayers by the use of paramagnetic quenching, *Biochim. Biophys. Acta* **552**:519–530.

Luly, P., and Shinitzky, M., 1979, Gross structural changes in isolated liver cell plasma membranes upon binding insulin, *Biochemistry* **18**:445–450.

Lyte, M., and Shinitzky, M., 1979, Cholesteryl-phosphoryl-choline in lipid bilayers, *Chem. Phys. Lip.* **24**:45–55.

Maeda M., Tanaka, Y., and Akamatsu, Y., 1981, Presence of phospholipid methylation pathway in mammalian cultured cells, *Biochim. Biophys. Acta* **663**:578–582.

Mahoney, E. M., Hamill, A. L., Scott, W. A., and Cohn, Z. A., 1977, Response of endocytosis to altered fatty acyl composition of macrophage phospholipids, *Proc. Natl. Acad. Sci. USA* **74**:4895–4899.

Mailer, C., Taylor, C. P. S., Schreier-Muccillo, S. and Smith, I. C. P., 1974, The influence of cholesterol on molecular motion in egg lecithin bilayers. A variable-frequency ESR study of a cholestane spin probe, *Arch. Biochem. Biophys.* **163**:671–678.

Marinetti, G. V., and Cattieu, K., 1982, Asymmetric metabolism of phosphatidylethanolamine in the human red cell membrane, *J. Biol. Chem.* **257**:245–248.

Marshall, J. D., Heiniger, H.-J., and Waymouth, C., 1979, High affinity concanavalin A binding to sterol-depleted L cells, *J. Cell Physiol.* **100**:539–550.

Matayoshi, E. D., 1980, Distribution of shape-changing compounds across the red cell membrane, *Biochemistry* **19**:3414–3422.
Mayer, L. D., and Nelsestuen, G. L., 1981, Ca^{++}- and prothrombin-induced lateral phase separation in membranes, *Biochemistry* **20**:2457–2463.
Maziere, C., Maziere, J. C., Mora, L., and Polonovski, J., 1981, Enhancement of phospholipid methylation in cultured hamster cells by viral transformation, *FEBS Lett.* **129**:67–69.
McCaleb, M. C., and Donner, D. B., 1981, Affinity of the hepatic insulin receptor is influenced by membrane phospholipids, *J. Biol. Chem.* **256**:11051–11057.
McConnell, H. M., and McFarland, B. G., 1972, The flexibility gradient in biological membranes, *Ann. N.Y. Acad. Sci.* **195**:207–217.
McGookey, D. J., Fagerberg, K., and Anderson, R. G. W., 1983, Filipin-cholesterol complexes form in uncoated vesicle membrane derived from coated vesicles during receptor-mediated endocytosis of low density lipoprotein, *J. Cell Biol.* **96**:1273–1278.
McLean, L. R., and Phillips, M. C., 1981, Mechanism of cholesterol and phosphatidylcholine exchange or transfer between unilamillar vesicles, *Biochemistry* **20**:2893–2900.
Mely-Goubert, B., Calvo, F., and Rosenfeld, C., 1979, Study of platelet membrane proteins through fluorescence polarization of diphenylhexatriene, *Biomedicine* **31**:155–156.
Minetti, M., Aducci, P., and Viti, V., 1979, Interaction of neutral polysaccharides with phosphatidylcholine multilamellar liposomes. Phase transitions studied by binding of fluorescein-conjugated dextrans, *Biochemistry* **18**:2541–2548.
Molotkovsky, J. G., Manevich, Y. M., Gerasimova, E. N., Mololkovskaya, I. M., Polessky, V. A., and Bergelson, L. D., 1982, Differential study of phosphatidylcholine and sphingomyelin in human HDL with lipid-specific probes, *Eur. J. Biochem.* **122**:573–579.
Montesano, R., Perrelet, A., Vassalli, P., and Orci, L., 1979, Absence of filipin-sterol complexes from large coated pits on the surface of culture cells, *Proc. Natl. Acad. Sci. USA* **76**:6391–6395.
Moore, B. M., Lentz, B. R., and Meissner, G., 1978, Effects of sarcoplasmic reticulum Ca^{++}-ATPase on phospholipid bilayer fluidity: Boundary lipid, *Biochemistry* **17**:5248–5255.
Moroni, M., Capsoni, F., and Caredda, F., 1976, *Boll. Inst. Siesotes Milana* **55**:317 (Abstract).
Morse, P. D., Ruhlig, M., Snipes, W., and Keith, A. D., 1975, A spin-label study of the viscosity profile of sarcoplasmic reticular vesicles, *Arch. Biochem. Biophys.* **168**:40–56.
Moskal, J. R., Emaus, R. K., and Holland, J. F., 1978, β-Paranaric acid as a membrane probe in cultured human epidermoid carcinoma (KB) cells, *Fed. Proc.* **37**:1597 (abstract 1805).
Nelson, G. J., 1973, The lipid composition of plasma lipoprotein density classes of sheep, ovis aries, *Comp. Biochem. Physiol.* **46**:81–91.
Nelson, G. J., 1972, Blood lipids and lipoproteins: Quantitation, composition, metabolism (G. J. Nelson, ed.), Wiley-Interscience, New York.
Nelson, D. H., 1980, Corticosteroid-induced changes in phospholipid membranes as mediators of their action, *Endocr. Rev.* **1**:180–199.
Okuda, C., and Ogli, K., 1979, Fluorometric study with aminonaphtholsulfonic acid on membrane surface charge changes induced by inhalation anesthetics, *FEBS Lett.* **101**:399–402.
Op den Kamp, J. A. F., 1979, Lipid asymmetry in membranes, *Annu. Rev. Biochem.* **48**:47–71.
Opella, S. J., Yesinowski, J. P., and Waugh, J. S., 1976, Nuclear magnetic resonance description of molecular motion and phase separations of cholesterol in lecithin dispersions, *Proc. Natl. Acad. Sci. USA* **73**:3812–3815.
Orci, L., Miller, R. G., Montesano, R., Perrelet, R., Armherdt, M., and Vassalli, P., 1980, Opposite polarity of filipin-induced deformations in the membrane of condensing vacuoles and zymogen granules, *Science* **210**:1019–1020.
Pagano, R. E., Ozato, K., and Ruysschaert, J. M., 1977, Intracellular distribution of lipophilic fluorescent probes in mammalian cells, *Biochim. Biophys. Acta* **465**:661–666.

Pagano, R. E., Martin, O. C., Schroit, A. J., and Struck, D. K., 1981, Formation of asymmetric phospholipid membranes via spontaneous transfer of fluorescent lipid analogues between vesicle populations, *Biochemistry* **20**:4920–4927.

Pal, R., Barenholz, Y., and Wagner, R. R., 1981, Depletion and exchange of cholesterol from the membrane of VSV by interaction with serum lipoproteins or PVP complexed with bovine serum albumin, *Biochemistry* **20**:530–539.

Parasassi, T., Conti, F., and Gratton, E., 1984, Study of heterogeneous emission of parinaric acid isomers using multifrequency phase fluorometry, *Biochemistry* **23**:5660–5664.

Patel, K. M., Sklar, L. A., Currie, R., Pownall, H. J., Morriset, J. D., and Sparrow, J. T., 1979, Synthesis of saturated, unsaturated, spin-labeled, and fluorescent cholesteryl esters: Acylation of cholesterol using fatty acid anhydride and 4-pyrrolidinopyridine, *Lipids* **14**:816–818.

Patzer, E. J., Wagner, R. R., and Barenholz, Y., 1978, Cholesterol oxidase as a probe for studying membrane organization, *Nature* (Lond.) **274**:394–395.

Peifer, J. J., 1968, Disproportionally higher levels of myocardial docosahexaenoate and elevated levels of plasma and liver arachidonate in hyperthyroid rats, *J. Lipid Res.* **9**:193–199.

Perrelet, A., Garcia-Segura, L., Singh, A., and Orci, L., 1982, Distribution of cytochemically detectable cholesterol in the electric organ of *Torpedo marmorata*, *Proc. Natl. Acad. Sci. USA* **79**:2598–2602.

Perret, B., Chap, H. J., and Dovate-Blazy, L., 1979, Asymmetric distribution of arachidonic acid in the plasma membrane of human platelets, *Biochim. Biophys. Acta* **556**:434–446.

Peters, R., Peters, J., Tews, K. H., and Bahr, W., 1974, A microfluorimetric study of translational diffusion in erythrocyte membranes, *Biochim. Biophys. Acta* **367**:282–294.

Petri, W. A., Jr., Pal, R., Barenholz, Y., and Wagner, R. R., 1981, Fluorescence studies of dipalmitoylphosphatidylcholine vesicles reconstituted with the glycoprotein of vesicular stomatitis virus, *Biochemistry* **20**:2796–2800.

Pike, M. C., Kredich, N. M., and Synderman, R., 1979, Phospholipid methylation in macrophages is inhibited by chemotactic factors, *Proc. Natl. Acad. Sci. USA* **76**:2922–2926.

Platner, W. S., Patnayak, B. C., and Chaffee, R. R. J., 1972, A comparison of magnesium deficiency, cold acclimation and thyroxine administration on mitochondrial fatty acid composition, *Proc. Soc. Exp. Biol. Med.* **140**:857–861.

Poznansky, M., and Lange, Y., 1978, Transbilayer movement of cholesterol in phospholipid vesicles under equilibrium and non-equilibrium conditions, *Biochim. Biophys. Acta* **506**:256–264.

Poznansky, M., and Lange, Y., 1978, Transbilayer movement of cholesterol in dipalmitoyl-lecithin-cholesterol vesicles, *Nature* (Lond.) **259**:420–421.

Prendergast, F. G., Haugland, R. P., and Callahan, P. J., 1981, 1-[4-(trimethylamino)phenyl]-6-phenylhexa-1,3,5-triene: Synthesis, fluorescence properties, and use as a fluorescence probe of lipid bilayers, *Biochemistry* **20**:7333–7338.

Presti, F. T., and Chan, S. I., 1982, Cholesterol-phospholipid interaction in membranes. 1. Cholestane spin-label studies of phase behavior of cholesterol-phospholipid liposomes, *Biochemistry* **21**:3821–3830.

Presti, F. T., Pace, R. J. and Chan, S. I., 1982, Cholesterol-phospholipid interaction in membranes. 2. Stoichiometry and molecular packing of cholesterol-rich domains, *Biochemistry* **21**:3831–3835.

Puri, J., Shinitzky, M., and Lonai, P., 1980, Concomitant increase in antigen binding and in T cell membrane lipid viscosity induced by lymphocyte activating factor, LAF, *J. Immunol.* **124**:1937–1942.

Radda, G. K., and Vanderkooi, J., 1972, Can fluorescent probes tell us anything about membranes? *Biochim. Biophys. Acta* **265**:509–549.

Rahwan, R. G., Piascik, M. F., and Witiak, D. T., 1979, The role of calcium antagonism in the therapeutic action of drugs, *Can. J. Physiol. Pharmacol.* **57**:443–460.

Rama Sastry, B. V., Statham, C. N., Meeks, R. G., and Axelrod, J., 1981, Changes in phospholipid methyltransferases and membrane microviscosity during induction of rat liver microsomal cytochrome P-450 by phenobarbital and 3-methylcholanthrene, *Pharmacology* **23**:211-222.

Rando, R. R., Orr, G. A., and Bangerter, F. W., 1979, Threshold effects on the concanavalin A-mediated agglutination of modified erythrocytes *J. Biol. Chem.* **254**:8318-8323.

Rando, R. R., and Bangerter, F. W., 1979, Threshold effects on the lectin-mediated aggregation of synthetic glycolipid-containing liposomes, *J. Supramol. Struct.* **11**:295-309.

Rando, R. R., Bangerter, F. W., and Alecio, M. R.. 1982, The synthesis and properties of a functional fluorescent cholesterol analog, *Biochim. Biophys. Acta.* **684**: 12-20.

Renooij, W., and Van Golde, L. M. G., 1977, The transposition of molecular classes of phosphatidylcholine across the rat erythrocyte membrane and their exchange between the red cell membrane and plasma lipoproteins, *Biochim. Biophys. Acta* **470**:465-474.

Renooij, W., and van Golde, L. M. G., 1979, Asymmetry in the renewal of molecular classes of phosphatidylcholine in the rat erythrocyte membrane, *Biochim. Biophys. Acta* **558**:314-319.

Rigaud, J. L., and Lablanc, G., 1980, Effect of membrane cholesterol on action of phospholipase A_2 in *M. mycoides* var. Capri, *Eur. J. Biochem.* **10**:77-87.

Rimon, G., Hanski, E., and Levitzki, A., 1980, Temperature dependence of β-receptor, adenosine receptor, and sodium fluoride stimulated adenylate cyclase from turkey erythrocytes, *Biochemistry* **19**:4451-4460.

Rintoul, D. A., and Simoni, R. D., 1977, Incorporation of a naturally occurring fluorescent fatty acid into lipids of cultured mammalian cells, *J. Biol. Chem.* **252**:7916-7918.

Rintoul, D. A., Sklar, L. A., and Simoni, R. D., 1978, Membrane lipid modification of chinese hamster ovary cells: Thermal properties of membrane phospholipids, *J. Biol. Chem.* **253**:7447-7452.

Rittenhouse, H. G., and Fox, C. F., 1974, Concanavalin A mediated hemagglutination and binding properties of LM cells, *Biochem. Biophys. Res. Commun.* **57**:323-331.

Rittenhouse, H. G., Williams, R. E., Wisnieski, B., and Fox, C. F., 1974, Alterations of characteristic temperatures for lectin interactions in LM cells with altered lipid composition, *Biochem. Biophys. Res. Commun.* **58**:222-228.

Rivnay, B., Berman, S., Shinitzky, M., and Globerson, A., 1980, Correlations between membrane viscosity, serum cholesterol, lymphocyte activation and aging in man, *Mech. Aging Dev.* **12**:119-126.

Robenek, H., Schopper, C., Fasske, E., Fetting, R., and Themann, H., 1982, Tissue connections in transplantable virus-producing sebaceous adenoma of the mouse, II. A freeze-fracture study in conjunction with filipin, *J. Cancer Res. Clin. Oncol.* **102**:215-226.

Roberts, J., and Quastel, J. H., 1963, Particle uptake by polymorphonuclear leucocytes and Ehrlich ascites-carcinoma cells, *Biochem. J.* **89**:150-156.

Robinson, J. M., and Karnovsky, M. J., 1980, Evaluation of the polyene antibiotic filipin as a cytochemical probe for membrane cholesterol, *J. Hist. Cytochem.* **28**:161-168.

Robinson, N. C., Strey, F., and Talbert, L., 1980, Investigation of the essential boundary layer phospholipids of cytochrome C oxidase using Triton X-100 delipidation, *Biochemistry* **19**:3656-3661.

Rogers, J., Lee, A. G., and Wilton, D. C., 1979, The organisation of cholesterol and ergosterol in lipid bilayers based on studies using non-perturbing fluorescent sterol probes, *Biochim. Biophys. Acta* **552**:23-37.

Romer, J., and Bickel, M. H., 1979, Interactions of chlorpromazine and imipramine with artificial membranes investigated by equilibrium dialysis, dual-wavelength photometry, and fluorimetry, *Biochem. Pharmacol.* **28**:799-804.

Rothblat, G. H., and Phillips, M. C., 1982, Mechanism of cholesterol efflux from cells: Effects of acceptor structure and concentration, *J. Biol. Chem.* **257**:4775-4782.

Rothman, J. E., and Lenard, J., 1977, Membrane asymmetry, *Science* **195**:743–753.
Rouser, G., Nelson, G. J., Fleischer, S., and Simon, G., 1968, Lipid composition of animal cell membranes, organelles, and organs, in: *Biological Membranes*, (D. Chapman, ed.), Academic Press, New York.
Salesse, R., and Garnier, J., 1979, Effects of drugs on pigeon erythrocyte membrane and asymmetric control of adenylate cyclase by the lipid bilayer, *Biochim. Biophys. Acta* **554**:102–113.
Salesse, R., Garnier, J., Leterrier, F., Daveloose, D., and Viret, J., 1982, Modulation of adenylate cyclase activity by the physical state of pigeon erythocyte membrane. 1. Parallel drug-induced changes in bilayer fluidity and adenylate cyclase activity, *Biochemistry* **21**:1581–1586.
Sandra, A., and Pagano, R. E., 1978, Phospholipid asymmetry in LM cell plasma membrane derivatives: Polar head group and acyl chain distributions, *Biochemistry* **17**:332–338.
Schachter, D., Cogan, U., and Abbot, R. E., 1982, Asymmetry of lipid dynamics in human erythrocyte membranes studied with permeant fluorophores, *Biochemistry* **21**:2146–2150.
Schlager, S. I., 1982, Relationship between cell-mediated and humoral immune attack on tumor cells, *Cellular Immunol.* **66**:300–316.
Schmidt, C. F., Barenholz, Y., Huang, C., and Thompson, T. E., 1978, Monolayer coupling in sphingomyelin bilayer systems, *Nature* (Lond.) **271**:775–777.
Schroeder, F., 1978a, Differences in fluidity between bilayer halves of tumor cell membranes, *Nature* **275**:528–530.
Schroeder, F., 1978b, Isothermal regulation of membrane fluidity in murine fibroblasts with altered phospholipid polar head groups, *Biochim. Biophys. Acta* **511**:356–376.
Schroeder, F., 1980a, Fluorescence probes as monitors of surface membrane fluidity gradients in murine fibroblasts, *Eur. J. Biochem.* **112**:293–307.
Schroeder, F., 1980b, Regulation of aminophospholipid asymmetry in murine plasma membranes by choline and ethanolamine analogues, *Biochim. Biophys. Acta* **599**:254–270.
Schroeder, F., 1981a, Use of a fluorescent sterol to probe the transbilayer distribution of sterols in membranes, *FEBS Lett.* **135**:127–130.
Schroeder, F., 1981b, Altered phospholipid composition affects endocytosis in cultured LM fibroblasts, *Biochim. Biophys. Acta* **649**:162–174.
Schroeder, F., 1982a, Phagosomal membrane lipids of LM fibroblasts, *J. Membrane Biol.* **68**:141–150.
Schroeder, F., 1982b, Hormonal effects on fatty acid binding and physical properties of rat liver plasma membranes, *J. Membrane Biol.* **68**:1–10.
Schroeder, F., 1982c, Phospholipid polar head group manipulation modulates Concanavalin A agglutinability of LM fibroblasts, *Biochemistry* **21**:6782–6790.
Schroeder, F., 1983, Lipid domains in plasma membranes from rat liver, *Eur. J. Biochem.*, **132**:509–516.
Schroeder, F., 1984, Fluorescence probes in metastatic B-16 melanoma membranes, *Biochim. Biophys. Acta*, **776**:299–312.
Schroeder, F., 1984, Fluorescent sterols: Probe molecules of membrane structure and function, *Progr. Lipid Res.* **23**:97–113.
Schroeder, F., 1984, Fluorescence probes in metastatic B-16 melanoma membranes, *Biochim. Biophys. Acta* **776**:299–312.
Schroeder, F., 1985, Role of membrane lipid asymmetry in aging. *Neurobiol. Aging* **5**:323–333.
Schroeder, F., and Bieber, L. L., 1975, Studies on hypo- and hypercholesterolemia induced in insects by filipin, *J. Insect. Biochem.* **5**:201–221.
Schroeder, F., and Goh, E. H., 1979, Regulation of VLDL interior core physicochemical properties, *J. Biol. Chem.* **254**:2464–2470.

Schroeder, F., and Kinden, D. A., 1980, Differences in fluidity between bilayer halves of cell plasma membranes, *Nature* **287**:255–256.

Schroeder, F., and Kinden, D. A., 1983, Measurement of phagocytosis using fluorescent latex beads, *J. Biochem. Biophys. Meth.*, **8**:15–27.

Schroeder, F., and Soler-Argilaga, C., 1983, Ca^{++} modulates fatty acid dynamics in rat liver plasma membranes, *Eur. J. Biochem.*, **132**:517–524.

Schroeder, F., and Thompson, T. E., 1985, Lateral interaction of fluorescent sterols with sphingomyelin, *Biochemistry*, to be submitted.

Schroeder, F., Holland, J. F., and Bieber, L. L., 1971, Fluorometric evidence for the binding of cholesterol to the filipin complex, *J. Antibiot.* **24**:846–849.

Schroeder, F., Holland, J. F., and Bieber, L. L., 1972, Fluorometric investigations of the interactions of polyene antibiotics with sterols, *Biochemistry* **11**:3105–3111.

Schroeder, F., Holland, J. F., and Bieber, L. L., 1973, Reversible interconversions of sterol-binding and sterol-nonbinding forms of filipin as determined by fluorimetric and light scattering properties, *Biochemistry* **12**:4785–4789.

Schroeder, F., Holland, J. F., and Vagelos, P. R., 1976a, Use of β-parinaric acid, a novel fluorimetric probe to determine characteristic temperatures of membranes and membrane lipids from cultured animal cells, *J. Biol. Chem.* **251**:6739–6746.

Schroeder, F., Holland, J. F., and Vagelos, P. R., 1976b Physical properties of membranes isolated from tissue culture cells with altered phospholipid composition, *J. Biol. Chem.* **251**:6747–6756.

Schroeder, F., and Holland, J. F., 1978, Fluorescence probes and the structure of mammalian membranes, in *Biomolecular Structure and Function* (P. F. Agris, R. N., Loeppky, and B. D. Sykes, Eds.), pp. 137–145, Academic Press, New York.

Schroeder, F., Goh, E. H., and Heimberg, M., 1979a, Investigation of the surface structure of the very low density lipoprotein using fluorescence probes, *FEBS Lett.* **97**:233–236.

Schroeder, F., Goh, E. H., and Heimberg, M., 1979b, Regulation of the surface physical properties of the very low density lipoprotein, *J. Biol. Chem.* **254**:2456–2463.

Schroeder, F., Holland, J. F., and Doi, O., 1979c, Physical properties of murine fibroblasts with altered acyl chain unsaturation, *Arch. Biochem. Biophys.* **194**:431–438.

Schroeder, F., Fontaine, R. N., Feller, D. J., and Weston, K. G., 1981, Drug-induced surface membrane phospholipid composition in murine fibroblasts, *Biochim. Biophys. Acta* **643**:76–88.

Schroeder, F., Fontaine, R. N., and Kinden, D. A., 1982a, LM fibroblast plasma membrane subfractionation by affinity chromatography on con A-sepharose, *Biochim. Biophys. Acta* **690**:231–242.

Schroeder, F., Wilcox, H. G., Keyes, W. G., and Heimberg, M., 1982c, Effects of thyroid status on the structure of the VLDL secreted by the perfused liver, *Endocrinology* **110**:551–562.

Schroeder, F., Dempsey, M. E. and Fischer, R. T., 1985a, Fluorescence properties of sterol and squalene carrier protein interactions with fluorescent $\Delta^{5,7,9(11)}$-cholestatrien-3β-ol, *J. Biol. Chem.*, **260**:2904–2911.

Schroeder, F., Barenholz, Y. and Thompson, T. E., 1985b, Fluorescence of dehydroergosterol and cholestatrienol in phosphatidylcholine bilayer vesicles, *Biochemistry*, to be submitted.

Schroit, A. J., and Gallily, R., 1979, Macrophage fatty acid composition and phagocytosis: Effect of unsaturation on cellular phagocytic activity, *Immunology* **36**:199–205.

Seed, T. M., and Kreier, J. P., 1972, Plasmodium–gallinaccum–erythrocyte–membrane alterations and associated plasma changes induced by experimental infections, *Proc. Helminthol. Soc. Wash.* **39**:387–411.

Seigneuret, M., Zachowski, A., Hermann, A. and Devaux, P. F., 1984, Asymmetric lipid fluidity

in human erythrocyte membrane: New spin label evidence, *Biochemistry,* **23**:4271–4275.
Sekiya, T., Kitajima, Y., and Nozawa, Y., 1979, Effects of lipid-phase separation on the filipin action on membranes of ergosterol-replaced *Tetrahymena* cells, as studied by freeze-fracture electron microscopy, *Biochim. Biophys. Acta* **550**:269–278.
Sessions, A. and Horwitz, A. F., 1981, Myoblast aminophospholipid asymmetry differs from that of fibroblasts, *FEBS Lett.* **134**:75–78.
Sheetz, M. P., Febbroriello, P., and Koppel, D. E., 1982, Triphosphoinositide increases glycoprotein lateral mobility in erythrocyte membranes, *Nature* **296**:91–93.
Sheetz, M. P., and Singer, S. J., 1974, Biological membranes as bilayer couples. A molecular mechanism of drug-erythrocyte interactions, *Proc. Natl. Acad. Sci. USA* **71**:4457–4461.
Sheetz, M. P., and Singer, S. J., 1976, Equilibrium and kinetic effects of drugs on the shapes of human erythrocytes, *J. Cell Biol.* **70**:247–251.
Sheetz, M. P., Painter, R. G., and Singer, S. J., 1976, Biological membranes as bilayer couples, III. Compensatory shape changes induced in membranes. *J. Cell Biol.* **70**:193–203.
Sherman, I. W., 1979, Biochemistry of *Plasmodium* (malarial parasites), *Microbiol. Rev.* **43**:453–495.
Shinitzky, M., and Inbar, M., 1974, Difference in microviscosity induced by different cholesterol levels in the surface membrane lipid layer of normal lymphocytes and malignant lymphoma cells, *J. Mol. Biol.* **85**:603–615.
Shinitzky, M., and Rivnay, B., 1977, Degree of exposure of membrane proteins determined by fluorescence quenching, *Biochemistry* **16**:982–987.
Shinitzky, M., and Souroujon, M., 1979, Passive modulation of blood-group antigens, *Proc. Natl. Acad. Sci. USA* **76**:4438–4440.
Shinitzky, M., Dianoux, A.-C., Gitler, C., and Weber, G., 1971, Microviscosity and order in the hydrocarbon region of micelles and membranes determined with fluorescent probes. I. Synthetic micelles, *Biochemistry* **10**:2106–2113.
Shinitzky, M., Skornick, Y., and Haran-Ghera, N., 1979, Effective tumor immunization induced by cells of elevated membrane-lipid microviscosity, *Proc. Natl. Acad. Sci. USA* **76**:5313–5316.
Shlatz, G. S., and Marinetti, G. V., 1972, Hormone-calcium interactions with the plasma membrane of rat liver cells, *Science* **176**:175–177.
Shukla, S. D., and Hanahan, D. J., 1982, Identification of domains of phosphatidylcholine in human erythrocyte plasma membranes, *J. Biol. Chem.* **257**:2908–2911.
Silberkang, M., Havel, C. M., Friend, D. S., McCarthy, B. J., and Watson, J. A., 1983, Isoprene synthesis in isolated embryonic drosophila cells. I. Sterol-deficient eukaryotic cells, *J. Biol. Chem.* **258**:8503–8511.
Sillerud, L. O., and Barnett, R. E., 1982, Lack of transbilayer coupling in phase transitions of phosphatidylcholine vesicles, *Biochemistry* **21**:1756–1760.
Silverstein, S. C., Steinman, R. M., and Cohn, Z. A., 1977, Endocytosis. *Annu. Rev. Biochem.* **46**:669–722.
Singer, S. J., and Nicolson, G. L., 1972, The fluid mosaic model of the structure of membranes. *Science* **175**:720–731.
Skipski, V. P., Barclay, M., Barclay, R. K., Fetzer, V. A., Good, J. J., and Archibald, F. M., 1967, Lipid composition of human serum lipoproteins, *Biochem. J.* **104**:340–352.
Sklar, L. A., Hudson, B. S., and Simoni, R. D., 1975, Conjugated polyene fatty acids as membrane probes: Preliminary characterization, *Proc. Natl. Acad. Sci. USA* **72**:1649–1654.
Sklar, L. A., Hudson, B. S., and Simoni, R. D., 1976, Conjugated polyene fatty acids as fluorescent membrane probes: Model system studies, *J. Supramol. Struct.* **4**:449–465.
Sklar, L. A., Hudson, B. S., and Simoni, R. D., 1977a, Conjugated polyene fatty acids as fluorescent probes: Binding to BSA, *Biochemistry* **16**:5100–5108.

Sklar, L. A., Hudson, B. S., Petersen, M., and Diamond, J., 1977b, Conjugated polyene fatty acids as fluorescent probes: Spectroscopic characterization, *Biochemistry* **16**:813–819.

Sklar, L. A., Hudson, B. S., and Simoni, R. D., 1977c, Conjugated polyene fatty acids as fluorescent probes: Synthetic phospholipid membrane studies, *Biochemistry* **16**:819–828.

Sklar, L. A., Miljanich, G. P., and Dratz, E. A., 1979a, Phospholipid lateral phase separation and the partition of *cis*-parinaric acid and *trans*-parinaric acid among aqueous, solid lipid, and fluid lipid phases, *Biochemistry* **18**:1707–1716.

Sklar, L. A., Miljanich, G. P., Bursten, S. L., and Dratz, E. A., 1979b, Thermal lateral phase separations in bovine retinal rat outer segment membranes and phospholipids as evidenced by parinaric acid fluorescence polarization and energy transfer. *J. Biol. Chem.* **254**:9583–9591.

Sklar, L. A., Miljanich, G. P., and Dratz, E. A., 1979c, A comparison of the effects of Ca^{++} on the structure of bovine retinal rod outer segment membranes, phospholipids, and bovine brain phosphatidylserine, *J. Biol. Chem.* **254**:9592–9597.

Sklar, L. A., Doody, M. C., Gotto, A. M., and Pownall, H. J., 1980, Serum lipoprotein structure: Resonance energy transfer localization of fluorescent lipid probes, *Biochemistry* **19**:1294–1301.

Slama, J. S., and Rando, R. R., 1980, Lectin-mediated aggregation of liposomes containing glycolipids with variable hydrophilic spacer arms, *Biochemistry* **19**:4595–4600.

Slavik, J., 1982, ANS as a probe of membrane composition and function, *Biochim. Biophys. Acta* **694**:1–25.

Smith, R. J. M., and Green, C., 1974a, Fluorescence studies of protein–sterol relationships in human plasma lipoproteins, *Biochem. J.* **137**:413–415.

Smith, R. J. M., and Green, C., 1974b, The rate of cholesterol "flip-flop" in lipid bilayers and its relation to membrane sterol pools, *FEBS Lett.* **42**:108–111.

Steck, T. L., Fairbanks, G., and Wallach, D. F. H., 1971, Disposition of the major proteins in the isolated erythrocyte membrane. Proteolytic dissection. *Biochemistry* **10**:2617–2624.

Strittmatter, W. J., Hirata, F., Axelrod, J., Mallorga, P., Tallman, J. F., and Henneberry, R. C., 1979, Benzodiazepine and β-adrenergic receptor ligands independently stimulate phospholipid methylation, *Nature* **282**:857–859.

Strittmatter, W. J., Hirata, F., and Axelrod, J., 1981, Regulation of the β-adrenergic receptor by methylation of membrane phospholipids, in: *Advances in Cyclic Nucleotide Research*, Vol. 14 (J. E. Dumont, P. Greengard and G. A. Robison, eds.), pp. 83–91, Raven Press, New York.

Stryer, L., 1978, Fluorescence energy transfer as a spectroscopic ruler, *Annu. Rev. Biochem.* **47**:819–846.

Sun, A. Y., 1976, Aging and *in vivo* norepinephrine-uptake in mammalian brain, *Exp. Aging Res.* **2**:207–219.

Surewicz, W. K., and Leyko, W., 1982, Interaction of local anaesthetics with model phospholipid membranes. The effect of pH and phospholipid composition studied by quenching of an intramembrane fluorescent probe, *J. Pharm. Pharmacol.* **34**:359–363.

Tanaka, K.-I., and Ohnishi, S. I., 1976, Heterogeneity in the fluidity of intact erythrocyte membrane and its homogenization upon hemolysis, *Biochim. Biophys. Acta* **426**:218–231.

Tank, D. W., Wu, E.-S., and Webb, W. W., 1982, Enhanced molecular diffusibility in muscle membrane blebs: Release of lateral constraints, *J. Cell Biol.* **92**:207–212.

Tecoma, E. S., Sklar, L. A., Simoni, R. D., and Hudson, B. S., 1977, Conjugated polyene fatty acids as fluorescent probes: Biosynthetic incorporation of parinaric acid by *E. coli* and studies of phase transitions, *Biochemistry* **16**:829–835.

Teissie, J., 1981, Interaction of cytochrome C with phospholipid monolayers. Orientation and penetration of protein as functions of the packaging density of film, nature of phospholipids, and ionic content of aqueous phase, *Biochemistry* **20**:1554–1560.

Thulborn, K. R., Tilley, L. M., Sawyer, W. H., and Treloar, F. E., 1979, The use of n-(9-anthroyloxy) fatty acids to determine membrane fluidity and polarity gradients in phospholipid bilayers, *Biochim. Biophys. Acta* **558**:166–178.

Tombaccini, D., Ruggieri, S., Fallani, A., and Mugnai, G., 1980, Concanavalin A-mediated agglutinability of BALB/c 3T3 cells grown in media supplemented with different phosphatidylcholines, *Biochem. Biophys. Res. Commun.* **96**:1109–1115.

Trauble, H., and Sackman, E., 1972, Studies of the crystalline-liquid crystalline phase transition of lipid model membranes. III. Structure of a sterol–lecthin system below and above the lipid-phase transition, *J. Am. Chem. Soc.* **94**:4499–4510.

Vadas, E. B., Melancon, P., Braun, P. E., and Galley, W. C., 1981, Phosphorescence studies of the interaction of myelin basic protein with phosphatidylserine vesicles, *Biochemistry* **20**:3110–3116.

van Deenen, L. L. M., 1979, Structural organization and dynamics of phospholipids in red cell membranes, *Normal and Abnormal Red Cell Membranes* **1**:451–456.

Vanderkooi, J., and McLaughlin, A., 1975, Use of fluorescent probes in the study of membrane structure and function, in: *Biochemical Fluorescence: Concepts*, Vol. II (R. F. Chen and H. Edelhoch, eds.), pp. 737–765, Marcel Dekker, New York.

van Dijck, P. W. M., van Zoelen, E. J. J., Seldenrijk, R., van Deenen, L. L. M., and de Gier, J., 1976, Calorimetric behavior of individual phospholipid classes from human and bovine erythrocyte membranes, *Chem. Phys. Lip.* **17**:336–343.

van Meer, G., Poorthuis, B. J. H. M., Wirtz, K. W. A., Op den Kamp, J. A. F., and van Deenen, L. L. M., 1980, Transbilayer distribution and mobility of phosphatidylcholine in intact erythrocyte membranes. A study with phosphalidylcholine exchange protein, *Eur. J. Biochem.* **103**:283–288.

Vincent, M., de Foresta, B., Gallay, J., and Alfsen, A., 1982, Nanosecond fluorescence anisotropy decays of n-(9-anthroyloxy) fatty acids in dipalmitoylphosphatidylcholine vesicles with regard to isotropic solvents, *Biochemistry* **21**:708–716.

Vaughan, D. J., and Keough, K. M., 1974, Changes in phase transitions of phosphatidylethanolamine- and phosphatidylcholine-water dispersions induced by small modifications in the headgroup and backbone regions, *FEBS Lett.* **47**:158–161.

Wade, C. G., Baker, D. E., and Bartholomew, J. C., 1978, Selective fluorescence quenching of benzo[a]pyrene and a mutagenic diol epoxide derivative in mouse cells, *Biochemistry* **17**:4332–4337.

Waggoner, A. S., 1979, Dye indicators of membrane potential, *Annu. Rev. Biophys. Bioeng.* **8**:47–68.

Wampler, J. E., 1976, Fluorescence spectroscopy with on-line computers: Methods and instrumentation, in: *Modern Fluorescence Spectroscopy*, Vol. 1, (E. L. Wehry, ed.), pp. 1–44, Plenum Press, New York.

Ware, W. R., 1971, Transient luminescence measurements, in: *Creation and Detection of the Excited State*, Vol. 1A (A. A. Lamola, ed.), pp. 213–302, Marcel Dekker, New York.

Waring, A. J., Glatz, P., and Vanderkooi, J. M., 1979, Subzero temperature study of the inner mitochondrial membrane and related phospholipid membrane systems with the fluorescent probe, trans-parinaric acid, *Biochim. Biophys. Acta* **557**:391–398.

Watts, A., Volotovski, I. D., and Marsh, D., 1979, Rhodopsin-lipid association in bovine rod outer segment membranes. Identification of immobilized lipid by spin-labels, *Biochemistry* **18**:5006–5013.

Webb, W. W., Barak, L. S., Tank, D. W., and Wu, E.-S., 1981, Molecular mobility on the cell surface, *Biochem. Soc. Symp.* **46**:191–205.

Weber, G., 1953, Rotational brownian motion and polarization of the fluorescence of solutions, *Adv. Protein Chem.* **8**:415–460.

Weber, G., 1978, Limited rotational motion: recognition by differential phase fluorometry, *Acta Phys. Polon.,* **A54**:859–865.

Weber, G., 1981, Resolution of fluroescence lifetimes in a heterogeneous system by phase and modulation measurements, *J. Phys. Chem.* **85**:949–953.

Weltzien, H. U., 1975, Lysolecithin induced membrane alterations in thymocytes, *Z. Naturforsch. C. Biosci.* **30C**:785–792.

White, D. A., 1973, The phospholipid composition of mammalian tissues, in: *Form and Function of Phospholipids,* Chapter 16 (G. B. Ansell, J. N. Hawthore, and R. M. C. Dawson, eds.), pp. 441–482, Elsevier Scientific Publishing Company, New York.

Whitin, J. C., Clark, R. A., Simons, E. R., and Cohen, H. J., 1981, Effects of the myeloperoxidase system on fluorescent probes of granulocyte membrane potential, *J. Biol. Chem.* **256**:8904–8906.

Wise, G. E., 1982, Isolation of human erythrocyte inside-out vesicles alters their molecular architecture, *The Anatomical Record* **202**:317–324.

Wisnieski, B. J., and Iwata, K. K., 1977, Electron spin resonance evidence for vertical asymmetry in animal cell membranes, *Biochemistry* **16**:1321–1326.

Wolber, P. K., and Hudson, B. S., 1981, Fluorescence lifetime and time-resolved polarization anisotropy studies of acyl chain order and dynamics in lipid bilayers, *Biochemistry* **20**:2800–2810.

Wolf, D. E., Henkart, D., and Webb, W. W., 1980, Diffusion, patching, and capping of stearolylated dextrans on 3T3 cell plasma membranes, *Biochemistry* **19**:3893–3904.

Yeagle, P. L., Bensen, J., Greco, M., and Arena, C., 1982a, Cholesterol behavior in human serum lipoproteins, *Biochemistry* **21**:1249–1254.

Yeagle, P. L., Bensen, J., Boni, L., and Hui, S. W., 1982b, Molecular packing of cholesterol in phospholipid vesicles as probed by dehydrocholesterol, *Biochim. Biophys. Acta* **692**:139–146.

Yuli, I., Wilbrandt, W., and Shinitzky, M., 1981, Glucose transport through cell membranes of modified lipid fluidity, *Biochemistry* **20**:4250–4256.

Zwaal, R. F. A., Roelofsen, B., Comfurius, P., and van Deenen, L. L. M., 1975, Organization of phospholipids in human red cell membranes as detected by the action of various purified phospholipases, *Biochim. Biophys. Acta* **406**:83–96.

Chapter 3

Functional Aspects of Gram-Negative Cell Surfaces

Volkmar Braun, Eckhard Fischer, Klaus Hantke, Knut Heller, and Heinz Rotering

Department of Microbiology
University of Tübingen
D-7400 Tübingen, Federal Republic of Germany

1. OUTER MEMBRANE PROTEINS OF *ESCHERICHIA COLI*

1.1. Introduction

Escherichia coli, in particular *E. coli* K-12, and the bacteriophages that multiply in *E. coli* became the principal model organisms of molecular biology. Since the genetic material can be exchanged easily by conjugation or introduced by phage infection, a great deal of basic knowledge on the nature and organization of genes and their regulation and expression, and about metabolic pathways in the cytoplasm and in membranes was gained and is still obtained with this organism. It was certainly wise to concentrate on one system to unravel basic features of life on the level of single cells. The theories developed led also to fruitful concepts for eukaryotes. However, the world of microbiology does not consist solely of *E. coli* K-12. A wealth of properties expressed by various bacteria is not contained in *E. coli*. But even when we only consider *E. coli* there are many features that cannot be studied with the K-12 strain. The K-12 strain was isolated in 1922 and has been kept in the laboratory since then. During these six decades it has lost properties that were once essential for survival in the natural habitat. *E. coli* is usually a harmless inhabitant of the gut but some strains cause diarrhea, some are invasive and lead to enterocolitis, certain strains distribute in the urinary tract or the kidney, or appear in the blood and elicit severe septicemia, others invade the brain and cause meningitis. Investigations aimed at unraveling the pathogenic properties of *E. coli* defined up to now 164 O (lipopolysaccharide), 103 K (capsular), and 56 H (flagellar) antigens. All these different antigens are exposed at the cell surface. Therefore, *E. coli* is a very complex species with regard to its contact with the surroundings. The antigen armory enables the organism to adhere to certain tissues and to avoid cellular and humoral defense reactions of the host. In contrast to the situation in molecular biology where insights obtained with one organism apply in principle to most organisms, ecological parameters are tailored to a single organism in a certain environment. The question of why an *E. coli* strain lives in a certain environment is not only of medical but also of immense biological interest. For example, studies on resistance to antibiotics opened the field to what we now call gene technology (Cohen, 1976). Recombinant DNA techniques are presently being applied to separate the various virulence factors that contribute to an infectious disease, to study them separately, and then to put them together again in various combinations with the aim of understanding the causes of a disease.

For the scope covered by this publication series, we selected a few topics, which have been studied biochemically in some detail. We mainly focus on the structure and function of surface proteins because in this field the greatest progress has been made in recent years. This report is by no means comprehensive and we refer the reader to reviews in which certain aspects are more fully covered: the structure and function of capsular polysaccharides (K antigens) at the cell surface (Jann, 1983); structure, function, and assembly of lipopolysaccharides (entotoxins, O antigens) (Rietschel *et al.*, 1983; Lüderitz *et al.*, 1983; Tallayama *et al.*, 1983); the O and K antigens of *E. coli* (Orskov *et al.*, 1977); the surface structures of photosynthetic bacteria (Weckesser *et al.*, 1979); the enterobacterial common antigen (Mäkelä and Mayer, 1976); the primary structure, conformation, and biosynthesis of surface carbohydrates (Troy, 1979); surface compounds as phage and colicin receptors (Braun and Hantke, 1981; Koninsky, 1982); fimbriae as adhesins (Gaastra and De Graaf, 1982); biosynthesis and assembly of proteins in the outer membrane (Di Rienzo *et al.*, 1978; Osborn and Wu, 1980; Lugtenberg and van Alphen, 1983); outer membrane constituents that affect sensitivity to complement (Taylor, 1983). Additional reviews are cited in the following sections in which selected aspects are treated.

1.2. The Outer Membrane

To discuss functions related to the cell surface we briefly introduce the structure of the outer membrane. The components at the surface are embedded in the outer membrane and the way they function can only be understood when the whole membrane is considered. Moreover, the outer membrane is not a separate layer disconnected from the metabolism taking place mainly in the cytoplasm and the cytoplasmic membrane. It becomes increasingly evident that the whole cell envelope, comprising the outer membrane, the murein (peptidoglycan) layer, the periplasm and the cytoplasmic membrane (Figure 1), has to be regarded as one functional entity. The outer membrane consists of proteins, phospholipids, and lipopolysaccharide (glycolipid). The major proteins are present in the order of 10^5 copies per cell. The number of lipopolysaccharide molecules (Figure 2) roughly equals the total number of protein molecules (about 10^6), whereas the amount of phospholipids is five times

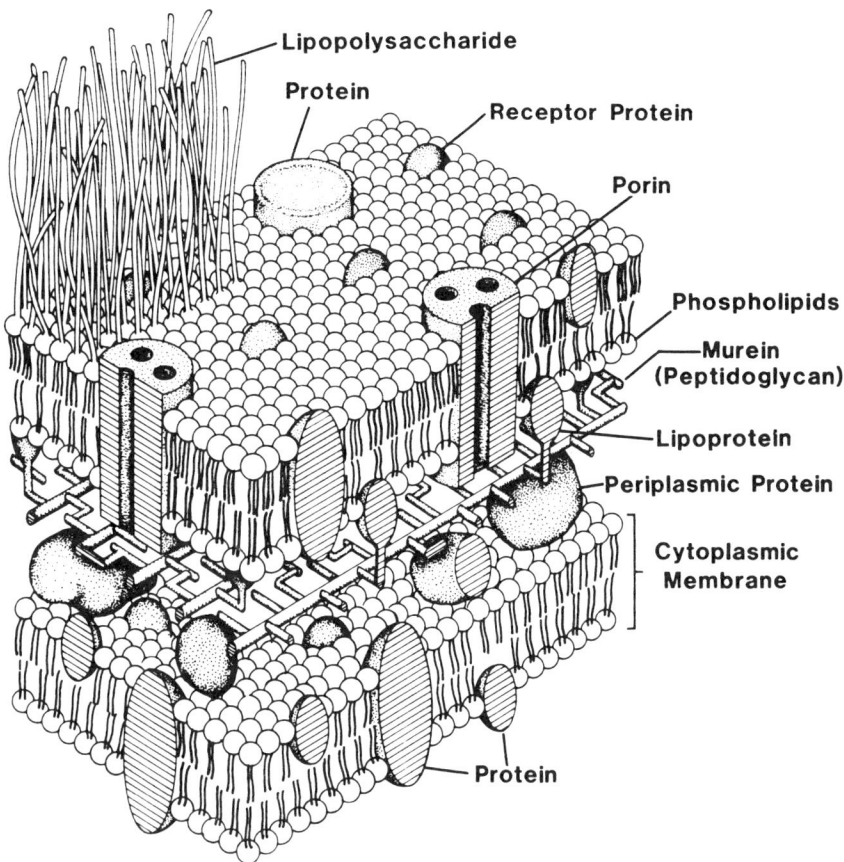

FIGURE 1. Model of the cell envelope of *E. coli* consisting of the cytoplasmic membrane, the periplasm, the murein (peptidoglycan), and the outer membrane. For the sake of clarity the individual components were not drawn to scale, and some components (e.g., lipopolysaccharide, proteins in the periplasm) were restricted to certain areas of the picture. (Drawn by H. Frank.)

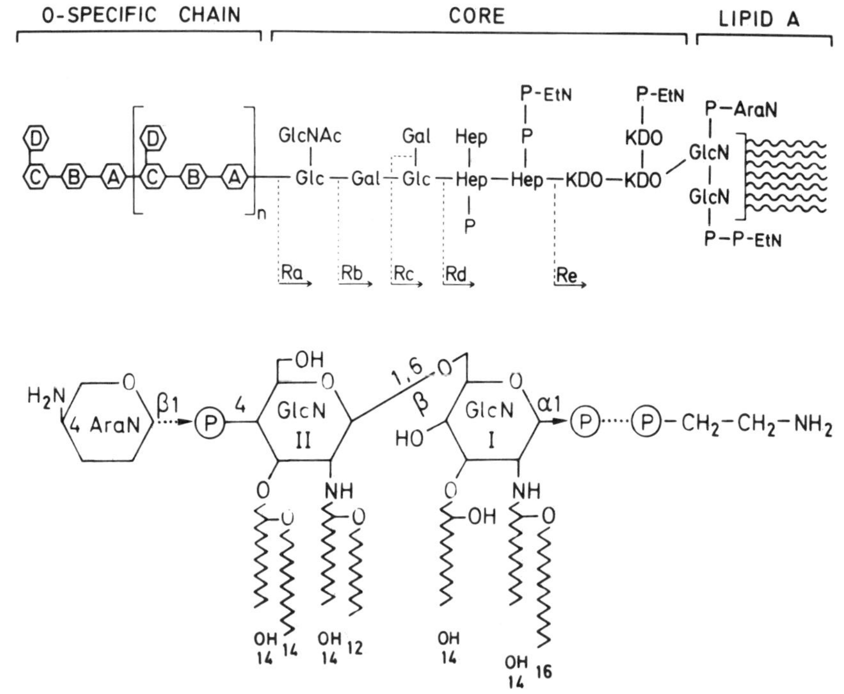

FIGURE 2. General structure of lipopolysaccharide of *Salmonella*, according to Lüderitz, Rietschel, Wollenweber, Kusumoto, and Redmont *et al.* (1983). Recent results show that only 1 KDO residue is located within the main chain and the other 2 residues form an $\alpha 2,4$ oligosaccharide linked by an $\alpha 2,4$ bond to the KDO in the sugar chain (Brade and Rietschel, 1984). A, B, C, D mark sugar residues. Glc, D-glucose, Gal, D-galactose; GlcN, D-glucosamine; GlcNAc, N-acetyl-D-glucosamine; Hep, L-glycero-D-manno-heptose; KDO, 2-keto-3-deoxy-D-manno-octonate; AraN, 4-amino-L-arabinose; P, phosphate; EtN, ethanolamine; hydroxy- and nonhydroxy fatty acids; Ra to Re indicate the various rough forms of lipopolysaccharide. The lower figure represents the present day view of the structure of lipid A.

higher. Lipopolysaccharide occurs at the surface, whereas the phospholipids are mainly confined to the inner leaflet of the outer membrane. Cells freshly isolated from their natural habitat usually contain polysaccharides at the surface that sometimes form capsules thicker than the cell body. In addition, several hundred fimbriae composed of identical protein subunits extend from wild-type cells (Figure 3) with which they adhere to human, animal, and plant tissues. One cell can express more than one fimbrial structure and they mainly determine the tissue specificity.

The area between the outer membrane and the cytoplasmic membrane is called periplasm or periplasmic space. It contains proteins, hydrolases, binding

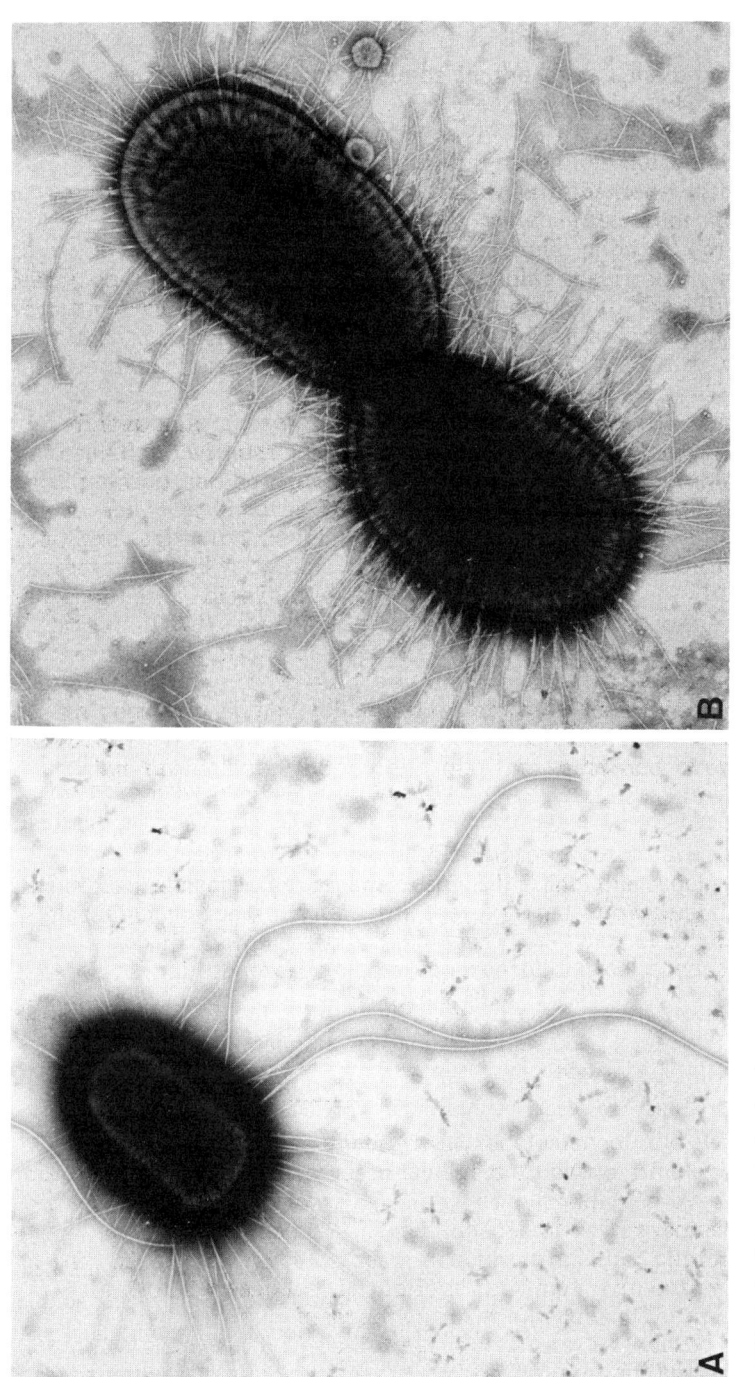

FIGURE 3. (A). *E. coli* showing fimbriae (short appendages) and flagellae (long). Magnification × 18,000. (B) *Proteus* showing fimbriae. Magnification × 30,000. (EM by H. Frank.)

proteins involved in transport and chemotaxis, and the murein (peptidoglycan) layer, whose basic structure seems to be essentially the same in all gram-negative bacteria.

1.3. Proteins in the Outer Membrane

The proteins firmly embedded in the outer membrane usually serve several functions. Some act as passive diffusion channels through which small hydrophilic substrates enter the periplasm. Others recognize substrates in a stereospecific manner. Most of them also serve as specific binding sites for certain phage and/or bacteriocins. In *E. coli* the preponderant secondary structure of the major outer membrane proteins is the β-structure. They are transmembrane proteins and those that form pores are assembled in trimers. In contrast, the major (murein) lipoprotein has a very high α-helical content, and there is no evidence for a transmembrane arrangement, nor have functions in the uptake of compounds or for binding of phage or colicins been discerned. It seems to be inserted into the inner half of the outer membrane and apparently serves a structural role.

Phage and colicins seemingly adapted to proteins that had beneficial functions for the cells. Although they usually kill their hosts, they have to be considered as constituents of a biological system in which many organisms live in a balanced relationship. Having phage and colicin receptors is detrimental for the individual strain but it is of advantage for the system as a whole. Moreover, transducing phages have the benefit of distributing genes among sensitive bacteria.

The protein composition of the outer membrane of gram-negative bacteria is unique in that a few species are present in large amounts. However, under certain growth conditions some of the minor proteins can also become major proteins. The number of proteins also varies, even within one species. For example, between one and four major proteins in the molecular weight range between 30,000 and 42,000 were identified electrophoretically in clinical isolates of *E. coli* (van Alphen *et al.*, 1982). *Shigella flexneri* contains a similar number of proteins. However, 14 additional polypeptides were found in invasive strains of *S. flexneri* that contained a 140-megadalton plasmid (Hale *et al.*, 1983). A similar plasmid was present in an enteroinvasive *E. coli* strain. Another extensive variation in the protein composition of the outer membrane of *Neisseria* related to pathogenicity is discussed in Section 2. The complex protein compositions frequently observed in bacteria freshly isolated from their natural environment should be kept in mind when the following rather simple situations in *E. coli* K-12 is described. This laboratory strain maintained primarily those properties that are required for growth in synthetic media.

1.4. *E. coli* K-12

E. coli K-12 when grown under standard conditions expresses four major outer membrane proteins, the major (murein-) lipoprotein, two proteins designated OmpC and OmpF (symbols of the structural genes *ompC, ompF*), and the OmpA protein *(ompA)*.

1.5. Lipoproteins

The major (murein-) lipoprotein was the first outer membrane protein and among the first membrane proteins whose primary structure was entirely determined (see Braun, 1975, where the early period of lipoprotein research is summarized). It was found as a constituent of the so-called rigid layer where it remained together with the murein (peptidoglycan) insoluble in boiling 4% sodium dodecyl sulfate. Later it was noticed that besides the murein-bound form twice as much lipoprotein is present in a free form (summarized by Di Rienzo *et al.*, 1978). It was also the first protein where a covalently linked lipid was proved by determination of the lipid structure and of the bond at the polypeptide chain (Hantke and Braun, 1973) (see Table I). The major lipoprotein is the most abundant protein in *E. coli* (about 7×10^5 copies per cell).

Mutants lacking the murein-lipoprotein survive but they shed off vesicles from the outer membrane, they are sensitive to EDTA, and they release proteins from the periplasm (Suzuki *et al.*, 1978). Double mutants lacking the lipoprotein and the OmpA protein (see later) grow only in the presence of high concentrations of Mg^{2+} and Ca^{2+} and form spheres (Sonntag *et al.*, 1978). These observations suggest that lipoprotein is important for outer membrane structure. The bound form plays a major role in the anchorage of the outer membrane to the underlaying murein. Mizushima and co-workers have shown in a series of publications describing reconstitution studies with isolated membrane components that lipoprotein bound to the murein facilitates the formation of ordered, regular hexagonal lattices on the murein layer composed of lipopolysaccharide and the OmpC protein, or the OmpF protein, or the LamB protein (Yamada *et al.*, 1981).

Five additional lipoproteins were discovered in the outer membrane of *E. coli* by labeling cells with [^{3}H]glycerol and [^{3}H]palmitate (Ichihara *et al.*, 1981). One of them, designated PAL, was found associated with murein. It is assumed from the labeling data that they contain the same lipid structure as the major lipoprotein. Their electrophoretic mobility suggests molecular weights between 16,000 and 52,000 so that they are larger than the major lipoprotein (mol. wt. 7200). They cannot be precipitated with antibodies to the major lipoprotein. However, a radioactive DNA probe containing 100 base-

Table I

Amino Acid Sequence of the Mature Form of the *E. coli* Major (Murein) Lipoprotein Compared with Those of *Serratia marcescens*, *Erwinia amylovora* and *Moraxella morganii*[a]

```
Fatty acid—O— CH₂
              |
Fatty acid—O— CH₂
              |
              S
              |
              CH₂
              |
Fatty acid-NH-  Cys Ser Asn Ala Lys Ile Asp Gln Leu Ser  Ser Asp Val Gln Thr Leu Asn Ala Lys Val Asp Gln Leu Ser Asn Asp Val Asn Ala   (30)
Serratia
Erwinia amylovora                                                          Thr
M. morganii                                 Phe                    Asp Asn                                              Lys
                                                                                                                                        (31)
E. coli            Met Arg Ser Asp Val Gln Ala Ala Lys Asp Asp Ala Ala Arg Ala Asn Gln Arg Leu Asp Asn Met Ala Thr Lys Tyr Arg Lys     (58)
Serratia                                                                                             Gln         His Ala
Erwinia amylovora  Ile                      Gln                                                      Gln         His
M. morganii        Ile              Ala                 Gln Glu                                      Gln Val Arg Ser         Lys        Thr
```

[a] In the latter cases only the amino acid replacements are given. For references consult the text.

pairs of the structural gene *lpp* of the major lipoprotein hybridized with more than seven *Hind*III fragments of the chromosomal DNA indicating structural similarities between several DNA regions and the lipoprotein gene (Huang *et al.*, 1982). It will be interesting to see whether the homologous sequences code for lipoproteins.

Labeling experiments with radioactive glycerol and palmitate indicate that the lipid residue of the major lipoprotein of *E. coli* is also present in the membrane-bound forms of certain penicillinases of gram-positive bacteria (Table II) that are secreted into the medium after further processing (Smith *et al.*, 1981; Lai *et al.*, 1981; Nielsen *et al.*, 1981; Nielsen and Lampen, 1982). Glyceryl cysteine sulfone released by hydrolysis from the performic acid oxidized lipoproteins was a further indication for a lipid structure of the *E. coli* type. The amino acid sequence around the modified cysteine residue is in the part of the signal peptide strikingly similar in all organisms whereas the structures extending to the *C*-terminal end (the mature forms of the lipoproteins) vary more extensively (Table II). The individual steps in lipid biosynthesis have mainly been studied by Henry Wu (Wu *et al.*, 1982).

Fatty acids covalently bound in ester linkage to hydroxyamino acids or in amide linkage to *N*-terminal amino acids of viral and nonviral membrane proteins were also found in eukaryotic cells (Maggee and Schlesinger, 1982; Carr *et al.*, 1982; Henderson *et al.*, 1983).

Comparison of the amino acid sequence of the major *E. coli* lipoprotein with the structure derived from DNA sequencing of other Enterobacteria indicate that the structure has largely been conserved (Table I). Most substitutions do not change the physicochemical properties of the amino acids. The major lipoprotein is commonly found in Enterobacteriaceae. In addition to the strains listed in Figure 4 an *E. coli* m-RNA probe of the *lpp* gene hybridized with *Hind*III endonuclease-generated fragments of the DNA of *Shigella dysenter-*

Table II
Comparison of the Amino Acid Sequences of the Proform of Lipoproteins Around the Lipid Attachment Site at the Cystein Residues Marked by Stars.[a]

E. coli	Leu Leu Ala Gly Cys Ser Ser Asn Ala Lys Ile Asp
Serratia	His Ser Ala Gly Cys Ser Ser Asn Ala Lys Ile Asp
Erwinia amylovora	Leu Leu Ala Gly Cys Ser Ser Asn Ala Lys Ile Asp
M. morganii	Leu Leu Ala Gly Cys Ser Ser Asn Ala Lys Phe Asp
P. mirabilis	Cys Ser Ser Asn Lys Asn Asp Asp
B. licheniformis 794/C	Ala Leu Ala Gly Cys Ala Asn Asn Glu Thr Asn
S. aureus PC1	Val Leu Ser Ala Cys Asn Ser Asn Ser Asn Ser

[a] For references see text.

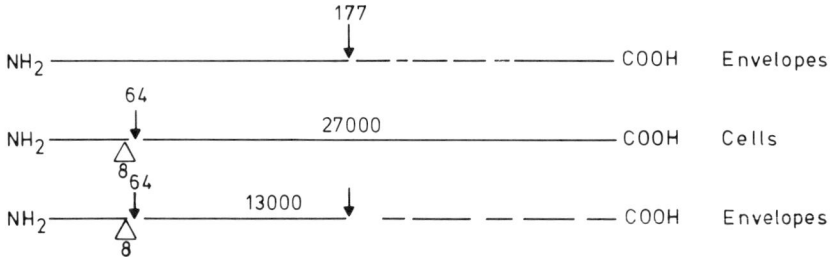

FIGURE 4. Cleavage site at amino acid residue number 177 of Pronase in isolated cell envelopes of wild-type (upper lane), and in cells and envelopes (middle and lower lane), respectively, of a mutant OmpA protein with an insert of eight amino acids.

iae, Salmonella typhimurium, Citrobacter freundii, Enterobacter aerogenes, and *Edwardsiella tarda* (Nakamura *et al.*, 1979). No hybridization was obtained with *Proteus mirabilis* but a lipoprotein similar in size to the one in *E. coli* was identified (Gruss *et al.*, 1975; Katz *et al.*, 1978). Moreover, a larger lipoprotein was isolated from *P. mirabilis* (Mizuno, 1981), and one of the major outer membrane proteins in *Pseudomonas aeruginosa* is a lipoprotein with basically the same structural characteristics (Figure 5) as the *E. coli* major lipoprotein (Mizuno and Kageyama, 1979). Protein covalently bound to murein was also found in *Legionella pneumophila* (Amano and Williams, 1983). The sample studied contained substantial amounts of fatty acids, suggesting that lipoprotein(s) may also be present in this genus.

1.6. Porins

The major outer membrane proteins OmpC and OmpF are called porins because they form water-filled channels through which small hydrophilic substrates diffuse into the periplasm from where they are usually actively taken

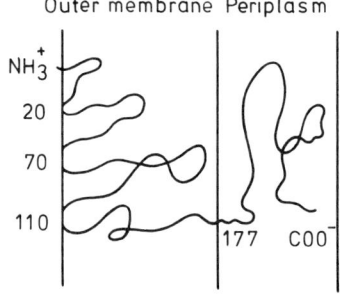

FIGURE 5. Model of the arrangement of the OmpA protein of Enterobacteria, based on the results obtained after Pronase hydrolysis and with mutants affected in phage binding (see citation in the text).

up by transport systems across the cytoplasmic membrane. Their transmembrane structure was derived from two observations. They serve as binding sites for phages and they are found associated with the murein when most of the constituents of the envelope are dissolved in 2% SDS at 60°C. The pore-forming ability of these proteins from *E. coli* and *Salmonella typhimurium* was directly demonstrated by inserting these proteins into planar lipid bilayers and into liposomes consisting of phospholipids and lipopolysaccharide. Incorporation into black lipid membranes resulted in the formation of ion-permeable pores (Benz *et al.*, 1978, 1979; Schindler and Rosenbusch, 1978).

Conductance measurements were extended to planar lipid bilayers formed from outer membrane vesicles (Schindler and Rosenbusch, 1981). Determination of the diffusion rate of sugars, β-lactam antibiotics, amino acids, peptides, nucleosides, and inorganic ions through reconstituted liposomes (Nikaido and Rosenberg, 1983) and into spheroplasts revealed that mainly three properties of the compounds are of importance: their size, the degree of hydrophilicity, and the charge. Basically, the porins allow the diffusion of all hydrophilic substrates up to a molecular weight of about 600 with preference for neutral and cationic compounds (Nikaido and Nakae, 1979). The porins are acidic proteins that may impede diffusion of anions by repulsion between the negative charges. There seems to be no stereochemical specificity of the porins for substrates. The porins form trimers and each of the subunits is apparently able to form a channel (Schindler and Rosenbusch, 1979, 1981). However, the three channels seem to combine into one at the inner side of the outer membrane (J. Rosenbusch, personal communication).

In addition to the OmpC and OmpF porins, encapsulated *E. coli* strains bearing acidic polysaccharides express a protein termed K, which can be separated electrophoretically from the OmpC, F proteins. The amino-terminal sequences of the K proteins, shown in Table III, indicate that they are largely homologous to the porins (Paakkanen *et al.*, 1979).

1.7. Outer Membrane Proteins That Are Part of Specific Transport Systems

In addition to the channels formed by the porins, the outer membrane contains a set of proteins that exhibit specificity for structurally related substrates, maltodextrins, nucleosides, and others that recognize a single substrate [iron(III) complexes and vitamin B_{12}]. The discovery of a selective transfer of substrates across the outer membrane via receptorlike proteins was unexpected. The outer membrane had rather been considered to be a mere permeability barrier for hydrophobic substances and for biopolymers that protects the cell against the bile salts and hydrolytic enzymes in the gut.

Table III
Comparison of the Amino-Terminal Sequences of the Outer Membrane Proteins OmpC, OmpF, PhoE and K of *E. coli*

	1												13
OmpC	Ala	Glu	Val	Tyr	Asn	Lys	Asp	Gly	Asn	Lys	Leu	Asp	Leu
OmpF		Ile					Val						
PhoE		Ile											
K						Ser				Tyr*			

	14				22				
OmpC	Tyr	Gly	Lys	Val	Asp	Gly	Leu	His	Tyr
OmpF			Ala	Val		Met			
PhoE			Lys	Ala					
K			Asn	Ala	X				

*a*Only the amino acid replacements related to the OmpC protein are indicated. The two other K proteins contain Asp instead of Tyr at the residue marked by a star. For references see text.

1.7.1. LamB Protein

The system studied in most detail is that of maltose uptake and catabolism. The LamB protein is encoded in the *malB* operon and expression of the *lamB* gene is induced by maltose (see reviews by Hall and Silhavy, 1981; Hengge and Boos, 1983). Maltodextrins bind to cells containing the LamB protein (Ferenci *et al.*, 1980). Stereochemical specificity of the LamB protein for maltose was shown by a liposome swelling assay (Luckey and Nikaido, 1980). Maltose diffused 40 times faster than sucrose, 10 times faster than lactose, and 8 times faster than cellobiose into liposomes containing the LamB protein. This clearly indicates that the LamB protein is able to form a specific transmembrane channel. It serves as receptor for phage lambda and is found associated with peptidoglycan like the porins, which indicates its transmembrane arrangement. However, many other solutes (e.g., monosaccharides) can diffuse through the LamB channel into liposomes faster than maltose. In whole cells there is a preference for maltose. This is explained by the maltose-binding protein (MalE) in the periplasm, which takes over maltose from the LamB protein. In fully induced cells the number of LamB trimers equals the number of molecules of the maltose-binding protein. The functional coupling of both proteins becomes essential for growth on larger maltodextrins as carbon source. Physical interaction of both proteins was shown when the LamB protein was adsorbed to a column loaded with binding protein covalently linked to Sepharose 6B beads. The importance of the complex of the two proteins for transport of maltodextrins revealed transport mutants with altered LamB and MalE proteins that exhibited weak binding to each other (Bavoil and Nikaido, 1981; Wandersmann and Schwartz, 1982; Bavoil *et al.,* 1983). This system had seemingly been developed for growth on low concentrations of maltose and on maltodextrins for which starch is an abundant source.

1.7.2. PhoE Protein

Synthesis of the PhoE protein is derepressed when cells are grown in low-phosphate medium (Overbeeke and Lugtenberg, 1980). It forms a channel that exhibits some specificity for anions, especially for inorganic and organic phosphate compounds (reviewed in Tommassen and Lugtenberg, 1982).

1.7.3. Tsx Protein

The Tsx protein, the product of the *tsx* gene, facilitates the diffusion of nucleosides and deoxynucleosides except cytidine and deoxycytidine through the outer membrane of *E. coli* K-12 (Hantke, 1976). The observation that expression of the *tsx* gene is controlled by *cytR* and *deoR* like the nucleoside uptake systems strongly supports the notion that the Tsx protein is part of the nucleoside transport systems (Krieger-Brauer and Braun, 1980).

1.7.4. Comparison of the Pore-Forming Proteins

The primary structure of the matrix protein of *E. coli* B that corresponds to the OmpF protein of *E. coli* K-12 has been determined (Chen *et al.,* 1982). The amino acid sequence of the latter was deduced from the nucleotide sequence of the *ompF* gene (Inokuchi *et al.,* 1982). Comparison of these two nearly identical proteins (three amino acid exchanges Gln/Glu) with the sequence derived from the *ompC* gene (Mizuno *et al.,* 1983) and the *phoE* gene (Overbeeke *et al.,* 1983) exhibits large homologies between these proteins. The mature OmpC, PhoE, and OmpF proteins consist of 346, 330, and 340 amino acids, respectively. The site where the signal sequence is cleaved off the proforms is identical for the three proteins (Ala-Ala-Glu) (see also Table IV in Section 4.) There are 210 identical amino acid residues between OmpC and PhoE and between OmpC and OmpF. On the DNA level the identity is also over 60%. There is a cluster of basic amino acids in the PhoE protein that is not present in the OmpC, OmpF proteins, which may explain the preference of the PhoE channel for anions. The diameters of the pores they form have been estimated to be 1.3 nm for OmpC, 1.2 nm for PhoE, and 1.4 nm for OmpF.

Antibodies raised against these porins cross-reacted with the heterologous proteins (Overbeeke *et al.,* 1980). They also reacted with porins isolated from other Enterobacteria, *Salmonella typhimurium, Klebsiella aerogenes, Enterobacter cloacae,* and *Proteus mirabilis.* They were, however, inactive against the LamB protein, which has an entirely different amino acid sequence (Clement and Hofnung, 1981). The diameter of the pore formed by the LamB protein in black lipid membranes was estimated to be 1.6 nm (Boehler-Kohler *et al.,* 1979) and thus somewhat larger than those formed by the porins.

1.8. The Receptor Proteins of Iron Uptake Systems

Proteins in the molecular weight range between 74,000 and 83,000 are strongly expressed when cells of *E. coli* or *S. typhimurium* are grown at iron-limiting concentrations (see reviews by Braun, 1985; Braun and Hantke, 1982; Neilands, 1982). These proteins serve as receptors for iron ligands, called siderophores, and are absolutely essential for the uptake of iron via siderophores. The FhuA (formerly TonA) protein (mol. wt. 78,000) was assigned to ferrichrome uptake, the FhuE protein (mol. wt. 76,000) to the uptake of ferric coprogen and ferric rhodotorulate (Hantke, 1983), the FepA protein (mol. wt. 81,000) to ferric enterochelin uptake, the FecA protein (mol. wt. 80,000) to ferric dicitrate uptake, and the Iut protein (mol. wt. 74,500) to ferric aerobactin uptake. In addition, two other proteins, Cir (mol. wt. 74,000) and Fiu (mol. wt. 83,000) are expressed in a low-iron medium. Hitherto, they were not related to an iron siderophore uptake system. For these proteins, except for FecA, Fiu, and FhuE, phages and/or colicins were found that use these proteins as receptors, indicating that they are exposed at the cell surface. All iron uptake systems require the product of the *tonB* gene, which is assumed to be involved in the energy-dependent release of the iron siderophores from the receptor proteins during the uptake process (Hancock and Braun, 1976; Hantke and Braun, 1978). The DNA sequence of the *tonB* gene (Postle and Good, 1983) suggests a signal peptide at the N-terminal end that is only found in proteins that are translocated across the cytoplasmic membrane. On the other hand, the *tonB* gene product was identified in the cytoplasmic membrane (Plastow and Holland, 1979). These data can be reconciled with the concept that the TonB protein is bound to the cytoplasmic membrane and extends into the periplasm. Such an arrangement would also meet the proposed function of the TonB protein as coupling factor between outer membrane receptors and the energized cytoplasmic membrane.

1.9. The BtuB Protein

The BtuB protein was the first for which it was shown that it is a constituent of a transport system. Mutants in the *btuB* gene are unable to take up vitamin B_{12}. They can easily be obtained by selecting mutants resistant for colicins E2 or E3, which use the BtuB protein as receptor. Growth in the presence of vitamin B_{12} represses synthesis of the receptor. When the synthesis of the BtuB protein was inhibited, cells became insensitive first to colicin E3, then to phage BF23, and only much later inactive in vitamin B_{12} transport. This indicates that the multifunctional properties of the receptor can be lost stepwise. It is assumed that the surroundings of the BtuB protein gradually change during growth of the cells and that the immediate ambient of the "old" BtuB protein becomes unfavorable for its function as phage and colicin receptor (Kadner and Bassford, 1978).

1.10. The OmpA Protein

Another multifunctional major outer membrane protein with about 10^5 copies per cell is the OmpA protein encoded by the *ompA* gene. It has the typical characteristic that its electrophoretic mobility decreases on heating to 100°C in 2% SDS. It serves as receptor for the phages TuII*, K3, and Ox2, and it is required for sensitivity of cells to colicins K and L. Only in *S. typhimurium* it is the receptor for a bacteriocin termed 4-59 (Stocker *et al.*, 1979). For F-mediated conjugation OmpA protein has to be present in the recipient so that stable aggregates are formed between donor and recipient cells (see Section 4). Interaction with lipopolysaccharide is a prerequisite for OmpA to be functional. Mutants lacking the OmpA protein also show a reduced rate of ferrichrome uptake (Coulton and Braun, 1979).

The primary structure of the OmpA protein has been determined by sequence analysis of the amino acids (Chen *et al.*, 1982) as well as by establishing the nucleotide sequence of the *ompA* gene (Beck and Bremer, 1980; Movva *et al.*, 1980). The mature polypeptide consists of 325 amino acids and is synthesized with a signal peptide of 21 amino acids. The C-terminal half of the molecule (148 amino acids) can be cleaved off by treating isolated cell envelopes (but not cells) with Pronase (Figure 4) (Bremer *et al.*, 1982). The sequence of the linkage between the Pronase-sensitive and the Pronase-resistant portion consists of four repeats of the sequence Pro-Ala, which exhibits analogy to the hinge region of immunoglobulins, the preferred cleavage site of papain (Chen *et al.*, 1980). These data suggest that the amino-terminal half of the molecule is embedded in the outer membrane and thus protected against Pronase, whereas the carboxy-terminal portion is accessible to Pronase and extends into the periplasm (Figure 5). Cleavage of the carboxy-terminal portion does not affect the receptor activity for the phages, which supports the proposed arrangement of the protein. A similar conclusion was derived from the receptor activity of an artificial hybrid protein in which 96 amino acid residues at the carboxy-terminal end were replaced by five residues of a cloning vector sequence (Movva *et al.*, 1980). Moreover, truncated OmpA polypeptides derived of cloned partial genes show that more than 133 but less than 193 of the amino-terminal amino acid residues are required to exhibit all OmpA-related functions, that is phage and colicin L sensitivity, and formation of stable mating aggregates (Bremer *et al.*, 1982). In contrast to the native protein, a mutant OmpA protein with eight additional amino acid residues (Figure 4) was cleaved when whole cells were incubated with Pronase. Pronase cuts the protein between residues 64 and 65. This region must therefore be exposed at the cell surface of the mutant OmpA protein. Additional results obtained with different experimental approaches provide a coherent model of the arrangement of the OmpA protein. Resistant mutants against the phages TuII*, K3h1, and Ox2 were isolated. Cross-resistance among the phage, insensitivity to colicins K and L, and impairment of conjugation were determined (Manoil, 1983). The nucleotide sequence of mutants of *E. coli* (Henning's group) and of other Enterobacteria (Freudl and Cole, 1983) were determined. They show that the functional sites at the polypeptide overlap considerably. The sequence around amino acid residue number 70 must be at the surface and serves as binding site for the phages K3 and TuII*. It also participates in the uptake of colicin L. The region around residue 25 is required for binding of phage Ox2, for conjugation, and for the uptake of colicin K. The region comprising residue 110 forms the binding site for the bacteriocin 4-50 in *S. typhimurium* (Freudl and Cole, 1983). For all these functions the amino-terminal portion of the molecule is sufficient. However, comparison of the primary structures of the OmpA protein among various Enterobacteria indicates that the carboxy-terminal portion remained more constant than the amino-terminal region, suggesting an important yet unknown function of the carboxy-terminus. The relation deduced from the OmpA sequences of the Enterobacteria shows that *Shigella* is closest to *E. coli* and *Serratia* is most distantly related to *E. coli* (G. Braun, S. Cole, U. Henning, private communication).

1.11. The TolC Protein

The TolC protein is a minor protein in the outer membrane of *E. coli* (Morona *et al.*, 1983) but lack of this protein has severe functional consequences. *tolC* mutants are insensitive to colicin E1, they are hypersensitive to detergents, dyes, and antibiotics; and they lack the OmpC and OmpF proteins in the outer membrane. Since they exert little effect on the transcription of the OmpF outer membrane protein, it is thought that the TolC protein influences posttranscriptional processing steps that several membrane proteins have in common (Morona and Reeves, 1982). Its primary structure has been deduced from the nucleotide sequence of the *tolC* gene (Hackett and Reeves, 1983). TolC is certainly an interesting protein because of the pleiotropic effects *tolC* mutants have on the formation and function of the outer membrane.

1.12. Enzymes in the Outer Membrane

Up to now mostly hydrolytic enzymes were found associated with the outer membrane. Some of them were in addition localized in the cytoplasmic membrane. Hitherto, a functional necessity for these enzymes in the cell's physiology was not established. In most cases it remains uncertain whether artificial redistribution of the enzymes occurs during preparation and separation of the outer and cytoplasmic membrane. Outer membrane protease was found that released the A and B subunits of nitrate reductase from the isolated cytoplasmic membrane (Mac Gregor *et al.*, 1979). The FepA receptor protein in isolated cell envelopes was degraded by protein a (Fiss *et al.*, 1982) but hydrolysis played no physiological role since mutants lacking protein a took up iron(III) via enterochelin with unchanged rate. It was also shown that cleavage of colicin I was not required for colicin activity (Bowless and Konisky, 1980). It is also questionable whether proteolytic cleavage of the proforms of the periplasmic alkaline phosphatases (Inouye and Beckwith, 1977), of the M13 coat protein (Wicker, 1980), and of additional exported proteins to the mature polypeptides by signal peptidase(s) has to occur by the enzyme(s) in the outer membrane in addition to the same enzyme(s) located in the cytoplasmic membrane. Comparison of different publications does not permit deduction of how many proteases are genuine in the outer membrane (Regnier, 1981; Pacaud, 1982).

The first well-characterized hydrolase to be found in the outer membrane was the phospholipase A of *E. coli* (Scandella and Kornberg, 1971), whose structural gene, *pldA*, has been cloned into high copy vectors (de Geus *et al.*, 1983).

It is feasible that hydrolases in the outer membrane gain access to substrates too large to be taken up, so that they have to be hydrolyzed into fragments that are then transported into the cells.

2. THE OUTER MEMBRANE OF *NEISSERIA*

Neisseria gonorrhoeae is the cause of a rather commonly reported communicable disease, so there is considerable interest into factors contributing to its pathogenicity. It is a gram-negative bacterium with an outer membrane that shares many characteristics with the outer membrane of other gram-negative organisms. The overall architecture is assumed to be similar to the well-studied *E. coli* outer membrane. The outer leaflet of the membrane contains mainly lipopolysaccharide, the inner leaflet phospholipids. Only very few protein species make up the 60% protein content of this membrane: a porin-type protein I, one or more type II proteins, which show a heat-modifiable mobility in SDS gels, and a protein III (Blake and Gotschlich, 1983). All these proteins appear in relatively high copy numbers and are accompanied by only a few minor proteins. In addition, fresh isolates are covered with pili. All these surface components seem to contribute to the pathogenic properties of the gonococci.

Pili certainly have a vital role in keeping the *Neisseria* at the surface of epithelia, which are constantly washed by mucus and other secretions. This is confirmed by the fact that most freshly isolated gonococci are piliated. On further culturing under nonselective conditions, unpiliated forms also appear. It was demonstrated that these unpiliated forms are less pathogenic than the piliated forms. Moreover, a comparison *in vitro* showed that the piliated forms adhere much better to human epithelia than the unpiliated forms (Schoolnik *et al.*, 1983; Watt and Ward, 1980). However, in recent years this simple picture has become more complicated. The unpiliated organisms can revert again to piliated ones. An investigation of Lambden *et al.*, (1980) showed that a new pilus type may appear that is different from the first one. The example studied showed that the two pili types had different affinities. The α-type pili strongly bound to buccal epithelial cells in a pH-dependent manner while the β-type pili showed a rather low, pH-independent binding. Pili from the α-type could also be distinguished from the β-type by the molecular weight of the pilin subunits (19,500 and 20,500, respectively). The involvement of sialic acid as part of the receptor for the α-pili was demonstrated. Recently, the same strain was shown to be capable of even further pilus variation. From subcutaneous chambers implanted into guinea pigs two new pilus-type variants designated γ and δ were isolated that differed from α and β by the molecular weights, the amino acid composition, and the isoelectric points of the subunits (Lambden *et al.*, 1982). Within the species *Neisseria gonorrhoeae* not only are there more than 50 serologically different pilus types (Brinton, cited in Meyer *et al.*, 1982) but even a single clone is able to elaborate four different pilus types. The next years will certainly show if this is an upper limit, or if a single clone is able to vary even further its pilus structure. It is assumed that this variation enables gonococci to escape the defense of their host and it may also allow adaptation to changing environments.

Although so many different pilus types can be differentiated serologically, it seems that all gonococcal pili have a common antigenic determinant, which, however, is immunogenic only in some pili (Brinton *et al.*, 1978). The pilin proteins from four serologically different types of pili had an identical N-terminus up to residue number 29 (Hermodson *et al.*, 1978), which was identical to the N-terminus of a pilin isolated from *Neisseria meningitidis*. Further comparison revealed also a high homology to the N-terminal sequence of pilin isolated from *Moraxella nonliquefaciens* and *Pseudomonas aeruginosa* (Schoolnik *et al.*, 1983). The similarity of the N-terminal ends of serologically different *N. gonorrhoeae* pili was also demonstrated by characterization of the three peptides CNBr1 (residue 1-7, *N*-terminal), CNBr2 (residue 8 to about 84) and CNBr3 (*C*-terminal end) obtained by cleavage with cyanogen bromide (CNBr) (Schoolnik *et al.*, 1983). The peptides CNBr2 and CNBr3 were compared for their ability to inhibit agglutination of human erythrocytes by homologous and heterologous pili. The fragment CNBr2 was able to inhibit hemagglutination of both types of pili. CNBr2 fragments of two different pili had a highly similar tryptic peptide map since 13 from 14 and 16 peptides, respectively, seemed to be identical. It was concluded that the fragment CNBr2 contained the region that determines pilus binding to erythrocytes and that this binding domain is highly conserved in pili from different gonococcal strains. These conclusions could be further substantiated by comparing antisera prepared against different CNBr2 fragments. The antibodies bound to homologous and heterologous CNBr2 fragments, and interestingly also to homologous and heterologous pili. In contrast, antisera against the different *C*-terminal peptides CNBr3 reacted only with the pilus from which the peptide was isolated. Furthermore, a tryptic peptide map of two different CNBr3 fragments showed only 10 out of 20 and 22 peptides respectively to be identical.

The following picture for a gonococcal pilus subunit was suggested (Figure 6). The *N*-terminal part with hydrophobic amino acid residues may have a role in pilin polymerization, the middle part carries a highly conserved domain for receptor binding and is immunorecessive, while the highly variable *C*-terminal part is immunodominant. There is some hope of developing a vaccine against *N. gonorrhoeae* based on the constant receptor binding site.

This picture has possibly to be modified to explain the observed specificities of some pili for certain tissues (Lambden *et al.*, 1980). In addition, further investigation of the pilin structure seems to be necessary to clarify the position

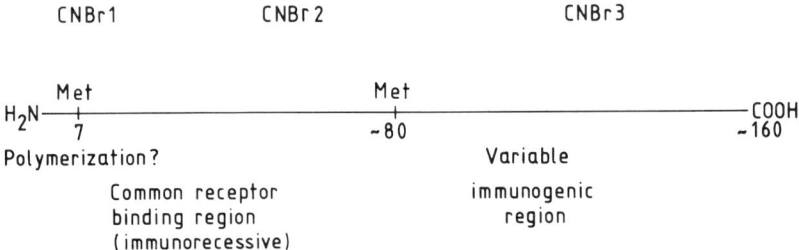

FIGURE 6. Functional and antigenic domains of a gonococcal pilus subunit (according to Schoolnick *et al.*, 1983). Three fragments are obtained from a pilus subunit after cleavage with cyanogen bromide (CNBr). The *N*-terminal part is possibly necessary for interaction of the subunits to form the pilus structure. The CNBr2 fragment contains a receptor-binding domain that is common to all gonococcal pilus types and that is not, or is only weakly, immunogenic in the pilus. Fragment CNBr3 carries the type-specific immunodeterminant of the pilus.

and function of the galactose residues that have been reported to be bound to pilin (Robertson et al., 1977) and to influence binding to host cell receptors (Gubish et al., 1982).

The colony forms of *N. gonorrhoeae* are highly variable, and one factor determining this variation is the change between piliated and unpiliated forms. The change from piliated to unpiliated forms occurs at a frequency of about 2×10^{-3} per colony-forming unit and per generation while the reverse seems to be less frequent (Mayer, 1982). For a better understanding of this genetic instability, pilus genes were cloned into *E. coli* and the DNA of the gene was used as a probe in different colonial morphology types (Meyer et al., 1982). It turned out that the conversion from the pilus positive to the pilus negative type was accompanied by a chromosomal rearrangement of pilus genes in the gonococcal chromosome. It is unknown which mechanisms bring about these gene rearrangements, but at first sight one is reminded of the phase variation of *Salmonella* strains, or the change of host range in bacteriophage Mu by the G loop.

Not only pilus production but also opacity of the colonies is a highly variable character of gonococci. Colony opacity changed from opaque to transparent and from transparent to opaque at a frequency of about 2×10^{-3} per colony-forming unit and per generation time (Mayer, 1982). This change in opacity is correlated with a change in the type or content of protein II in the outer membrane. Five different types of protein II have been differentiated in one strain, whereby one and more different protein II species were demonstrated in one colony type (Lambden et al., 1979; Swanson, 1982). Mostly these different protein II species were associated with a certain degree of opacity so that there was a graduation and not a clear-cut difference between opaque and transparent colony types (Swanson, 1982).

The variation of protein II species seems to be important for virulence, serum sensitivity, and adherence (Lambden et al., 1979) since the same strain isolated from different locations of one patient (Draper et al., 1982) or at different times in the menstruation cycle showed typical differences in opacity and protein II content (James and Swanson, 1978). Pili and often protein II variations were also found in gonococci isolates from infected consorts of the female cervix and urethra and the male partner (Duckworth et al., 1983).

One-third of the antibodies made against a seemingly unpiliated strain was directed against protein II while the rest was against LPS ($\frac{1}{3}$) and protein I ($\frac{1}{3}$) (Diaz and Heckels, 1982). It is assumed that the different protein II species observed form a protein family that varies mainly in the exposed hydrophilic parts of the protein (Heckels, 1981) and therefore elicits rather different antibodies. The genetic mechanism of protein II variation may be similar or connected to that observed for the pili (Meyer et al., 1982).

Protein I has also been implicated to be a major factor determining serotype diversity of gonococci (Johnston, 1980). But only three major types of protein I have been differentiated according to their molecular weight (Swanson, 1979). A very interesting hypothesis on the function of protein I during infection has been forwarded by Blake and Gotschlich (1983). Protein I is the major porin in the outer membrane of gonococci. It creates hydrophilic channels with preference for cations (Greco et al., 1980). The pore diameter seems to be larger than that of the E. coli porins (Douglas et al., 1981).

It has been observed that the protein was transferred from gonococci into red blood cells or into artificial lipid bilayers. The asymmetric protein in these membranes had a reverse orientation compared to the one in the gonococcal outer membrane. This was elegantly demonstrated by the different proteolytic sensitivity of the protein in the membrane depending on its orientation (Blake and Gotschlich, 1983). It may be that the generation of ion-permeable pores near the attachment site of the gonococcus elicits ion movements across the membrane of the epithelial cell. This may be a signal that helps the gonococcus to be taken up by the epithelial cell. However, up to now no data are available that substantiate this hypothesis (Blake and Gotschlich, 1983).

In conclusion, several outer membrane constituents of *N. gonorrhoeae*, pili, protein II, and also possibly protein I are highly variable between different strains and even within a single strain. This variability seems to be an adaptation to its highly specialized parasitic life since humans are the only natural host.

Another aspect of this specialization is possibly the way in which *N. gonorrhoeae* and *N. meningitidis* acquire iron. It has been found that iron is one of several factors that influence the outcome of an infection. Some *Neisseria* seem to have evolved a rather direct way to acquire the necessary iron from their host. These strains are able to use transferrin, the major iron carrier in the serum (Mickelsen and Sparling, 1980; Archibald and DeVoe, 1979). Transferrin is a protein with a molecular weight of 78,000; it exerts a high iron affinity and it is usually 30% saturated with iron. This results in a very low free-iron concentration in the serum, which is thought to be the reason for its bacteriostatic properties. In a comprehensive study 29 gonococci and 21 meningococci were able to use transferrin as an iron source. However, of the 45 nonpathogenic *Neisseria* tested only 10 were able to use transferrin (Mickelsen and Sparling, 1980).

A similar iron-binding protein is lactoferrin, which has an even higher iron-binding affinity and retains its iron-binding capacities below pH 6 in contrast to transferrin. Lactoferrin is mainly found in secretory fluids and in secondary granules of polymorphonuclear leucocytes and is assumed to have an antimicrobial effect by lowering the available iron for microorganisms. A protein with a similar function is conalbumin, found in the hen egg white. In a survey, Mickelsen *et al.* (1982) found that all 15 meningococci, 53% of 59 gonococci, and only 24% of 33 commensal *Neisseria* could use iron bound to lactoferrin for growth. However, conalbumin was not able to supply iron to any of the 95 *Neisseria* strains tested (Mickelsen and Sparling, 1981). This indicates that the removal of iron from transferrin and lactoferrin is a rather specific mechanism. *Salmonella paratyphi B* is also able to remove iron from transferrin *in vivo*, but this process is enterochelin-dependent (Tidmarsh and Rosenberg, 1981). In contrast, the mechanism of iron removal by meningococci seems to be different. A direct contact between transferrin and the cells is necessary (Simson *et al.*, 1982). Further elucidation of this rather unusual iron uptake system for bacteria will be interesting. In addition, it was claimed that *N. gonorrhoeae* and *N. meningitidis* produce iron-chelating hydroxamates (Yanccy and Finkelstein, 1981), although these iron uptake systems do not seem to work very efficiently, since they were not detected by another group (Archibald and DeVoe, 1980). In *N. gonorrhoeae* some outer membrane proteins in the molecular weight range of 80,000 were induced under iron-limiting conditions, indicating the presence of iron uptake systems organized similar to those of other gram-negative bacteria (Norqvist *et al.*, 1978).

3. THE OUTER MEMBRANE OF *PSEUDOMONAS AERUGINOSA*

Enteric bacteria like *E. coli* and *S. typhimurium* have relatively narrow pores in their outer membranes that accommodate only substances with a molecular weight up to 600. As pointed out by Nikaido (1979), this may be an adaptation to the environment of the enteric bacteria and not typical for porins of gram-negative outer membranes. In fact, the porins of *Pseudomonas aeruginosa* allow diffusion of molecules with a molecular weight up to 6000. These results, obtained with outer membrane vesicles or with porin-reconstituted vesicles, were further substantiated by the incorporation of the purified porin protein F into black lipid bilayers and measurement of the conductance increase.

From the high conductance steps it was calculated that the pores have a diameter of 2.2 nm, which is larger than that of the *E. coli* OmpF porin (1.4). However, most of the pores seem to be closed (100–300 times less open pores formed by protein I in *P. aeruginosa* than open porins in *E. coli*) (Benz and Hancock, 1981). It is assumed that this is also the case *in vivo* and it may be the reason for the relatively high resistance of *P. aeruginosa* against a wide range of antibiotics (Yoshimura and Nikaido, 1982; Angus *et al.*, 1982).

Permeability of the outer membrane was studied *in vivo* by measuring the rate with which β-lactams and phosphate esters were cleaved by periplasmic enzymes of intact cells. The permeability was 100-fold lower than that observed in *E. coli*. A second barrier shielding the periplasmic enzymes seems improbable, since the apparent K_M values for the uptake of various carbon source compounds were above 20 μM in *P. aeruginosa*. In *E. coli* the K_M values are mostly below 1 μM. One exception was the apparent K_M below 1 μM observed for glucose (Yoshimura and Nikaido, 1982). This is explained by the fact that glucose induced a protein designated D1, which functions as a porin (Hancock and Carey, 1980). It is assumed that protein D1 serves together with an inducible glucose-binding protein in glucose transport. It is an open question whether there are growth conditions at which *P. aeruginosa* opens more than 1 out of 400 pores—for example, when it grows in distilled water from hospitals on air-borne contaminants (Favero *et al.*, 1971); perhaps it affords to have a less permeable outer membrane because it is able to grow on a rather broad spectrum of 70 to 80 carbon sources.

Under phosphate-limiting conditions the porin protein P is found in the outer membrane of *P. aeruginosa*, which is coregulated with alkaline phosphatase and phospholipase C. The channel has been calculated to have a diameter of 0.7 nm and is clearly anion specific, indicating that it should have a role in phosphate uptake (Hancock *et al.*, 1982). In *E. coli* the PhoE protein is similarly regulated and has similar functions. However, the pore diameter of 1.2 nm is much larger, suggesting that the pore has been adapted to polyphosphates (see also Section 1).

The genus *Pseudomonas* is divided into two groups: one is able to produce fluorescent pigments under low-iron growth conditions and the other is not. The structures of pseudobactin (Teintze *et al.*, 1981) and pyochelin (Cox *et al.*, 1981) and the partially characterized structures of ferribactin and pyoverdine (Maurer *et al.*, 1968; Philson and Llinas, 1982) are totally different from the known siderophores used by the enteric bacteria such as *E. coli* and *Salmonella*. From a nonfluorescent *Pseudomonas*, nocardamine (ferrioxamine E) was isolated (Meyer and Abdallah, 1980).

Some studies indicate that also in these organisms proteins in the molecular weight range of 70,000–80,000 are induced under iron limitation and in one case one of these proteins seems to be a receptor for pyocin S2 and for a phage (Ohkawa *et al.*, 1980).

Specific binding of ^{59}Fe pyochelin to outer membranes of *Pseudomonas aeruginosa* was demonstrated (Sokol and Woods, 1983). The receptor protein had a molecular weight of 14,000 although in the strain used, proteins in the molecular weight range of 80,000 were induced. This low molecular weight compound is rather unusual since all receptors for siderophores identified in outer membranes have a molecular weight in the range of 70,000–80,000.

The production of pseudobactin has been implicated in the protective effect of certain *Pseudomonas* strains in so-called disease suppressive soils. In these soils the vascular wilt disease from *Fusarium* sp. and the take-all disease

(Gaeumannomyces graminis) are suppressed and it is assumed that the bacterial siderophores complex iron(III) efficiently, rendering it unavailable to the pathogens (Kloepper *et al.*, 1980).

4. EXPORT OF PROTEINS

4.1. Introduction

Prokaryotes such as gram-negative bacteria do not contain distinct organelles. However, the gram-negative cell envelope forms four separate compartments: the cytoplasm, the cytoplasmic membrane, the outer membrane, and the aqueous space between both membranes, called the periplasm. Each compartment contains a unique set of proteins (Osborn and Wu, 1980). Since protein synthesis takes place in the cytoplasm, mechanisms must exist for translocating newly synthesized proteins through membranes and distributing them to the different compartments or to the surrounding medium.

During recent years a wealth of information concerning the export process has been accumulated, not least by particular genetic strategies applied in isolating mutants defective in protein export. For detailed discussion of such strategies and for comprehensive analyses of protein export, the reader is referred to several recent reviews (Michaelis and Beckwith, 1982; Inouye and Halegoua, 1980; Osborn and Wu, 1980; Garwin and Beckwith, 1982; Hall and Silhavy, 1981; Randall and Hardy, 1982, Silhavy *et al.*, 1983).

4.2. Models for Protein Transport

Many exported proteins are synthesized in a precursor form containing amino-terminal signal peptides (or signal sequences), which are not present in the mature proteins. According to the signal hypothesis of Blobel and Dobberstein (1975), a nascent polypeptide destined for export binds to the membrane via the signal sequence. Binding is suggested to initiate formation of a complex of membrane proteins and ribosomes. This complex constitutes a membrane pore through which the growing polypeptide is cotranslationally extruded. At the noncytoplasmic face of the membrane the signal sequence is proteolytically cleaved off to generate the mature protein.

An alternative model, the membrane trigger hypothesis, has been proposed by Wickner (1979). In this model the signal sequence is suggested to promote folding of the precursor into a conformation that is soluble in the cytoplasm. On interaction with the membrane, a conformational change of the protein is triggered that allows insertion of the protein into the lipid bilayer. Cleavage of the signal sequence renders the process irreversible. In contrast to the signal hypothesis, this model does not require membrane-bound polysomes, nor ordered vectorial export, nor any cellular export system.

4.3. Signal Sequence

4.3.1. Structure of the Signal Sequence

Precursor proteins containing a signal sequence show a higher molecular weight than the mature species. Thus, they can be detected by means of sodium dodecylsulfate polyacryamide gel electrophoresis. Because of their short half-life *in vivo*, they can only be seen if cells are pulse-labeled for a few seconds. The rapid and quantitative chase into the mature proteins clearly establishes their nature as true intermediates in the maturation of exported proteins. Using techniques such as *in vitro* protein synthesis, phage infection of irradiated cells, plasmid-directed protein synthesis in minicells, and protein sequencing together with DNA sequencing, many bacterial proteins were shown to be synthesized via precursor forms (for a review see Michaelis and Beckwith, 1982).

As far as their sequences have been determined, signal peptides from different proteins show only little homology (Table IV). Exceptions are the lipoprotein and the arabinose binding protein, whose signal peptides are very similar (Wilson and Hogg, 1980). Interestingly, both proteins are localized in different compartments, suggesting that the signal sequence does not determine the final localization of a protein. Nevertheless, signal peptides do exhibit striking similarity in their degree of polarity. They all start with two to eight positively charged, basic amino acids at the amino-terminus, followed by 15–19 uncharged, mainly hydrophobic amino acids (Michaelis and Beckwith, 1982).

According to Austen (1979) the amino acid sequences indicate highly ordered helical conformations with very little random coiling for the signal peptides.

To relate the structure of the signal peptides to their function, Inouye and Halegoua (1980) propose in their loop model that the positively charged, basic amino-terminal segment binds to the inner face of the cytoplasmic membrane by ionic interaction with the negatively charged surface. Insertion of the hydrophobic segment into the lipid bilayer forms a loop structure that initiates protein export.

4.3.2. Role of the Signal Sequence

The essential role of the signal sequence for protein export has been established mainly by genetic studies. The powerful tool of gene fusion (Casadaban, 1976) has been used to demonstrate that the cytoplasmic protein β-galactosidase can be directed to two different compartments of the cell envelope. Cer-

Table IV
Signal Sequences of Prokaryotic Proteins

Proteins	Sequences	References
Periplasmic Proteins		
Alkaline phosphatase	MetLysGlnSerThrIleAlaLeuAlaLeuLeuProLeuLeuPheThrProValThrLysAlaArg′	Inouye et al., 1982a
ampC β-lactamase	MetPheLysThrThrLeuCysSerAlaLeuLeuIleThrAlaSerCysSerThrPheAlaAla′	Jaurin and Grundstrom, 1981
TEM β-lactamase	MetSerIleGlnHisPheArgValAlaLeuIleProPhePheAlaAlaPheCysLeuProValPheAlaHis′	Suttcliffe, 1978
Histidine-binding protein	MetLysLysLeuAlaLeuSerLeuSerLeuValLeuAlaPheSerSerAlaThrAlaAlaPheAlaAla′	Higgins and Ames, 1981
Isoleucine–valine-binding protein	MetAsnIleLysGlyLysAlaLeuLeuAlaGlyCysIleAlaLeuAlaPheSerAsnMetAlaLeuAla′	Landick and Oxender, 1982
Leucine-binding protein	MetLysAlaAsnAlaLysThrIleIleAlaGlyMetIleAlaLeuAlaIleSerHisThrAlaMetAlaAsp′	Oxender et al., 1980
Lysine–arginine–ornithine-binding protein	MetLysLysThrValLeuAlaLeuSerLeuLeuIleGlyLeuGlyAlaThrAlaAlaSerTyrAlaAla′	Higgins and Ames, 1981
Maltose-binding protein	MetLysIleLysThrGlyAlaArgIleLeuAlaLeuSerAlaLeuThrThrMetMetPheSerAlaSerAlaLeuAlaLys′	Bedouelle et al., 1980
Arabinose-binding protein	MetLys ? ThrLysLeuValLeuGlyAlaValIleLeuThrAlaGlyLeuSer ? GlyAla ? AlaGlu	Wilson and Hogg, 1980

Outer Membrane Proteins

Protein	Sequence	Reference
Lipoprotein	MetLysAlaThrLysLeuValLeuGlyAlaValIleLeuGlySerThrLeuLeuAlaGly'Cys	Nakamura and Inouye, 1979
Lipoprotein (*S. marcescens*)	MetAsnArgThrLysLeuValLeuGlyAlaValLeuGlyAlaValIleLeuGlySerHisSerAlaGly'Cys	Nakamura and Inouye, 1980
Lipoprotein (*E. amylovora*)	MetAsnArgThrLysLeuValLeuGlyAlaValIleLeuGlySerThrLeuLeuAlaGly'Cys	Yamagata *et al.*, 1981
LamB protein	MetMetIleThrLeuArgLysLeuProLeuAlaValAlaAlaAlaGlyValMetSerAlaGlnAlaMetAlaVal	Hedgpeth *et al.*, 1980
OmpA protein	MetLysLysThrAlaIleAlaIleAlaValAlaLeuAlaGlyPheAlaThrValAlaGlnAla'Ala	Movva *et al.*, 1980
OmpA protein (*S. dysenteriae*)	MetLysLysThrAlaIleAlaIleThrValAlaLeuAlaGlyPheAlaThrValAlaGlnAla'Ala	Braun and Cole, 1982
OmpC protein	MetLysValLysValLeuSerLeuLeuValProAlaLeuLeuValAlaGlyAlaAlaAsnAla'Ala	Mizuno *et al.*, 1983
OmpF protein	MetMetLysArgAsnIleLeuAlaValIleValProAlaLeuLeuValAlaGlyThrAlaAsnAla'Ala	Mutoh *et al.*, 1982
PhoE protein	MetLysLysSerThrLeuAlaLeuValValMetGlyIleValAlaSerAlaSerValGlnAla'Ala	Overbeeke, 1982
TolC protein	MetGlnMetLysLysLeuLeuProIleLeuIleGlyLeuSerLeuSerGlyPheSerSerGlnAla'Glu	Hackett *et al.*, 1983

Toxins

Protein	Sequence	Reference
Heat-labile toxin, subunit B (*eltB*)	MetAsnLysValLysCysTyrValLeuPheThrAlaLeuLeuSerSerLeuTyrAlaHisGly'Ala	Dallas and Falkow, 1980
Heat-labile toxin, subunit B (*toxB*)	MetAsnLysValLysCysTyrValLeuPheThrAlaLeuLeuSerSerLeuCysAlaTyrGly'Ala	Yamamoto and Yokata, 1983
Heat-labile toxin, subunit A	MetLysAsnIleThrPheIlePhePheIleLeuLeuAlaSerProLeuTyrAlaAsn	Spicer *et al.*, 1981

a The positively charged areas are underlined. The dashed line indicates the cleavage sites between the signal sequences of the mature polypeptides.

tain LamB–LacZ hybrid proteins are exported to the outer membrane (Silhavy et al., 1977), which is the normal location of the LamB protein. Fusions of β-galactosidase to periplasmic proteins such as maltose-binding protein and alkaline phosphatase result in the direction of β-galactosidase to the cytoplasmic membrane but not to the periplasm (Bassford et al., 1979; Michaelis et al., 1983a). In these cases, export is probably initiated but then aborted because the amino acid sequence of β-galactosidase interferes with transport through the membrane.

The unusual properties of certain fusion-bearing strains allowed the isolation of signal sequence mutations in *lamB* (Emr et al., 1978; Emr and Silhavy, 1980), *malE* (Bassford and Beckwith, 1979) and *phoA* (Michaelis et al., 1983b). All mutations that had serious effects on export were located in the hydrophobic segment of the signal peptide: they changed a hydrophobic or weakly hydrophilic amino acid into a charged one (Bedoueile et al., 1980; Emr et al., 1980; Michaelis et al., 1983b) or they deleted several amino acids from the hydrophobic region (Emr and Silhavy, 1980). Two pseudorevertants isolated from these deletions each contained an additional point mutation that led to an amino acid substitution within the signal sequence (Emr and Silhavy, 1983). Analyses of the secondary structures suggest that the secondary mutants restored export by allowing the formation of a stable α-helical conformation in the central, hydrophobic region of the signal peptide (Emr and Silhavy, 1983). Mutants with alterations in the positively charged, basic amino-terminal segment of the signal peptide have recently been isolated for LamB (Schwartz et al., 1981) and lipoprotein (Inouye et al., 1982; Vlasuk et al., 1983b). Most of these mutations seem to affect translation rather than export, which may reflect a tight coupling of translation and export (Hall and Schwartz, 1982; Hall et al., 1983). For lipoprotein mutants, where the amino-terminus was negatively charged, a drastic effect on export was observed (Inouye et al., 1982b; Vlasuk et al., 1983).

The results so far clearly establish the essential role of the signal sequence for protein export. The cytoplasmic enzyme β-galactosidase can be directed to membranous compartments when provided with an intact signal sequence of exported proteins. Certain mutations in the signal sequences of exported proteins cause accumulation of the precursor forms of these proteins, indicating that translocation cannot be initiated. The initiation of export seems to be the true function of the signal peptide. The final localization of a protein appears not to be determined by the signal peptide, and therefore should be determined by sequences of the mature protein.

4.4. Role of the Sequence of the Mature Protein

From a catalogue of exported proteins it appears that periplasmic and outer membrane proteins are initially synthesized via precursor forms, whereas integral inner membrane proteins appear not to be made as precursors (Michaelis and Beckwith, 1982). This has been shown for lactose permease (Buchel *et al.*, 1980; Ehring *et al.*, 1980), for subunits b and c of the F_0 complex of ATP synthase (Nielsen *et al.*, 1981), and for histidine permease (Higgins *et al.*, 1982). On the other hand, two inner membrane penicillin-binding proteins, PB5 and PB6 (Pratt *et al.*, 1981), two inner membrane lipoproteins (Ichihara *et al.*, 1981), and the inner membrane Tsr protein (Boyd *et al.*, 1983) are apparently synthesized in precursor form. These differences among inner membrane proteins may be due to different procedures in preparing inner membranes, since it is well known that different separation procedures of *E. coli* envelopes yield different protein profiles (Dassa *et al.*, 1978). Another interpretation would be that inner membrane proteins synthesized as precursors exert their functions in the periplasm at the outer face of the inner membrane. Thus, the bulk of their polypeptides would have to be translocated through the inner membrane, while a short hydrophobic carboxy-terminal region anchors them to the membrane. Such a membrane disposition has been demonstrated for penicillin-binding proteins in gram-positive organisms (Waxman and Strominger, 1981a,b). Integration into the inner membrane of proteins synthesized without signal sequence is probably due to the existence of long stretches of hydrophobic amino acids and may occur simply by partition into the lipid bilayer. Thus, for integral inner membrane proteins the amino acids sequence of the mature protein determines the final location. However, since the signal sequence of periplasmic and outer membrane proteins appears not to determine the final localization, as mentioned above, there must exist information within the mature protein relevant for translocation and final deposition.

From experimental data it has become obvious that the carboxy-terminus does not play any active role in export. This has been shown by certain fragments of OmpA (Bremer *et al.*, 1982), MalE (Ito and Beckwith, 1981), β-lactamase (Koshland, 1982), and arginine-binding protein (Celis, 1981), which are all translocated to the proper cell compartment. A passive role of the carboxy-terminus is possible, especially in hybridproteins, where certain amino acid sequences may be incompatible with export (Von Heijne and Blomberg, 1979).

However, for efficient direction to and stable incorporation into the membrane of outer membrane proteins a considerable portion of the protein seems to be necessary (Bremer *et al.*, 1982; Hall *et al.*, 1982; Moreno *et al.*, 1980). Benson and Silhavy (1983) propose two different regions within the mature LamB protein required for export and incorporation. The first region, corresponding to the amino acids 15–70, is required for transfer, localization, and processing. The second region, around amino acid 240, is needed for efficient translocation.

An internal protein region may also be essential for export of colicin E1

(Yamada et al., 1982a). This region is homologous to the signal sequences of E. coli lipoprotein and Bacillus licheniformis β-lactamase (Yamada et al., 1982b). Conditions that inhibit lipoprotein translocation also block translocation of colicin E1, suggesting that by the aid of the internal signal sequence colicin E1 is exported at the same site as lipoprotein (Yamada et al., 1982a). The internal region is located near the carboxy-terminus (Yamada et al., 1982b). However, since colicin E1 forms an ion-permeable channel in the cytoplasmic membrane of bacteria, and since this activity is located in a carboxy-terminal fragment of colicin E1 (Konisky, 1982), the internal sequence may be involved in the bactericidal action of colicin E1 rather than in export (Cleveland et al., 1983).

4.5. Cellular Components of the Export System

The signal hypothesis demands that membrane and ribosomal proteins form a complex through which proteins are cotranslationally exported. Thus, ribosomal and/or membrane components, essential for protein transport, should be detectable by either biochemical or genetic techniques.

A general involvement of the cytoplasmic membrane in protein export has been demonstrated by several investigators. Under a variety of conditions that affect the physiological state of the membrane, accumulation of precursor forms of several proteins has been shown (for a review see Michaelis and Beckwith, 1982). The results show that an energized membrane state and a certain membrane fluidity are required for protein export and processing.

Evidence that inner membrane proteins are required for protein export has been obtained by Smith (1980). He studied co-translational *in vitro* secretion of alkaline phosphatase and diphtheria toxin into inverted inner membrane vesicles of *E. coli*. Pretreatment of the cytoplasmic face of the vesicles with protease prior to initiation of translation had no effect on translation but inhibited export and processing. The involvement of further cellular components in protein export has been demonstrated by genetic techniques. Mutants in the signal sequence that prevented export made it possible to select for suppressor mutations that restore export. By this approach mutations at three unlinked loci have been obtained (Emr et al., 1981). They were called *prlA, prlB*, and *prlC* (standing for *pr*otein *l*ocalization). *prlA* suppresses signal sequence mutations in LamB (Emr et al., 1981), MalE (Emr and Bassford, 1982), and PhoA (Michaelis et al., 1983b). *prlB* only suppresses *lamB* mutations, whereas *prlC* suppresses *lamB* and *malE* mutations (Emr and Bassford, 1982). In contrast to *prlA* mutants, proteins are incorrectly processed in *prlB* and *prlC* mutants (Emr and Bassford, 1982). This may indicate that in *prlB* and *prlC* mutants proteins are exported via an unusual route, while *prlA*-mediated protein export occurs via the normal route.

prlA maps very close to the *spc* ribosomal protein operon (Emr *et al.,* 1981) but it does not constitute a ribosomal protein (Cerretti *et al.,* 1983). Another mutation, *ts215,* isolated by Ito and co-workers (1982) after localized mutagenesis of the *spc* ribosomal gene cluster drastically reduced the maturation of MalE, OmpA, and OmpF. *ts215,* a conditional lethal mutation, maps in a similar position to *prlA*.

Two further mutations have been described that apparently affect the secretory apparatus of the cell. *secA* (standing for *sec*retion defective), a temperature-sensitive lethal mutation, maps at 2.5 min in a cluster of genes involved in cell division and envelope biosynthesis (Oliver and Beckwith, 1981; Oliver *et al.,* 1982). At the nonpermissive temperature this mutant accumulates precursors of many proteins, including MalE, PhoA, OmpF, and LamB. The *secA* gene product appears to be a 90K protein of the cytoplasmic membrane (Oliver and Beckwith, 1982). *secB* mutants, which are not conditional lethal, also have pleiotropic effects and accumulate precursors of MalE and OmpF. *secB* maps at min 80 to the *E. coli* chromosome. There are several other mutations described such as *tpo, perA, cpxA, cpxB,* and *expA,* which reduce the amount of exported proteins in the envelope but do not accumulate precursors (for review see Michaelis and Beckwith, 1982). Since it is not clear whether they play a role in export or not, they will not be discussed further.

The mutations described in this section all have pleiotropic effects, accumulating the precursors of more than one exported protein. Similar effects have been described for a certain MalE–LacZ fusion. For this fusion, transport appears to be initiated but then the hybrid protein becomes stuck in the cytoplasmic membrane, leading to the accumulation of the periplasmic proteins MalE and PhoA and the outer membrane proteins OmpA, OmpF, and lipoprotein (Ito *et al.,* 1981). The simplest explanation for the pleiotropic effects is that there exist specific sites, as predicted by the signal hypothesis, in the cytoplasmic membrane that are shared by several periplasmic and outer membrane proteins in their translocation through the inner membrane.

4.6. Processing

As envisioned by both the signal and membrane trigger hypotheses a strong correlation exists between export and processing, processing being the cleavage of the signal peptide from the precursor. This may indicate that processing defines an essential step in protein translocation. However, since unprocessed precursors of lipoprotein (Lin *et al.,* 1980a), LamB (Emr *et al.,* 1981), and FhuA (Fecker and Braun, 1983; Hoffmann, private communication) are exported into the outer membrane, processing appears not to be essential for export. It seems to be essential for the binding of lipoprotein to the peptidoglycan (Lin *et al.,* 1980b), but a requirement of processing for the activity of

LamB does not seem to exist since unprocessed LamB has been described to restore LamB functions in an otherwise phenotypically *lamB* strain (Emr et al., 1981; Hall and Silhavy, 1981; Emr and Bassford, 1982). Furthermore, processing seems not to be necessary for export or enzymatic activity of β-lactamase, whereas it seems to be required for release of β-lactamase from the outer face of the cytoplasmic membrane into the periplasm (Koshland et al., 1982).

The tight coupling of export and processing suggests that processing occurs co-translationally rather than posttranslationally. As shown by Josefsson and Randall (1981a,b) both mechanisms may be realized even for the same protein. Interestingly, β-lactamase, specified by *ampC,* appears to be processed entirely co-translationally, whereas TEM β-lactamase appears to be processed entirely posttranslationally. For most co-translationally exported proteins processing is a relatively late event, taking place after about 80% of the total polypeptide have been synthesized (Josefsson and Randall, 1981a). With a similar approach, Randall (1983) has shown that translocation of nascent chains of the periplasmic maltose- and ribose-binding proteins occurs after the entire structural domains of the proteins have been synthesized. These data are inconsistent with a strict coupling of translocation to the elongation of these proteins.

The processing enzymes are most likely located at the outer face of the inner membrane, since precursors, accumulating in the cytoplasm under certain conditions, are not processed (Oliver and Beckwith, 1981; Kumamoto and Beckwith, 1983; Emr and Bassford, 1982).

A processing function has been demonstrated in the inner and outer membrane of *E. coli* (Zwisinski et al., 1981). This function, called leader peptidase, has been shown to process the precursors of several exported proteins *in vitro* (Zwisinski et al., 1981). By screening through a library of cloned *E. coli* DNA, a plasmid has been identified, bearing the gene for leader peptidase (Date and Wickner, 1981). Using this plasmid, the essential character of leader peptidase could be demonstrated with a novel technique (Date, 1983). Plasmids bearing the intact or inactivated gene-encoding leader peptidase were transformed into a *polA* strain of *E. coli*. In *polA* strains plasmid-determined antibioticum resistance can only be expressed after recombination of the plasmid into the chromosome, since the *polA* mutation prevents replication of the plasmid. Transformation with plasmids bearing the intact gene for leader peptidase yielded resistant colonies while transformation with plasmids bearing the defective gene yielded none. With the same genetic approach the gene locus *(lep)* for leader peptidase has been mapped at 54–55 min on the *E. coli* chromosome (Silver and Wickner, 1983).

The signal peptidase responsible for processing of lipoprotein has been shown to be different from the *lep* leader peptidase (Tokunaga et al., 1982a). It maps at about min 0 near the *rpsT* locus (Yamagata et al., 1983). Processing occurs only after glyceride modification of the lipid-free prolipoprotein (Tokunaga *et al.*, 1982b) and is a prerequisite of final acylation to yield the free form of lipoprotein (Tokunaga *et al.*, 1982b). Recently it has been demonstrated that the prolipoprotein signal peptidase only perfoms cleavage of the signal peptide as a whole from the protein to be matured, while further hydrolysis of the signal peptide is performed by an additional enzyme, the leader peptide peptidase (Hussain *et al.*, 1982a,b).

4.7. Translocation into or through the Outer Membrane

The signal hypothesis as well as the membrane trigger hypothesis provide some suggestions on how a protein may be translocated from the site of synthesis into or through a single membrane. However, the situation is more complex in gram-negative bacteria, since there are two membranes surrounding the cytoplasm. Thus, for outer membrane proteins and those proteins excreted into the surrounding medium both hypotheses are applicable only to the step of translocation through the inner membrane.

Indeed, it appears that many proteins that have to cross both membranes, such as toxins, hemolysins, and subunits of adhesins, cross the inner membrane with the help of a signal peptide and are transiently found in the periplasmic space (Dallas and Falkow, 1980; Spicer *et al.*, 1981; So and McCarthy, 1980; Wagner *et al.*, 1983; Mooi *et al.*, 1983).

Accumulation in the periplasm also seems to occur for some bacteriocins (Pugsley and Rosenbusch, 1981; Oudega *et al.*, 1982). However, their translocation through the inner membrane appears to be different from the translocation of periplasmic and outer membrane proteins, since they are synthesized without an amino-terminal signal peptide. Finally, the bacteriocins seem to be released by partial cell lysis (Jakes and Model, 1979; Pugsley and Rosenbusch, 1981; Varenne *et al.*, 1981). Protein functions appear to be necessary for translocation of hemolysin (Wagner *et al.*, 1983) and of the K88ab fimbrial subunit (Mooi *et al.*, 1983) from the periplasm through the outer membrane. In the latter case a rather complicated mechanism requiring two periplasmic and one outer membrane protein seems to function. According to the model proposed by Mooi *et al.* (1983), the outer membrane protein may function as a pore for the subunit and as an anchor for the assembled adhesin. One periplasmic protein is assumed to form a complex with the fimbrial subunit, thus preventing it from polymerizing or from being degraded in the periplasm. The second periplasmic protein possibly initiates enzymatically protrusion of the fimbrial subunit through the outer membrane pore.

An intermediate accumulation in the periplasm of integral outer membrane proteins appears not to occur. This may not be surprising, at least for the porins of *E. coli* and *S. typhimurium*, because several lines of evidence suggest that these proteins are assembled into the membrane via regions of adhesion between cytoplasmic and outer membrane (Boyd and Holland, 1980; Smit and Nikaido, 1978; for a review of adhesion sites see Bayer, 1979). However, this seems not to be the route by which LamB is assembled into the outer membrane. Ryter *et al.* (1975) have shown that newly synthesized LamB appears at the septal region of dividing cells, and Ohki (1979) has demonstrated a cell cycle dependent synthesis of LamB. Since these results clearly distinguish LamB from the porins, Hall and Silhavy (1981) have presented a model in which LamB transport to the outer membrane occurs by a vesicle intermediate. This model is purely speculative but at least it offers an explanation of how a large portion of β-galactosidase in a certain LamB–LacZ fusion (Silhavy *et al.*, 1977) can get into the outer membrane without even passing the inner membrane. In addition, this model shows how the carboxyterminus of LamB may become exposed on the cell surface (Gabay *et al.*, 1983). This orientation again distinguishes LamB from another outer membrane protein, OmpA, for which the *C*-terminus is exposed to the periplasm (Bremer *et al.*, 1982).

4.8. Concluding Remarks

During recent years many data have been accumulated that provide a better understanding of protein export in prokaryotic systems. The enormous progress in this field was mainly based on the ready accessibility of the prokaryotic genome to genetic manipulations. The protein fusion technique in particular has been proved to be a powerful tool in this respect. This technique will also help to clarify the biochemical aspects of protein export in prokaryotes since it allows raising of specific antisera against components of the export machinery. The discovery of a possible prokaryotic equivalent to the eukaryotic signal recognition particle (SRP) (Walter and Blobel, 1980) by means of antibodies directed against the *secA* gene product (Liebke, personal communication) demonstrates this. At the moment the data suggesting a bacterial SRP are preliminary. However, if they turn out to be correct, many data concerning the role of the signal sequence will have to be reconsidered. The role of the positively charged, basic amino-terminal segment of the signal sequence, as proposed in the loop model, will then especially require a new interpretation. A major part of studies on protein export in gram-negative bacteria concerns translocation through the cytoplasmic membrane. It appears that proteins at this stage are predominantly translocated as domains and not by linear extrusion. The significance of processing is not clear.

There are relatively few studies on translocation of proteins into or through the outer membrane, so this step remains a mystery. Even for the well-studied *E. coli* it is not yet clear if truly secretory proteins do exist. Most of those proteins finally found in the surrounding medium seem to be released by (partial) cell lysis. One candidate for a secretory protein in *E. coli* may be hemolysin, but also in this case the available data do not firmly establish hemolysin as a secretory protein.

5. COMPONENTS OF THE OUTER MEMBRANE INVOLVED IN THE UPTAKE OF DNA

5.1. Introduction

Because of its exposition, the outer membrane of gram-negative bacteria not only represents a protection barrier, but senses environmental parameters, perceives signals from the surrounding medium, and performs receptor and transport functions. Membranes of bacterial surfaces rapidly react upon interactions with macromolecules. For example, virus attachment immediately stimulates membrane changes that affect ion retention, membrane energization, active transport, and macromolecule synthesis (Bayer *et al.,* 1982a). The passage across the bacterial membrane of macromolecules such as proteins and especially DNA is an intriguing problem, being virtually contradictory to the true membrane function, namely, impermeability for macromolecules accumulated in the cytosol. In this chapter the uptake of DNA was chosen to demonstrate the cooperative action of membrane components constituting the efficient and versatile bacterial organelle, the outer membrane.

The transfer of DNA in populations of microorganisms represents an obvious and easily detectable means of intercellular communication. Not only does the exchange of biological information via DNA transfer occur among members of the same species or genus, but it has also been observed among bacteria (e.g., *Escherichia* and *Vibrio*) belonging to different taxonomic families (e.g., Enterobacteriaceae and Spirillaceae). In the crown gall disease of most dicotyledonous plants, even a prokaryote *(Agrobacterium tumefaciens)* genetically determines the infected eukaryotic host to express functions that benefit the survival of the parasite (for a review see Zambryski *et al.,* 1983). According to the different nature of the DNA donor vehicle, three naturally occurring types of DNA transfer are distinguished (for a review see Low and Porter (1978): (1) the direct transfer of DNA between cell fusions or mating aggregates of donor and recipient bacteria (conjugation), (2) the transfer of DNA from a phage capsid into the host cell (transduction), and (3) the uptake of free DNA from the surrounding medium without participation of a donor vehicle (transformation). Depending on the mode of uptake, the invading DNA often is replicated independently of cell division and replication of the chromosome. If the acquired DNA is stably inherited, the host cell may be endowed with useful additional properties (Willetts and Skurray, 1980).

The capability to take up DNA is not distributed statistically over the entire cell surface. Rather, it is restricted to small, scattered patches corresponding to only 5–10% of the cell surface. These zones (about 100 per cell) are characterized by high biological activity. The sites of synthesis of major membrane constituents (LPS) and of energy-dependent uptake of macromolecules appear to be located here. They function as export sites of cell wall components and serve as infection sites for bacteriophage. Morphologically, these sites can be distinguished as connections between the outer and the cytoplasmic membrane of gram-negative bacteria (adhesion sites, fusion sites, Bayer junctions; Bayer *et al.*, 1979; Bayer *et al.*, 1982b).

This section is intended to summarize briefly the participation of outer membrane constituents in the uptake of DNA. Emphasis is put on recent developments in the field.

5.2. Proteins

5.2.1. Proteins in Conjugation

Conjugation in bacteria is mediated by DNA transfer systems that are encoded by plasmids (for a review of F-plasmid conjugation see Willetts and Skurray, 1980). Cells harboring such conjugative plasmids are donor cells because of the autonomous plasmid transfer to other cells (recipients). Donor cells express few copies of sex pili (e.g., F-plasmids: 1–3). The pili coded for by the F-plasmid protrude from the outer membrane of *E. coli*. They are filamentous aggregates (up to 20 μm long) of an α-helical arrangement of pilin, a single peptide subunit ($\sim$10K), which may contain phosphate and carbohydrate groups. Based on blending experiments with wild-type plasmid *(Flac)* and a mutant that does not elaborate pili (F *lac tra*), Sowa *et al.*, (1983) proposed that pilin subunits are assembled transiently into pili. The pilin subunits then may be conserved by retraction and made available again for subsequent reassembly.

Pili can be distinguished according to the incompatibility of the coding plasmids (e.g., *Inc*F-, *Inc*I-pili), serologically, and by three different types of morphology: thin, flexible (I-, B-, K-pili); thick, flexible (*Inc*D-, *Inc*T-, F-pili); and straight, rigid (P-, W-, M-pili). These morphological variants were found within all incompatibility groups of *E. coli* K-12. No serological cross-reactions were found between the pili of the three different morphological groups (Bradley, 1980). The conjugative pili of *Pseudomonas* resemble those of *E. coli*. Rigid and flexible morphological variants have been identified by electron microscopy (Bradley, 1983). Thirteen groups of incompatibility (Pl–P13) of conjugative plasmids have been distinguished. Pili generate the initial contact between mating partners and appear to be required for DNA transfer. In conjugal transfer of F-plasmids in *E. coli* the pilus tip contacts receptor regions of the recipient, consisting of LPS or of aggregates of LPS and the outer membrane protein OmpA. This contact reaction is inhibited between cells carrying the same or a closely related plasmid (donor–donor matings). In this case, the pilus receptor of donor cells is assumed to be blocked by proteins that are coded by the respective plasmid. This discrimination of donor–donor matings is denoted *surface exclusion.* For F-like plasmids, four specificity groups have been defined. Three genes, *traJ, traS,* and *traT,* out of at least 23 transfer genes coded by a 35-kb region of the F-plasmid, have been shown to be independently responsible for surface exclusion. The *traJ* gene codes for a 24K protein of the outer membrane. The translocation of the TraJ protein to the outer membrane is controlled by a chromosomal gene, *cpx* (Sambucetti *et al.,* 1982).

In the presence of the TraJ protein neither mating aggregation nor conjugative transfer of DNA is observed. The *traS* gene codes for an 18K protein component in the cytoplasmic membrane of the donor (Kennedy *et al.,* 1977). The TraS protein does not interfere with the formation of mating aggregates but inhibits triggering of conjugal DNA transfer. The *traT* gene codes for a 25K protein that exists in multimeric aggregates on the surface of the outer membrane of donor cells. The TraT protein seems to inactivate the recipient receptors for the F-pilus, so that donor–donor mating aggregates become less stable (Manning *et al.,* 1980). The expression of the *tra* operon genes is under positive control by *traJ* (Willetts and Skurray, 1980). Rashtchian *et al.,* (1983) have shown that reduced expression of *traJ* has much less effect on *traT* expression than on F-plasmid transfer. Even a 1000-fold reduction in *traJ* expression does not show a detectable change in *traT* synthesis, indicating that either *traT* has its own promoter or is translated much more efficiently. The TraT protein alone is not sufficient for full expression of surface exclusion (Achtman *et al.,* 1980). Normal amounts of TraT can be present in the cell envelope without producing normal levels of surface exclusion. Although *traT* is involved in surface exclusion, the *traS* gene product that is regulated by *traJ* may be more important (Rashtchian *et al.,* 1983).

In addition to the plasmid-coded mating functions, components of the recipient bacteria determine conjugation. Mutants defective as recipients for F-plasmids have been identified to be altered in lipopolysaccharide or in the heat-modifiable major outer membrane protein, the OmpA protein. Recipients defective in OmpA protein do not undergo tight binding to donor cells during F-plasmid conjugation. The OmpA protein not only binds cells together during conjugation but also facilitates uptake of receptor-bound colicin K and L and serves as receptor for phage Ox2. Two further classes of combined OmpA functions have similarly been defined by Manoil and Rosenbusch (1982): the receptor functions in phage TuII* and K3 infections, and the structural function in the outer membrane providing resistance to chelating agents and to the hydrophobic antibiotic novobiocin. The sites in OmpA protein catalyzing these three groups of functions are overlapping (see Section I), and different cellular amounts of OmpA protein are required (Manoil and Rosenbusch, 1982).

Once male and female bacteria are attached to each other, a series of events occurs to initiate the replication and the transfer of donor DNA into the recipient. Two classes of nonconjugative mutants that normally attach to donor bacteria but produce low recombinants (0.01–1%) have been selected by Ou and Yura (1982). These mutants show a modified pattern of outer and cytoplasmic membrane proteins. In the case of class I mutants, Ou and Yura propose that a channel across the outer membrane is made, but putatively is missing in the cytoplasmic membrane. Mutants of class II seemingly do not generate a mating signal to induce synthesis of donor DNA (Ou, 1982). Apparently, proteins of the cell membranes fulfill a key role in conjugation. They are involved in initiation and control of mating aggregation. Moreover, they seem to be involved in the induction of DNA synthesis in the donor, as well as in the subsequent transfer of DNA.

5.2.2. Proteins in Virus Infections

Proteins located on the surface of the outer membrane usually serve as receptors for bacteriophage (for a review see Braun and Hantke, 1981). Such receptors are readily accessible to phage and provide a stable attachment necessary for a successful transfer of phage DNA. The outer membrane protein FhuA, which is involved in the transport of ferrichrome, serves as a receptor for phage T5, T1, and Ø80. The maltodextrin pore, formed by the LamB protein, is recognized as a receptor by phage lambda. Phage BF23 adsorbs to the BtuB protein, which is involved in the uptake of vitamin B_{12}. The nonspecific porins OmpF (Ia) and OmpC (Ib) serve as receptors for phage T2 and TuIa, or TuIb and 434, respectively. The Tsx protein is used as a receptor by phage T6. Activity as phage receptors is not restricted to proteins that are constituents of transport systems. For example, the major heat-modifiable protein OmpA is the receptor for phage K3, Ox2, and TuII*.

In addition to the function of host outer membrane proteins as receptors, phage-borne proteins may act in recognition of the DNA uptake site, in DNA transport through the membrane, and in the control of DNA-related functions in the cytoplasm. In the case of phage ØX174 the protein coded by gene H (gpH) and in the case of phage M13 the protein coded by gene 3 (gp3) have been shown to participate in DNA transfer. Since a protein is involved in the attachment of DNA to the host cell and since it is a protein that leads the invading end of DNA into the cytoplasm, this type of proteins has been denoted pilot proteins (Kornberg, 1980). It is not clear whether the pilot protein becomes part of a pore, or enters the cell through a preexisting pore, prior to or after the DNA (Labedan and Goldberg, 1982).

The *Salmonella* phage P22 carries its own DNA pore, or a specific modifier of preexisting pores. Upon infection, the protein coded by gene 16 (gp16) appears to integrate as a hexamer into the outer membrane of *Salmonella* and hereby creates a pore for DNA uptake (Labedan and Goldberg, 1982).

5.2.3. Proteins in Transformation

The gram-negative bacteria *Haemophilus influenzae* and *Neisseria gonorrhoeae* are transformed predominantly by DNA originating from their own genus. DNA prepared from distantly related organisms does not bind to the DNA uptake system of naturally competent cells, nor does it compete with homologous DNA (Scocca *et al.*, 1974). Recognition of homologous DNA occurs at the cell surface. After cloning in *Escherichia coli* and reisolation, an 8.1-kb fragment of *Haemophilus* DNA was still efficiently taken up by competent *Haemophilus* cells. This finding implies that the recognition signal for uptake is not a modification pattern but a base sequence specific for *Haemophilus* DNA. Based on restriction enzyme digestion and uptake competition studies, Sisco and Smith (1979) estimated that the sequence conferring specificity occurs at high frequency in *Haemophilus* DNA (~ 600 copies; 1 per 4 kb), but at low frequency in heterologous DNA (~ 8 copies; 1 per 300 kb). It is a nucleotide sequence 11 base pairs long (Danner *et al.*, 1980) that is recognized by a restriction-type endonuclease, situated in the outer membrane of *Haemophilus* (for a review see Goodgal, 1982).

Haemophilus influenzae and *Neisseria gonorrhoeae* show no appreciable interaction with each other's DNA. Transformation by homologous DNA was not affected by up to 10-fold excess of heterologous DNA. *Haemophilus* and *Neisseria* apparently recognize different specificity determinants in the DNA during the uptake step of genetic transformation (Mathis and Scocca, 1982; Graves *et al.*, 1982).

In contrast to gram-positive cells, which hydrolyze one strand of the transforming DNA along with the uptake and thus internalize single-stranded DNA, *Haemophilus* and *Neisseria* take up double-stranded DNA. Although it seems likely that small amounts of the donor DNA are converted, at least transiently, to single-stranded DNA, the bulk of donor DNA is taken up in double-stranded form (Biswas and Sparling, 1982). The mechanism of uptake is poorly understood (Goodgal, 1982). *E. coli* does not develop competence spontaneously. If competence is induced through intermittent intervals of cooling and refrigeration in the presence of specific divalent cations (see below), the uptake system does not discriminate against foreign DNA. Differences in the structure of transforming DNA also have no effect since linear chromosomal DNA competes with the uptake of covalently closed circular plasmid DNA. Apparently, both kinds of DNA are taken up by the same mechanism (Brown *et al.*, 1981).

5.3. Lipopolysaccharides (LPS)

5.3.1. LPS in Conjugation

The inability of conjugation-deficient mutants *(con)* to form stable mating aggregates has been shown to be partly due to modification or absence of the OmpA protein (see above). A similar function is exerted by LPS in the transfer of I-like conjugative plasmids belonging to the incompatibility group *Inc*Iα. Duke and Guiney (1983) have shown that in the case of plasmid R64drdll (*Inc*Iα) the efficiency of transfer is dependent on the LPS structure of the recipient. While the F-like plasmids Flac (*Inc*FI) and Rldrd19 (*Inc*FII) transferred at equal frequencies either to smooth parent strains or to rough LPS mutant strains of *Salmonella minnesota*, the transfer of the I-like plasmid R64drdll was reduced to 4×10^{-4} in recipients with rough LPS. These results show that plasmids have evolved distinct conjugation systems that depend on different surface structures.

Purified LPS is known to block the adsorption of DNA-phage to the tip of the F-pilus (Kanegasaki *et al.*, 1978). Therefore, LPS is likely to form part of the recipient receptor prone to interact with the F-pilus tip. Concerning the inhibitory action of LPS on conjugation, inconsistencies are found in the literature. While some authors found that purified LPS by itself strongly inhibited conjugation, other authors showed that LPS only together with OmpA protein could produce a strong inhibition (for references see Willetts and Skurray, 1980).

This discrepancy is reconciled by the different requirements of the various conjugation systems for OmpA protein in the recipient. The transfer of plasmids ColV2 and R386 was markedly reduced with some *ompA* recipients, but the transfer of plasmids R100 and R136 was not. Mutants that failed to receive plasmid R100 were found uniformly defective in LPS structure (Havekes *et al.*, 1977).

5.3.2. LPS in Virus Infections

In analogy to outer membrane proteins, LPS can function as a receptor enabling tight attachment of phage to the bacterial surface (for a review see Lindberg, 1977). The *O*-antigenic polysaccharide moiety of LPS functions as a receptor for the *Salmonella* phage ε15 and P22. The base plate spikes of these phage having endoglycosidase or esterase activities promote the enzyme-substrate-like binding.

While the phage approaches the cell wall, oligosaccharides are released from the cell. Micellar complexes of purified entire LPS suffice to trigger the ejection of DNA *in vitro*. It is not known which part of the LPS finally triggers DNA ejection.

The phages T4, P1, and ØX174 adsorb to the LPS core oligosaccharide. Because the LPS core is embedded in the outer membrane, no enzymatic activity is required to enable the phage to approach the surface of the host cell. In contrast to LPS from *E. coli* B, LPS from *E. coli* K-12 is completely inactive as a receptor for phage T4. With K-12 LPS the presence of porin protein OmpC (formerly designated Ib) is required for phage T4 receptor activity (Mutoh *et al.*, 1978; Henning and Jann, 1979). This result suggests an interaction between LPS and protein OmpC, the latter enabling LPS of *E. coli* K-12 to act as T4 receptor.

In addition to its function as a true receptor, LPS can accelerate the adsorption of phage T5 to its protein receptor FhuA by means of a reversible interaction with the L-shaped tail fibers of this phage (Heller and Braun, 1979). The polymannose antigens 08 and 09 have been identified as the O-polysaccharide site interacting reversibly with the L-shaped tail fibers. The binding site resides in a trimannoside (Heller and Braun, 1982). Along with increasing amounts of 08 or 09 polymannose, the rate of adsorption of T5 phage is accelerated by a factor of 10. T5 mutants devoid of tail fibers do not show interaction. In this case, the rate of adsorption is reduced by increasing the polymannose component of the LPS.

Like LPS, exopolysaccharides of capsules and slimes function as receptors for phage (e.g., *E. coli* phage 29). In this case, the attachment of the phage occurs approximately eight phage diameters distant from the cell surface (Lindberg, 1977). The phage then approaches the bacterial cell wall by localized enzymatic lysis of the exopolysaccharide by its tail spikes. On contact with the cell wall, DNA ejection is triggered. The penetration of nucleic acid is supposed to take place at fusion sites between wall and cytoplasmic membrane.

5.3.3. LPS in Transformation

The presence or absence of the FhuA receptor protein for phage T1, T5, and Ø80 in the outer membrane was without influence on the transformation of $CaCl_2$-treated *E. coli* cells with T5 DNA. However, the efficiency of transfection with T5 DNA was low in strains altered in LPS structure. The Ca^{2+}-induced competence developed poorly in LPS mutants lacking the branch heptose in the LPS core or being deficient in the branch glucose and galactose residues (Taketo, 1977 and 1978).

5.4. Lipids

5.4.1. Lipids in Conjugation

It is evident that pili are involved in the early stage of mating-pair formation, although the precise role of pili in conjugal cell–cell interaction remains controversial. The outgrowth of conjugal pili is assumed to occur at adhesion sites between the cytoplasmic and outer membrane of the donor cell. Retraction of the sex pilus then may bring donor and recipient patches of membrane fusions into apposition. It has been proposed that, due to the particular structure of the fused lipid bilayers in these adhesion sites, a nucleic acid pore is formed that may be stablized by the transmembraneous OmpA protein (Willetts and Skurray, 1980).

5.4.2. Lipids in Viral Infections

Artificially contracted T4 phage particles, having a protruding tail core and representing most probably an intermediate stage in the course of infection, adsorbed to phospholipid liposomes by their core tips (Furukawa and Mizushima, 1982). This adsorption resulted in ejection of phage DNA. The DNA ejection was triggered by liposomes of either phosphatidylglycerol or cardiolipin but not by liposomes of phosphatidylethanolamine. While adsorption already occurred at 4°C, the process of DNA ejection was temperature dependent. A proton motive force (see the following discussion) was not required for these processes.

With uncontracted phage neither adsorption to liposomes nor DNA ejection was observed. This indicates that T4 phage particles are able to intereact with liposomes only after the protrusion of the core tip. The major part of ejected DNA was found to be sensitive to externally added DNAse. This suggests that liposomes provide a signal for the ejection of DNA. However, the transfer of DNA across the lipid bilayer probably requires additional functions.

When the outer membrane protein LamB of *Shigella,* which serves as the receptor for phage lambda, was integrated into unilamellar phospholipid vesicles, the DNA of phage lambda was injected into the liposomes. The phage head was found to be only partially empty when small liposomes were used, presumably because the vesicles were too small to contain all the DNA. Upon usage of larger liposomes, the heads of most bound bacteriophage appeared to be completely empty of DNA. The ejected DNA was not susceptible to DNAse unless the vesicle bilayer was disrupted. Apparently, high-molecular-weight

DNA can be entrapped within liposomes. This may possibly be used to facilitate gene transfer to eukaryotic cells (Roessner et al., 1983).

In this context it is interesting that the efficiency of delivery of liposome-entrapped SV40 DNA to African green monkey cells was also influenced by the nature of the lipids. Acidic phospholipids were found most effective in binding to cells and DNA delivery. The transfer of DNA by phosphatidylserine liposomes was stimulated by brief exposure of the recipient to glycerol. Because glycerol treatment causes the ruffling and vacuolization of the membrane, the enhanced uptake of liposomes probably occurs by an endocytosis-like process (Fraley et al., 1981).

5.4.3. Lipids in Transformation

The development of competence for DNA uptake by *E. coli* usually requires intermittent incubation at 0°C and elevated temperatures (25–42°C) in the presence of 50–100 mM CaCl$_2$ (Mandel and Higa, 1970). Under optimal conditions only very few cells become competent (10^{-3}). The optimal temperature during the heat shock incubation depends on the temperature applied for cultivation of the cells. If cells are grown at 22°C instead of 37°C, the temperature range to induce optimal competence is shifted by 5°C to lower temperatures (van Die et al., 1983a). Similarly, the optimal temperature to induce competence of fatty-acid-requiring mutants was decreased upon growth on oleic acid and increased upon growth on elaidic acid (van Die et al., 1983b). Incorporation of oleic acid yields a fluid membrane with low transition temperature, whereas incorporation of the stereoisomer elaidic acid results in a more rigid membrane with high transition temperature. These data indicate that lipid phase transition is a necessary requirement for the induction of competence in *E. coli*. The uptake of DNA occurs during the incubation at 0° after the heat shock. During temperature shift-down leakage of the periplasmic β-lactamase occurs.

This observation indicates that phase transitions of membrane lipids cause damage to the outer membrane that may be essential for the induction of competence.

The natural development of competence in *Haemophilus parainfluenzae* and *H. influenzae* appears to be a function of morphological alterations of the outer membrane. When thin sections of competent and noncompetent cells were compared, five-fold more membrane extrusions were present on the surfaces of competent cells than on noncompetent cells. These extensions disappeared on treatment with transforming DNA, while vacuole-like structures appeared in the cytoplasm (Kahn et al., 1982). Competent cells treated with radiolabeled DNA were shown to contain the majority of labeled DNA in membrane vesicles. The polypeptide composition of the vesicles was similar to that of the outer membrane. The generation of membrane extrusions by competent *Haemophilus* appears to be similar to the "blebs" found in *Neisseria meningitis* and *Bacteroides nodosus* (Kahn et al., 1982).

5.5. Divalent Cations

5.5.1. Divalent Cations in Conjugation

The formation of mating aggregates during conjugation is inhibited by Zn^{2+} ions at 1 mM concentration. Further steps in the mating process, such as mobilization, transfer, and integration of the donor DNA into the recipient, are not affected. It is likely that Zn^{2+} ions associated with LPS form part of the recipient receptor interacting with the F-pilus tip. Thus, an excess of Zn^{2+} ions in the medium may block the tips of the F-pili (Ou and Anderson, 1972). On the other hand, when recipient cells (F^-) are grown in the presence of Zn^{2+} ions, their ability to aggregate with donor cells (F^+) increases. Because of the higher incorporation of Zn^{2+} ions, receptor areas to bind the F-pilus tip may be multiplied in the cell envelope of the recipient (Ou, 1973).

5.5.2. Divalent Cations in Virus Infections

The infection by phage T5 depends on the presence of Ca^{2+} ions (Lanni, 1960). Complexing agents such as EDTA, EGTA, and citrate inhibit infection by T5. After treatment with Ca^{2+}-complexing agents, the susceptibility toward T5 is not restored immediately upon addition of even an excess of Ca^{2+} but is restored along with the incubation of the host bacteria in the presence of Ca^{2+}.

The inhibition of phage infection through chelating agents is not only effective with phage adsorbing to protein receptors. For example, adsorption of phage P1 to the basal core of LPS is inhibited by the addition of citrate. In protocols of generalized transduction this property is used to limit P1 infection to one growth cycle.

5.5.3. Divalent Cations in Transformation

The induction of competence in *E. coli* by heat pulse in the presence of Ca^{2+} ions is a technique frequently used for cloning experiments with DNA originating from a wide variety of prokaryotes and even eukaryotes.

Treatment with a heat pulse in the presence of Ca^{2+} ions perforates the permeability barrier of the outer membrane. Many gram-negative bacteria that are not naturally competent, such as *E. coli*, *Aerobacter aerogenes*, and *Salmonella* sp., develop competence upon this treatment. Even *Rhodopseudomonas sphaeroides*, a gram-negative facultative photoheterotrophic bacterium, was found susceptible (Fornari and Kaplan, 1982). The only modification these authors added to the original protocol of Mandel and Higa (1970) was a washing procedure with 500 mM Tris buffer prior to the CaCl$_2$ incubation and the completion of the transformation assay by 20% polyethyleneglycol 6000.

The order of effectiveness of divalent cations in transformation of *E. coli* with the replicative form of phage ØA double-stranded DNA was found to be Ba^{2+} > Ca^{2+} > Sr^{2+} > Mg^{2+}, but differs somewhat if single-stranded DNA is used in the transformation assay (Taketo, 1975). Zn^{2+} ions were found to be strongly inhibitory, even at concentrations as low as 0.5 mM. Monovalent cations would reduce competence only when present in concentrations similar to the competence inducing agent (Taketo and Kuno, 1974). Recently, a set of conditions has been described under which one in every 400 plasmid molecules produces a transformed *E. coli* cell (Hanahan, 1983). The mechanism of competence induction by heat pulse treatment in the presence of divalent cations is largely unknown. A possible explanation based on the phase transition of lipid bilayers is given in Section 5.4.1.

5.6. Energy Dependence

Metabolic energy of the bacterial cell is linked to transport processes by high-energy phosphates (ATP) and by means of the chemiosmotic potential (proton motive force). According to the chemiosmotic theory of Mitchell, the chemiosmotic potential Δp is composed of a chemical gradient, the pH, and a membrane potential $\Delta \psi$ (for a review see Harold, 1977). The dependence of DNA uptake on these energy components is shown in the following section.

5.6.1. Energy Dependence of Conjugation

As a consequence of the bidirectional activity of the membrane H$^+$/ATPase, the chemiosmotic potential can be used to drive the synthesis of ATP from ADP and P_i. Vice versa, hydrolysis of ATP by the H$^+$/ATPase can generate a proton motive force. Thus, experiments to determine the contribution of either Δp or ATP as an energy source in DNA transport make use of mutants *(uncA)* defective in the membrane H$^+$/ATPase. If such mutants are used as recipients in conjugation under anaerobic conditions (absence of Δp), mating pairs are formed but the frequency of recombination is lower than 1%. This result suggests that the proton motive force, rather than high-energy phosphate, is the source of energy for conjugal DNA transfer. In line with this, inhibition of oxidative phosphorylation by cyanide in recipients with proficient H$^+$/ATPase did not affect the transfer of the conjugative plasmid unless the intracellular ATP level was reduced by the addition of arsenate. However, if added in 60 mM concentration, conjugation proved to be susceptible to arsenate alone (Berzhinskene *et al.,* 1980). Because arsenate does not affect the chemiosmotic potential of the membrane, the presence of both Δp and ATP may be a prerequisite for the transfer of plasmid DNA during conjugation.

5.6.2. Energy Dependence of Virus Infection

The infection of *E. coli* K-12 by phage T4 has been shown to be strictly dependent on the membrane potential ($\Delta\psi$) if the ratio of phage to host is at least 5 (Labedan and Goldberg, 1982). Reduction of the membrane potential below -100 mV caused inhibition of infection. Reduction of $\Delta\psi$ had no effect if concomitantly the transmembranous pH gradient (ΔpH) was increased. By impairing the proton motive force the transfer of phage DNA was inhibited reversibly. These results presented by Kalasauskaitė *et al.* (1983) support the concept that phage T4 DNA uptake across the cell membrane occurs by expenditure of host cell metabolic energy.

However, if artificially contracted T4 phage having a protruding tail core were used, DNA injection into *E. coli* spheroplasts occurred independently of a membrane potential. Based on this experiment Furukawa *et al.* (1983) proposed a model in which a membrane potential was not required for DNA injection itself, but was necessary for the rapprochement of the cytoplasmic membrane to the outer membrane. Because of this energy-dependent step, the interaction of the cytoplasmic membrane with the core tip of the T4 tail may be facilitated.

Virtually independent from host metabolic energy is the transfer of phage T5 DNA across the cell envelope (Maltouf and Labedan, 1983). Addition of various metabolic inhibitors, uncouplers, cyanide, arsenate, and ionophores, separately or in combination, did not inhibit the transmembranous transfer of first-step DNA of T5.*

The first-step transfer remained indifferent to host cell energy when the phage capsid was removed and DNA was uncoiled prior to infection. Thus, the participation of the tension of DNA or of any energetic component of the phage capsid was ruled out.

Even the uptake of second-step DNA was independent from host metabolic energy, provided that proteins A_1 and A_2 were present in the cell. Based on these results, Maltouf and Labedan concluded that the traversal of the cell

---

*The DNA transfer upon infection with T5 and related phage naturally occurs in two steps (Lanni, 1968). After 10% of the genome [*first-step* (fst) DNA] has been threaded into the cytoplasm, further DNA transfer stops. As soon as proteins A_1 and A_2, coded by the fst segment of T5 DNA, have been synthesized, transfer of the remaining 90% of the T5 genome (second-step DNA) takes place.

membrane by the entire T5 DNA must occur by passive diffusion through protein channels.

5.6.3. Energy Dependence of Transformation

When *E. coli* cells were exposed to protonophores that collapse the chemiosmotic potential, $CaCl_2$-induced transformation by an ampicillin resistance plasmid was drastically reduced (Santos and Kaback, 1981). By alteration of the ambient pH or by addition of the ionophores valinomycin and nigericin, the efficiency of transformation was directly correlated with the magnitude of the membrane potential. Changes in the pH gradient had no significant effect.

Apparently, the membrane potential $\Delta\psi$ plays a critical role even in the largely unspecific uptake of DNA during the transformation of Ca^{2+}-induced gram-negative cells.

5.7. Summary

The transport of DNA across the gram-negative cell envelope is currently regarded as a manifold process. Virtually all components of the bacterial envelope, proteins, LPS, and lipids seem to operate in concert in the recognition, binding, and transport of DNA. Many features of the delivery of DNA to its site of uptake have been described. Labedan and Goldberg (1982) propose that the specificity of delivery and the sophistication of the delivery mechanism determine the efficiency of DNA transfer. Therefore, the ratio of productive transfer of DNA may vary from approximately 1 (phage T4) to orders of magnitude less (in $CaCl_2$-induced transformation). In view of the different delivery mechanisms it is not surprising that the requirements for the presence or structural integrity of specific constituents of the outer membrane vary within the initial stage of DNA uptake, which provides the contact of DNA with the cell surface. During membrane transfer of DNA, either proteins or lipids may create DNA channels. Both models, (1) creation of pores by protomers or oligomers of proteins and (2) fusion of lipids with protonated DNA or cation-aggregated DNA, seem reasonable to explain the uptake of linear DNA nucleotide by nucleotide or the penetration of covalently closed DNA circles.

The only feature of DNA uptake common to transformation, phage infection, and conjugation, namely, the dependence on the proton motive force, is abandoned by the diffusion-like mechanism in the uptake of phage T5 DNA. Whether this kind of DNA uptake is only the exception to the otherwise generally valid hypothesis (Grinius, 1980) that DNA transport is a function of the ion gradients across the membranes of metabolically active cells remains to be seen.

6. OSMOREGULATION

6.1. Introduction

Prokaryotes and many unicellular eukaryotes can acquire their nutrients only from solution, and thus they can grow only in direct contact with liquid water. This contact is mediated via the cell membrane, which forms the osmotic barrier of the cell. Osmosis is defined as the diffusion of water across differentiating membranes readily permeable to water and slowly permeable or impermeable to solutes. In gram-positive bacteria the peptidoglycan, offering only negligible resistance to the passage of water or solutes, is lined within by the differentially permeable cytoplasmic membrane. The water contained in the cytoplasm is "diluted," that is, its diffusion pressure is lowered by a high concentration of sugars and various other organic or inorganic substances in solution. In thermodynamic terms the water activity (a_w) is lowered. If the cell is transferred from medium into water, the "undiluted" water from without will tend to diffuse into the protoplast because of its higher diffusion pressure. The peptidoglycan cell wall, like the cover of a football, resists extension, and a pressure is built up within the cell as water diffuses in. At this point, it is important to recall that this turgor pressure increases the diffusion pressure of the water within the cell in the same way as would an externally applied pressure of the same magnitude. At the equilibrium, the diffusion pressure of the "diluted" water inside the cell equals that of the pure water outside.

In their natural habitats bacterial cells have to deal with situations in which the osmolarity slowly decreases or increases, and in addition they may be subjected to abrupt changes of environmental conditions. Enterobacteriaceae, like nonpathogenic and pathogenic strains of *E. coli,* multiply in the intestine of their hosts, but after excretion they must be able to survive in drying feces as well as in water to get ingested again by another host. This makes the problem evident: What are the molecular mechanisms that enable bacterial cells to survive and to grow under osmotic stress?

A microorganism responds to a modification of physicochemical parameters essentially in two stages. Stage I occurs when the bacterium, either growing or not, is exposed to new environmental conditions. Adaptation has to be comparatively rapid and decides whether the organism will survive or not. The second stage of adaptation occurs after the bacterium has adjusted thermodynamically to the new conditions. Stage II involves alterations in the organism's metabolism that may involve genetic induction or repression of enzyme activities. Therefore, the responses of stage II are much slower than those of stage I. If the bacterium has survived stage I, the limits of its environmental tolerance are set by the nature and extent of the metabolic alterations that can take place in stage II.

The discussion of osmoregulation in gram-negative bacteria will emphasize the stress associated with limiting water at high environmental osmolarity as well as the stress caused by an "excess" of water during growth at extremely low osmolarity. The optimal salt concentration for most gram-negative species is in the range of 0.2–0.3 M (Brown, 1976). Most gram-negatives will tolerate media without added sodium chloride, for example, peptone–water. On the other hand, gram-negative bacteria may also be able to grow in environments with salt concentrations raising up to 0.7 M, in some special cases even 1.4 M (Brown, 1976). An example for bacterial growth under extreme conditions are clinical isolates of *Pseudomonas aeruginosa* able to multiply in stored distilled water in cell densities up to 10^7 cells/ml. Organic compounds dissolved in minute amounts during storage are sufficient to enable growth (Favero *et al.*, 1971).

To understand the nature of the microbial cell's responses to osmotic stress one has to take into consideration a peculiarity of prokaryotic cells. In contrast to eukaryotes the enzymes of energy metabolism are concentrated in the cytoplasmic membrane of prokaryotes. This membrane at the same time mediates the cell's interactions with its environment, which is especially true for gram-positive bacteria, whereas gram-negatives possess an additional outer membrane. In osmotic terms this outer membrane can be regarded as a molecular sieve with an exclusion limit of 600 daltons in the case of Enterobacteriaceae, and higher for other species (Nikaido and Nakae, 1979). In any case, it is obvious that the cytoplasmic membrane has to be protected from inadequate pressure resulting from within or without not only to prevent the membrane from bursting or collapsing, but also to keep all enzymes and transport systems functioning by the maintenance of an optimal surface tension and density.

Throughout nature two mechanisms for protection of osmotically sensitive cell membranes have evolved: (1) the osmolarity of the cytoplasmic fluid can be adapted to the extracellular fluid by the uptake or excretion of water or osmotically active solutes, (2) if the volume of an extracytoplasmic fluid is fixed and does not exceed too much the volume of the system under consideration the osmolarity of the extracytoplasmic fluid can be regulated too.

As the cell wall of gram-negative bacteria is composed of two membranes, a second cell compartment—the periplasmic space—is established. Under certain conditions this space may comprise up to 40% of the total cell volume (Alemohammad and Knowles, 1974; Stock *et al.*, 1977). Smaller volumes have also been reported (Nikaido, 1979). We now may ask the question, is the periplasmic osmolarity regulated actively by the cell, or does the osmolarity of the periplasmic fluid reflect only the osmolarity of the medium in which the cells are growing?

In the following we shall show that gram-negative bacteria not only regulate the osmolarity in the cytoplasm and in the periplasm, but also the composition of the outer membrane, which, together with murein, resists the osmotic pressure of the cell under most conditions.

6.2. Osmoregulation in the Cytoplasm

6.2.1. Potassium Transport

In studies on the osmotic properties of *E. coli* the osmolarity of the medium has been varied and the responses of the cell observed. To simplify discussions of these responses, it is convenient to refer to abrupt increases and decreases in osmolarity of the medium as "upshock" and "downshock."

Ørskov (1948) found that the intracellular K^+ concentration increases with the osmolarity in the medium. Osmotic regulation of the intracellular potassium concentration also occurs in *Salmonella* (Christian, 1955) and may be characteristic of many genera of bacteria (Brown, 1976). Epstein and his group measured potassium uptake in *E. coli*. They showed that growing cells respond to sudden changes in the osmolarity of the medium with rapid changes of the K^+ content so that the osmotic difference was minimized (Epstein and Schultz, 1965). Two major K^+ transport systems are available in *E. coli* (Rhoads *et al.*, 1976): Under most growth conditions, potassium is taken up by the constitutive TrkA system, which has a high rate of transport and a low affinity for K^+. Its properties can be altered by mutations in the genes *trkB, trkC, trkD, trkE,* and *trkG* (Epstein and Laimins, 1980; Rhoads *et al.*, 1976; Epstein and Kim, 1971). The TrkB and TrkC functions seem to affect the efflux of potassium from the cell, and TrkE exerts some unknown regulatory functions on TrkA (Lubin and Kessel, 1960; Gunther and Dorn, 1966; Epstein and Kim, 1971; Rhoads *et al.*, 1976). Little is known about the mechanism of the Trk transport function, except that it is electrogenic (Bakker and Mangerich, 1981) and requires for activity cytoplasmic ATP as well as a transmembrane proton motive force (Rhoads and Epstein, 1977). At very low external K^+ concentrations, or when the Trk system has been inactivated by mutation, K^+ is taken up by the high-affinity Kdp system, which, in contrast to the Trk system, has a moderate rate of transport (Rhoads *et al.*, 1976). The ATP-driven Kdp potassium transport system is repressed at high K^+ concentrations in the medium, and it is associated with a membrane-bound K^+-stimulated ATPase (Rhoads and Epstein, 1977; Epstein *et al.*, 1978). The function of the Kdp system is dependent on the expression of four closely linked genes (Epstein and Davies, 1970). The *kdpD* gene codes for a positive regulator (Rhoads *et al.*, 1978) and the *kdpABC* genes form an operon coding for three inner membrane proteins (Laimins *et al.*, 1978).

Both Kdp and Trk systems carry out net uptake of K^+, either in K^+-

depleted cells or in response to an osmotic upshock. In the absence of a carbon source *E. coli* accumulated K^+ to an intracellular concentration of about 0.1 M, about half the level of actively metabolizing cells (Meury and Kepes, 1981). The intracellular K^+ pool can undergo K^+/K^+ exchange without changing the intracellular K^+ level. This exchange differs from net K^+ uptake in K^+ affinity, rate, and energy requirements (Rhoads and Epstein, 1977; Rhoads and Epstein, 1978). The efflux of K^+ is dependent on active metabolism and on the concentration of intracellular K^+ above a threshold level (Meury and Kepes, 1981). Therefore, it is suggested that the Trk system, active under conditions described by Meury and Kepes (1981), is not a "pump and leak" system but a device to maintain an intracellular K^+ concentration that depends on the osmolarity of the medium and the state of the cell metabolism. K^+ concentrations range from 0.1 M in very dilute media to 0.6 M in media with an osmotic strength of 1200 mosM (Epstein and Schultz, 1965). In cells with a functional Trk system, derepression of the Kdp operon will take place when the concentration of K^+ in the medium falls below 5 mM. In *trkA* mutants the threshold concentration is 10-fold higher (Rhoads *et al.*, 1976). This shows that expression of the Kdp system is not regulated by the absolute concentration of K^+ in the medium but by the potassium requirement of the cell. Therefore, the cell has to control Kdp expression. Using a strain of *E. coli* in which the *lacZ* gene had been fused into the *kdpA* gene of the Kdp operon, Epstein and co-workers (Laimins *et al.*, 1981) showed that the Kdp system is expressed by a decrease in turgor pressure caused by an elevated osmolarity in the growth medium. This was independent of the agent used to increase the osmolarity of the medium. Only glycerol (0.23 M) did not cause stimulation of Kdp expression as it easily diffuses into the cytoplasm. For the first time an example for the regulation of a bacterial operon by a mechanical force has been described! Epstein and co-workers (Laimins *et al.*, 1981) proposed a model in which the membrane-bound KdpD protein undergoes a conformational change following membrane relaxation due to lower turgor pressure. In the new conformation the KdpD protein is able to interact with the *kdp* promoter to stimulate Kdp expression.

As has been discussed, regulation of gene expression is rather slow, taking place in stage II of the cell's adaptation to osmotic stress. The primary event in osmoregulation of *E. coli* following an osmotic increase is an immediate activation of net K^+ uptake (Epstein and Schultz, 1965). The increase in the K^+ content starts within 10 sec of upshock and does not require protein synthesis. The activation of prexisting K^+ transport facilities endows the cell with the ability to respond immediately to a sudden increase of medium osmolarity. The stage I answer allows the cell to adapt thermodynamically to the new environmental situation, a prerequisite for protein synthesis that is required in stage

II. Nothing has been reported about the mechanisms of this rapid activation of the Trk potassium transport system. It does not reflect simply a decrease in K^+ efflux, since kinetic data provided by Epstein and Schultz (1965) show that the initial rate of net K^+ uptake following osmotic upshock is a saturable function of the extracellular K^+ concentration and is often more rapid than the steady-state influx before upshock.

To preserve electroneutrality and a constant pH in the cytoplasm, the uptake of K^+ has to be accompanied by the movement of other ions. K^+ cannot be compensated for by the extrusion of intracellular Na^+, since the cytoplasmic Na^+ concentration in *E. coli* is very low and does not vary with the osmolarity of the medium. In contrast to K^+, the internal Na^+ concentration is a function of the external Na^+ concentration (Epstein and Schultz, 1965). If, as suggested by these authors, H^+ excretion and retention of anionic organic metabolites normally secreted together with H^+ would balance the rapid uptake of K^+, the cytoplasmic pH should rise considerably. Munro *et al.* (1972) presented evidence that the rate of putrescine excretion from *E. coli* is dependent on the presence of K^+ in the medium and on the osmolarity of the medium. In fact, a linkage of K^+ uptake and putrescine excretion would be an appealing mechanism for the raising of cytoplasmic osmolarity via K^+ accumulation without raising the intracellular pH. The uptake of two osmotically active K^+ ions is compensated electrochemically by the loss of one molecule of putrescine into the medium. In contrast to spermidine, the putrescine content of *E. coli* varies inversely with the osmotic strength of the growth medium (Munro *et al.*, 1972). A variety of charged or uncharged solutes, but not glycerol, suddenly added to the growth medium, caused putrescine efflux from *E. coli*. Like Na^+, the spermidine concentration is not influenced by the osmolarity outside the cell. A comparison of kinetic data of K^+ uptake and putrescine efflux after an osmotic upshock suggests that putrescine efflux may be coupled to K^+ transport. Putrescine efflux is specifically dependent on the presence of K^+ in the medium. On the other hand, it seems that K^+ uptake does not require putrescine excretion. Like K^+ uptake, putrescine efflux is dependent on metabolic energy (Munro *et al.*, 1972).

6.2.2. Accumulation of Organic Compounds

It has been reported that the concentration of free amino acids, particularly glutamate, increases when *E. coli* is transferred to a medium of high osmotic strength (Munro *et al.*, 1972; Tempest *et al.*, 1970; Measures, 1975).

Since glutamate is a precursor in putrescine biosynthesis, a reduction in putrescine synthesis could account for part of the increase of intracellular glutamate. Pools of ornithine and arginine, which are direct precursors of putrescine, also increase under these conditions (Munro et al., 1972). Glutamate, which carries a negative charge at neutral pH, has to be accompanied by a cation to preserve electroneutrality. K^+ at a concentration of 0.5 M can activate glutamate dehydrogenase up to 10-fold (Measures, 1975).

Raised osmolarity of the medium may thus lead to a cyclic build up of glutamate, resulting from the osmotically induced increase in K^+ concentration. This increase activates glutamate synthesis, which then requires influx of new K^+ to balance the charge, thus further activating the enzyme. It is interesting to note that this mechanism seems to be confined to gram-negative bacteria, whereas most gram-positives accumulate proline in response to osmotic stress (Measures, 1975). As proline carries no charge, no electrochemical neutralization is necessary. For example, in *Bacillus subtilis* the intracellular K^+ concentration is 0.25 M, irrespective of the external osmolarity (Measures, 1975).

Although proline synthesis is not enhanced in gram-negative bacteria as a response to osmotic increase, addition of proline to the medium may well protect these bacteria from osmotic stress (Christian, 1955). Proline overproduction mutants of *Salmonella typhymurium* have been isolated and found to be osmotolerant and able to grow in 0.8 M media, which was not tolerated by a wild-type strain (Csonka, 1981).

This regulatory mutation is closely linked to the *proAB* genes. A very interesting result of this work is the regulation of the proline level, which increases up to 22-fold in response to 0.65 M NaCl in the growth medium (Csonka, 1981). Since the mutation responsible for the overproduction of proline was present on a F'-plasmid, it could be conjugally transferred into certain strains of *E. coli* and *Klebsiella pneumoniae,* there causing proline overproduction too (Le Rudulier and Valentine, 1982). Concomitantly, both organisms gain osmotolerance, and nitrogen fixation is drastically stimulated during osmotic stress (Le Rudulier and Valentine, 1982). This effect can be attenuated by exogenous application of 1 mM glycine betain (($CH_3)_3$-N^+-CH_2-COO^-). Already 10^{-5} M glycine betaine enables growth of wild-type *Klebsiella pneumoniae* in a medium containing 0.8 M NaCl, which otherwise strongly inhibits growth. Like proline, glycine betaine is accumulated to high concentrations in cells under osmotic stress, but not to a level where it completely balances the external osmotic pressure (Le Rudulier and Bouillard, 1983). This is mediated by a recently characterized high-affinity uptake system (Le Rudulier and Valentine, 1982).

An accumulation of glucose or arabinose in *E. coli* during growth at low osmolarity has also been reported (Roller and Anagnostopoulos, 1982). Carbohydrate accumulation is dependent on the carbon source used to increase the medium osmolarity. The intracellular glucose concentration is five-fold higher when salts are used instead of sucrose in a medium containing glucose as a carbon source (Roller and Anagnostopoulos, 1982).

6.2.3. Compatible Solutes

A variety of solutes can be accumulated in bacteria as a response to osmotic stress. Generally, such low-molecular-weight substances that increase to a high intracellular concentration are termed *compatible solutes* (Borowitzka and Brown, 1974). Their physicochemical properties in highly concentrated solutions have to be compatible with the environmental requirements of the cytoplasmic enzymes. They function partly as osmotically active agents and partly as protectors of enzyme activities under conditions at which the availability of water is limited. An enzyme shows optimal action at a definite ionic strength and/or concentration of organic low-molecular-weight compounds. An increase in solute concentration will cause competition for hydration water and above a certain level may even precipitate proteins. Because of its amphiphilic properties the zwitterionic proline molecule is able to interact with hydrophobic parts of a protein (Schobert, 1977). The increase in the number of hydrophilic groups, caused by binding of proline, results in an enhanced solubility of the protein (Schobert and Tschesche, 1978). The effect of proline seems to be nonspecific and not related to any particular protein. A substance possibly acting quite similarly is betaine (Schobert and Tschesche, 1978). Both substances may increase the cytoplasmic osmolarity and at the same time enhance the solubility of the proteins by increasing their water affinity. The compatible nature of K^+ compared to Na^+ may be ascribed to the protein-binding properties of both ions. K^+ has been observed to show a very low affinity for enzymes at sites important for activity (Brown, 1976). Therefore, K^+ does not disturb the structure of water bound to the protein. In contrast, Na^+ is a much more powerful inhibitor because of its tighter binding.

6.2.4. Osmotic Downshock

Many substances have been reported to be released from *E. coli* by a sudden decrease in medium osmolarity termed *osmotic downshock*. A depletion of the amino acid pools, as well as a rapid decrease in the K^+ concentration, adapts cells to the new condition (Brown, 1976; Britten and McLure, 1962). Since it appears that this loss is quite unspecific, it is supposed that it occurs through small holes or enlarged pores produced by the hydrostatic pressure developing within the cell (Brown, 1976). The leakage is a transient phenomenon since a wash with medium of low osmolarity for as little as 4 sec removes as much K^+ as do washes lasting 45 sec.

6.3. Osmoregulation in the Periplasm

6.3.1. Volume of the Periplasmic Space

It is a well-known phenomenon that strong increases in medium osmolarity cause efflux of water from bacterial cells. In gram-negative cells an increase of the periplasmic space at the expense of the cytoplasmic volume can easily be observed in the electron microscope. Cells that have been exposed to hyperosmolar fluids are termed *plasmolyzed*. This phenomenon is exhibited primarily in media like sucrose solutions in water where cells do not grow (Schleie, 1969). Under these conditions the cells cannot adapt to the new situation because of a lack of potassium and energy. Slight plasmolysis, which occurs primarily at the ends of the cells, is predominant in cells suspended in 0.2 M sucrose.

Extensive plasmolysis is most prevalent in 0.4 M sucrose where membrane separations appear along the sides of the cell. Severe plasmolysis in 1 M sucrose is characterized by a general cytoplasmic collapse along the axis of a cell, sometimes accompanied by a collapse of the cell. These observations of Schleie (1969) could indicate that the periplasmic space includes domains of different osmolarity. Stock *et al.* (1977) published measurements of the volumes of periplasm and cytoplasm under different osmotic conditions. In minimal medium the periplasmic space of *E. coli* and of *S. typhimurium* comprises 20–40% of the total cell volume. If the medium is made hypertonic by the addition of 1 M sucrose, which can penetrate only the outer membrane, the cell changes its total volume only slightly but the volume of the periplasmic space is increased to 80% of the total cell volume at the expense of the cytoplasmic volume. Sucrose raises the osmolarity of the periplasm above that of the cytoplasm (and medium). This causes an immediate water flux from the cytoplasm (and medium) into the periplasmic space, which thus gains in volume. If polyglutamate, which is unable to pass the outer membrane, is used to make the medium hypertonic, the cells shrink to about 60% of their original volume and the ratio of cytoplasmic to periplasmic volume remains constant. The increased osmolarity of the medium above that of the periplasm and cytoplasm results in water efflux from both compartments and causes shrinking of the cell. As no increase of the cytoplasmic volume at the expense of the periplasmic space could be observed, both cytoplasm and periplasm must lose equal proportions of their water content into the medium. This can only be explained if both fluids are isoosmolar. Ethanol, which easily penetrates both membranes, does not affect the cell volume and the ratio of the cell compartments. These observations have several important consequences (Stock *et al.*, 1977). (1) The barrier resisting the osmotic pressure of the gram-negative cell is formed by the outer membrane–peptidoglycan complex, which is termed exoskeleton by the authors. This exoskeleton is not a rigid structure. It collapses when subjected to increased osmotic pressure from without (sodium polyglutamate) and expands when the pressure is increased on the inside during growth in media of low osmolarity. (2) The osmolarity of the periplasm has to be regulated actively by the cell. As the outer membrane allows the permeation of hydrophilic substances with a molecular weight up to 600 daltons, maintenance of high osmolarity in the periplasm cannot be achieved by the accumulation of ions or of small organic substances. If cells of *S. typhimurium* are suspended in water, the resulting osmotic strength of the periplasm (and cytoplasm) is about 170 mosM. Therefore, the sum of the osmotically active material in the periplasm must be at least 170 mM. Solutes of high molecular weight cannot be present at this concentration in the periplasm. Assuming a molecular weight of 6000 daltons, the solute would have to be present in a concentration of 1 g/ml. Therefore, Stock *et al.* (1977) postulate the presence in the periplasm of polyvalent anions that are large enough to be retained within the cell. To preserve electroneutrality, small cations could be concentrated in the periplasm, thus maintaining high osmotic strength. As this would lead to an unequal distribution of cations and anions between periplasm and medium, a Donnan potential across the outer membrane should result. In fact, by measuring the distribution of radiolabeled Na^+ and Cl^- between periplasm and medium, Stock *et al.* (1977) calculated a potential of about 30 mV across the outer membrane of *S. typhimurium*. The validity of the results reported above is made doubtful by the fact that much smaller values for the size of the periplasmic space have been reported by Nikaido (1979). It is possible that this difference is due to the variations in the environment and in the energy state of the cells used in the experiments. Stock *et al.* (1977) obtained the high value (35% of cell volume) with cells suspended at high density under anaerobic conditions. Nikaido's measurements yielded 5% periplasmic space in normal, unplasmolyzed *S. typhimurium* and *E. coli* cells, except in starved cells or stationary phase cells, in which up to 13% of the cell volume became periplasmic space (Nikaido, 1979).

6.3.2. Regulation of the Periplasmic Osmotic Strength by Membrane-Derived Oligosaccharides (MDO)

The nature of the polyvalent anions postulated by Stock *et al.* (1977) was finally elucidated by Kennedy (1982). In 1973 van Golde *et al.* discovered that the turnover of membrane phospholipids in *E. coli* is related to the biosynthesis of the so-called membrane-derived oligosaccharides (MDO), which are constituents of the periplasmic space of gram-negative bacteria (Schulman and Kennedy, 1979).

Not being aware of the existence of MDO, Munro (1973) reported that the rate of turnover of phospholipids in *E. coli* was decreased when the osmolarity of the growth medium was high. This turnover was due mainly to the transfer of glycero-1-phosphate or phosphoenthanolamine residues to membrane-derived oligosaccharides (Schulman and Kennedy, 1977). *E. coli* MDO contains about 8–10 glucose units per molecule linked by β 1–2 and β 1–6 bonds, thus forming a highly branched structure (Schneider *et al.*, 1979). They are variously substituted with *sn* 1-phosphoglycerol and phosphoethanolamine in phosphodiester-linkage or succinate in *O*-ester linkage at the free C-6 atoms of the glucose residues (Kennedy *et al.*, 1976). The multiple substitution and the branched structure allow the MDO to maintain a high anionic charge per residue of glucose, and the molecular weight (2200–2500 daltons) prevents their loss through the porins of the outer membrane of *E. coli*. When Kennedy (1982) reported that the biosynthesis of MDO was regulated by the osmolarity of the medium, it became obvious that the properties of these periplasmic constituents of gram-negative bacteria ideally fulfill the requirements of the osmoregulatory polyanions postulated by Stock *et al.* (1977). When osmolarity is low, MDO may constitute up to 7% of the cell's dry weight (Kennedy, 1982); lower values (0.5–1%) have also been reported (Schulman and Kennedy, 1979). After a shift to high osmolarity, the biosynthesis of MDO is not arrested but the specific activity of the biosynthetic enzymes is reduced by dilution (Kennedy, 1982). From this observation Kennedy derived a hypothesis about osmoregulation in the periplasmic space of *E. coli* that has many features in common with the model presented by Epstein (1981) for the regulation of the Kdp operon expression. The function of MDO is the maintenance of high periplasmic osmolarity and the generation of a Donnan potential across the outer membrane. When the total osmolarity of the periplasm falls below that of the cytoplasm, caused by influx of water, the resulting difference in turgor pressure on the cytoplasmic membrane is detected by an osmotic sensor, presumably located in this membrane. The osmotic sensor generates a signal that induces the synthesis of some essential proteins, which may be MDO biosynthesis enzymes or may act as regulator(s) of these enzymes. Like the Kdp operon expression, induction of MDO biosynthesis at low osmolarity is a stage II response of the bacterial cell to osmotic stress. Again, gene expression is stimulated by a mechanical force.

Details of the biochemistry and genetics of MDO biosynthesis have still to be worked out. A proposal outlining a possible biosynthetic pathway has been presented (Goldberg *et al.*, 1981). This model involves UDP-glucose as cytoplasmic precursor and lipid carrier bound oligosaccharide as first acceptor for *sn* 1-phosphoglycerol. Oligosaccharide synthesis is mediated by a glucosyl transferase, which *in vitro* needs a "primer" (like the β-glucoside 2-*O*-β-D-glucopyranosyl α-D-glucopyranose) and UDP-glucose for the synthesis of branched β-linked glucose oligomers of the same size as MDO (Weissborn and Kennedy, 1983). The transfer of phosphoglycerol from phosphatidylglycerol to carrier-bound pre-MDO (and synthetic β-glucosides) is catalyzed by a membrane-bound transferase leading to membrane-bound pre-MDO-1-phosphoglycerol and 1,2-diglyceride (Jackson and Kennedy, 1983). Mutants of *E. coli* lacking digylceride kinase and thus blocked in the salvage of diglyceride accumulate the neutral lipid resulting from the biosynthesis of MDO (Raetz and Newman, 1978). Accumulation is low in media with high osmolarity and may rise up to 12% of total lipids in cells grown at low osmolarity (Raetz and Newman, 1979; Rotering and Raetz, 1983). A second periplasmic phosphoglycerol transferase has been partially purified and characterized (Goldberg *et al.*, 1981). This enzyme obviously transfers *sn* 1-phosphoglycerol residues from membrane-bound MDO to periplasmic MDO, previously released from the lipid carrier by a hydrolase. The transfer reaction leads to multiple substituted MDO molecules, each carrying up to three residues of *sn* 1-phosphoglycerol. The transferase used only membrane-bound pre-MDO, and no phosphatidylglycerol, as donor for phosphoglycerol residues. Nothing has been reported about the transfer of phosphoethanolamine and succinate to MDO. MDO might play a role in the transfer of phosphoethanolamine residues from phosphatidylethanolamine to the core oligosaccharide region of the lipopolysaccharides of *E. coli* since it has been shown by Hasin and Kennedy (1982) that phosphoethanolamine residues in lipopolysaccharide are derived from phosphatidylethanolamine, which has a large pool size in constrast to serine, phosphoserine, and CDP-ehtanolamine.

Another open question is what happens to periplasmic MDO if it is no longer required by the cell after an osmotic increase. Two periplasmic enzymes potentially able to degrade MDO are known in *E. coli*. An inducible periplasmic phosphodiesterase, able to cleave deacylated phospholipids, has been identified as part of *glp* regulon and genetically designated *glpQ* (Larson *et al.*, 1983). An enzyme with similar specificity, able to cleave MDO, has been isolated from *Aspergillus niger* (Schneider and Kennedy, 1978). The other appealing aspect of potential MDO degradation is the presence in *E. coli* of genetic information coding for periplasmic β-glucosidase that is part of the system for the utilization of β-glucosides (Reynolds *et al.*, 1981; Schaefler, 1976). The expression of this system can only be achieved by (spontaneous) mutation or by the insertion of a transposable element into the promoter region of the operon. Mutants of this type are able to ferment the β-glucosides arbutin and salicin, which, as mentioned earlier, may also act as phosphoglycerol acceptors for the transfer reaction catalyzed by the membrane-bound phosphoglycerol transferase involved in MDO biosynthesis. The presence of information in *E. coli* for two enzymes that potentially may degrade MDO could also indicate that under certain conditions the advantage of MDO-mediated osmoregulation can be sacrificed in favor of the utilization of substrates that require these enzymes to be metabolized. But this hypothesis has to be tested by future studies that especially should include the screening for MDO mutants. The use of such mutants should clarify the physiological functions of the Donnan potential across the outer membrane and confirm the indispensability of MDO for osmoregulation in the periplasmic space.

6.4. Osmolarity-Dependent Regulation of the Protein Composition of the Outer Membrane

Proteins OmpF and OmpC, the so-called porins of *E. coli* K-12, are major outer membrane proteins; they form transmembrane channels and interact strongly with the underlying peptidoglycan layer (Rosenbusch, 1974; Nikaido and Nakae, 1979). Both proteins are the products of two structural genes *ompF* and *ompC,* which map at 21 min and 47 min on the *E. coli* chromosomes, respectively (Ichihara and Mizushima, 1978; Hall and Silhavy, 1979; Sato and Yura, 1979; van Alphen *et al.,* 1979).

The relative amounts of the porins found in the outer membrane vary according to the growth conditions and are influenced by the osmolarity of the culture medium (van Alphen and Lugtenberg, 1977). The amounts of the OmpF and OmpC proteins in the outer membrane are affected differentially by high concentrations of solutes such as sucrose in the culture medium in such a manner that a decrease in the amount of the OmpF protein appears to be compensated by a reciprocal increase of the OmpC protein (Kawaji *et al.,* 1979). Sugars and dextrans of molecular weights greater than 600 switch the synthesis of OmpF to OmpC more effectively than those of lower molecular weight. Substances such as sucrose primarily control *ompF* gene expression, presumably via osmotic pressure, which in turn controls *ompC* gene expression on the transcriptional level (Ozawa and Mizushima, 1983). The *ompB* gene locus, mapping at 74 min on the *E. coli* chromosome (Sarma and Reeves, 1977), has been identified as the regulatory locus for porin protein synthesis (Hall and Silhavy, 1979; Hall and Silhavy, 1981). The *ompB* regulon codes for two products, the OmpR and EnvZ proteins. Mutations in *ompR* cause two phenotypes. One group loses both OmpF and OmpC proteins and the other constitutively synthesizes only OmpF. This indicates that there are two functional domains within the OmpR protein (Hall and Silhavy, 1981). Mutations in *envZ* are pleiotropic, affecting the production of OmpF and OmpC proteins as well as other outer membrane proteins and periplasmic proteins (Hall and Silhavy, 1981; Wandersman *et al.,* 1980; Wanner *et al.,* 1979). The *ompB* regulon has been cloned and sequenced by Inouye and co-workers (Wurtzel *et al.,* 1982; Mizuno *et al.,* 1982). A molecular weight of 32,500 daltons has been ascribed to the OmpR protein, which has an extreme basic sequence domain at its carboxy-terminal end. Sequences homologous to the DNA binding regions of the phage λ and P22 repressor proteins have been identified within the OmpR protein. The *envZ* gene product is a protein with a molecular weight of 44,000 daltons. A domain of 17 amino acid residues is strongly hydrophobic, indicating that the protein is membrane bound. Furthermore, the protein shows an extremely acidic domain within a stretch of 72 amino acid residues. In spite of this important insight into the nature of the *ompB* operon gene products, the role of the OmpR and EnvZ proteins in the osmolarity-dependent regulation of the porin genes *ompF* and *ompC* remains obscure. Inouye and co-workers (Wurtzel *et al.,* 1982; Mizuno *et al.,* 1982), however, point out a number of facts that suggest a model of *ompF/ompC* gene regulation similar to Epstein's hypothesis of osmolarity-dependent regulation of potassium transport (Laiminis *et al.,* 1981). OmpF protein synthesis is regulated by the *envZ* gene product as some mutations in *envZ* result in inhibition of OmpF synthesis and in constitutive, osmolarity independent production of OmpC. This is consistent with observations made by Ozawa and Mizushima (1983), who identified osmolarity-independent OmpC production in an *ompF* strain lacking OmpF, whereas introduction of an *ompC* mutation did not affect the sucrose-dependent profile of OmpF synthesis. Because of its hydrophobic sequences the EnvZ protein might be a cytoplasmic membrane protein that interacts via its acidic domain with the basic sequence of the OmpR protein. The latter is supposed to be a DNA-binding protein according to its amino acid sequence. Alterations in the state of the EnvZ protein, induced by osmolarity changes, could influence the binding of EnvZ to OmpR, which in this model would act as a cytoplasmic repressor for OmpF transcription. There are, however, several possible objections to this model. Hitherto, the location of EnvZ has not been determined. Cavard *et al.* (1982) suggest that EnvZ may be identical with the colicin A protease located at the external face of the outer membrane. The authors compared properties of *envZ* mutants with a mutant strain devoid of this protease, but they could not exclude the possibility that the mutation affects both functions in the same way. If the *envZ* gene product was located in the outer membrane, no direct interaction between EnvZ and OmpR could take place. Cloning of the DNA region coding for the OmpR protein on a multicopy plasmid does not cause repression of OmpF synthesis at low osmolarity (Mizuno *et al.,* 1982). This makes a stoichiometric interaction of EnvZ and OmpR as repressor of OmpF transcription improbable.

An understanding of the OmpF–OmpC regulation becomes even more complicated if one considers the role of cyclic AMP in the expression of both proteins. cAMP stimulates *ompF* expression (Scott and Harwood, 1980), whereas in an adenylate cyclase deficient mutant OmpC becomes constitutive, thus resembling the *envZ* phenotype (Mizuno *et al.,* 1982). Nikaido (1979) offers an interpretation for the physiological importance of porin regulation by osmotic strength and cyclic AMP:

> The salt-repressible porin (OmpF) appears to be the most effective in the diffusion of a lot of solutes and could form the basic, all purpose channels. . . . When enteric bacteria are thrown out of their protective environment, that is, the intestinal tract of warm-blooded animals, the salt concentration of the environment would increase as a result of the drying out of feces. This could act as a signal to tell *E. coli* cells of a change in the environment, so that they can shut down the most effective pores and survive in a state of minimal metabolic activity or with minimal exchange of material with the outside world. Under certain conditions, in contrast, the rise in cellular cyclic AMP concentrations tells *E. coli* that it is becoming starved and should expand its catabolic activities.

The increased production of OmpF protein under these conditions would endow the cell with optimal facilities to take in additional nutrients.

7. REFERENCES

Achtman, M., Manning, P. A., Kusecek, B., Schwauchow, S., and Willets, N., 1980, Genetic analysis of F sex factor cistrons needed for surface exclusion in *Escherichia coli*, *J. Mol. Biol.* **138**:779–795.

Alemohammad, M. M., and Knowles, C. J., 1974, Osmotically induced volume and turbidity changes of *Escherichia coli* due to salts, sucrose, and glycerol, with particular references to the rapid permeation of glycerol into the cell, *J. Gen. Microbiol.* **82**:125–142.

Amano, K., and Williams, J. C., 1983, Partial characterization of peptidoglycan-associated proteins of *Legionella pneumophila*, *J. Biochem.* **94**:601–606.

Angus, B. L., Carey, M. M., Caron, A. D., Kropinski, A. M. B., and Hancock, R. E. W., 1982, Outer membrane permeability in *Pseudomonas aeruginosa:* Comparison of a wild type with an antibiotic-supersusceptible mutant, *Antimicrob. Agents Chemother.* **21**:299–309.

Archibald, F. S., and DeVoe, J. W., 1979, Removal of iron from human transferrin by *Neisseria meningitidis*, *FEMS Microbiol. Lett.* **6**:159–162.

Archibald, F. S., and DeVoe, J. W., 1980, Iron acquisition by *Neisseria meningitidis* in vitro, *Infect. Immun.* **27**:322–334.

Austen, B. M., 1979, Predicted secondary structure of amino-terminal extension sequences of secreted proteins, *FEBS Lett.* **103**:308–313.

Bakker, E. P., and Mangerich, W., 1981, Interconversion of components of the bacterial proton motive force by electrogenic potassium transport, *J. Bacteriol.* **147**:820–826.

Bassford, P., and Beckwith, J., 1979, *Escherichia coli* mutants accumulating the precursor of a secreted protein in the cytoplasm, *Nature* **277**:538–541.

Bassford, P. J., Silhavy, T. J., and Beckwith, J. R., 1979, Use of gene fusion to study secretion of maltose-binding protein into *Escherichia coli* periplasm, *J. Bacteriol.* **139**:19–31.

Bavoil, P., and Nikaido, H., 1981, Physical interaction between the phage lambda receptor protein and the carrier-immobilized maltose-binding protein of *Escherichia coli*, *J. Biol. Chem.* **256**:11385–11388.

Bavoil, P., Wandersman, C., Schwartz, M., and Nikaido, H., 1983, A mutant form of maltose-binding protein of *Escherichia coli* deficient in its interaction with the bacteriophage lambda receptor protein, *J. Bacteriol.* **155**:919–921.

Bayer, M. E., 1979, The fusion sites between the outer membrane and cytoplasmic membrane of bacteria: Their role in membrane assembly and virus infection, in: *Bacterial Outer Membranes: Biogenesis and Function* (M. Inouye, ed.), John Wiley and Sons, Inc., New York, pp. 167–202.

Bayer, M. E., Bayer, M. H., Shephardson, S., Dall'Olio, A., Garrett, W., Ridley, R., and Scott, E., 1982a, Structure and function of cell membranes during growth and interaction with viruses, in: The Institute for Cancer Research Twenty-seventh Scientific Report (September 1981–September 1982), Fox Chase Cancer Center, Philadelphia, Pa., pp. 70–74.

Bayer, M. H., Costello, G. P., and Bayer, M. E., 1982b, Isolation and partial characterization of membrane vesicles carrying markers of the membrane adhesion sites, *J. Bacteriol.* **149**:758–767.

Beck, E., and Bremer, E., 1980, Nucleotide sequence of the gene *ompA* coding the outer membrane protein II of *Escherichia coli* K-12, *Nucleic Acid Res.* **8**:3011–3027.

Bedouelle, H., Bassford, P. J., Fowler, A. V., Zabin, I., Beckwith, J., and Hofnung, M., 1980, Mutations which alter the function of the signal sequence of the maltose-binding protein of *Escherichia coli, Nature* **285**:78–81.

Benson, S., and Silhavy, T., 1983, Information within the mature Lam B protein necessary for localization to the outer membrane of *Escherichia coli* K-12, *Cell* **32**:1325–1335.

Benz, R., and Hancock, R. E. W., 1981, Properties of the large ion-permeable pores formed from protein F of *Pseudomonas aeruginosa* in lipid bilayer membranes, *Biochim. Biophys. Acta* **646**:298–308.

Benz, R., Janko, K., Boos, W., and Läuger, P., 1978, Formation of large ion-permeable membrane channels by the matrix protein (porin) of *Escherichia coli, Biochim. Biophys. Acta* **511**:305–319.

Benz, R., Janko, K., and Läuger, P., 1979, Ionic selectivity of pores formed by the matrix protein (porin) of *Escherichia coli, Biochim. Biophys. Acta* **551**:238–247.

Berzhinskene, Y. A., Zizaite, L. Y., Baronaite, Z. A., and Grinyus, L. L., 1980, Investigation of energy supply for transfer of plasmic R 100-1 during conjugation of *Escherichia coli* cells, *Biokhimiya* (Engl. trans.) **45**:839–845.

Biswas, G. D., and Sparling, P. F., 1981, Entry of double-stranded deoxyribonucleic acid during transformation of *Neisseria gonorrhoeae, J. Bacteriol.* **145**:638–640.

Blake, M. S., and Gotschlich, E. C., 1983, Gonococcal membrane proteins: Speculation on their role in pathogenesis, *Progr. Allergy* **33**:298–313.

Blobel, G., and Dobberstein, B., 1975, Transfer of proteins across membranes, *J. Cell Biol.* **67**:835–851.

Boehler-Kohler, B. A., Boos, W., Dieterle, R., and Benz, R., 1979, Receptor for bacteriophage lambda of *Escherichia coli* forms larger pores in black lipid membranes than the matrix protein (porin), *J. Bacteriol.* **138**:33–39.

Borowitzka, L. J., and Brown, A. D., 1974, The salt relation of marine and halophilic species of the unicellular green algae *Dunaliella:* The role of glycerol as a compatible solute, *Arch. Microbiol.* **96**:37–52.

Bowles, L. K., and Konisky, J., 1980, Cleavage of Colicin Ia by the *Escherichia coli* K-12 outer membrane is not mediated by the Colicin Ia receptor, *J. Bacteriol.* **145**:668–671.

Boyd, A., and Holland, I. B., 1980, Intermediate location in the assembly of the matrix protein or porin into the outer membrane of *Escherichia coli, J. Bacteriol.* **143**:1538–1541.

Boyd, A., Kendall, K., and Simon, M., 1983, Structure of the serine chemoreceptor in *Escherichia coli, Nature* **301**:623–626.

Brade, H., and Rietschel, E.-T., 1984, α-2→4- interlinked 3-deoxy-D-manno-octulosonic acid disaccharide. A common constitutent of enterobacterial lipopolysaccharides, *Eur. J. Biochem.* **145**:231–236.

Bradley, D. E., 1980, Morphological and serological relationships of conjugative plasmids, *Plasmid* **4**:155–169.
Bradley, D. E., 1983, Specification of the conjugative pili and surface mating systems of Pseudomonas plasmids, *J. Gen. Microbiol.* **129**:2545–2556.
Braun, G., and Cole, S. T., 1982, The nucleotide sequence coding for major outer membrane protein Omp A of *Shigella dysenteriae*, *Nucleic Acid Res.* **10**:2367–2378.
Braun, V., 1975, Covalent lipoprotein from the outer membrane of *Escherichia coli*, *Biochim. Biophys. Acta* **415**:335–377.
Braun, V., 1985, The iron transport systems of *Escherichia coli*, in: *The Enzymes of Biological Membranes* (A Martonosi, ed.), Volume 3, Plenum Press, New York, **11**:617–652.
Braun, V., and Hantke, K., 1981, Bacterial cell surface receptors, in: *Organization of Procaryotic Cell Membranes*, Vol. II (B. K. Ghosh, ed.), CRC Press, Inc., Boca Raton, Fla., pp. 1–73.
Braun, V., and Hantke, K., 1982, Receptor-dependent transport systems in *Escherichia coli* for iron complexes and vitamin B_{12}, in: *Membranes and Transport*, Vol. 2 (A. Martonosi, ed.), Plenum Press, New York, pp. 107–113.
Bremer, E., Cole, S. T., Hindennach, I., Henning, U., Beck, E., Kurz, C., and Schaller, H., 1982, Export of a protein into the outer membrane of *Escherichia coli* K-12: Stable incorporation of the Omp A protein requires less than 193 amino-terminal amino-acid residues, *Eur. J. Biochem.* **122**:223–231.
Brinton, C. C., Bryan, J., Dillon, J. A., Guerina, N., Jacobson, L. J., Kraus, S., Labik, A., Lee, S., Levene, A., Lim, S., McMichael, J., Polen, S., Rogers, K., To, A. C. C., and To, S. C. M., 1978, Uses of pili in gonorrhea control: Role of pili in disease, Purification and properties of gonococcal pili and progess in the development of a gonococcal pilus vaccine for gonorrhea, in: *Immunobiology of Neisseria gonorrhoeae* (G. F. Brooks, E. C. Gotschlich, K. K. Holmes, W. D. Sawyer, and F. E. Young, eds.), American Society of Microbiology, Washington, pp. 155–178.
Britten, R. J., and McLure, F. T., 1962, The amino-acid pool in *Escherichia coli*, *Bacteriol. Rev.* **26**:292–335.
Brown, A. D., 1976, Microbial water stress, *Bacteriol. Rev.* **40**:803–846.
Brown, M. G. M., Saunders, J. R., and Humphreys, G. O., 1981, Lack of specificity in DNA binding and uptake during transformation of *Escherichia coli*, *FEMS Microbiol. Lett.* **11**:97–100.
Buchel, D. E., Gronenborn, B., and Müller-Hill, B., 1980, Sequence of the lactose permease gene, *Nature* **283**:541–545.
Carr, S. A., Biemann, K., Shoji, S., Parmelee, D. C., and Titani, K., 1982, *n*-Tetradecanoyl is the NH_2-terminal blocking group of the catalytic subunit of cyclic AMP-dependent protein kinase from bovine cardiac muscle, *Proc. Natl. Acad. Sci. USA* **79**:6128–6131.
Casadaban, M. J., 1976, Transposition and fusion of the *lac* genes to select promotors in *Escherichia coli* using bacteriophage lambda and mu, *J. Mol. Biol.* **104**:541–555.
Cavard, D., Pages, J. M., and Lazdunski, C. J., 1982, A protease as a possible sensor of environmental conditions in *Escherichia coli* outer membrane, *Mol. Gen. Genet.* **188**:508–512.
Celis, R. T. F., 1981, Chain terminating mutants affecting a periplasmic binding protein involved in the active transport of arginine and ornithine in *Escherichia coli*, *J. Biol. Chem.* **256**:773–779.
Cerretti, D. P., Dean, D., Davis, G. R., Bedwell, D. M., and Nomura, M., 1983, The *spc* ribosomal protein operon of *Escherichia coli:* Sequence and cotranscription of the ribosomal genes and a protein export gene, *Nucleic Acid Res.* **11**:2599–2616.
Chen, R., Krämer, C., Schmidmayr, W., Chen-Schmeisser, U., and Henning, U., 1982, Primary

structure of major outer membrane protein I (Omp F protein, porin) of *Escherichia coli* B/r, *Biochem. J.* **203**:33–43.

Christian, J. H. B., 1955, The influence of nutrition on the water relations of *Salmonella oranienburg*, *Aust. J. Biol. Sci.* **8**:75–82.

Clément, J. M., and Hofnung, M., 1981, Gene sequence of the lambda receptor, an outer membrane protein of *Escherichia coli* K-12, *Cell* **27**:507–514.

Cleveland, M. vB., Slatin, S., Finkelstein, A., and Levinthal, C., 1983, Structure–function relationships for a voltage-dependent ion channel: Properties of COOH-terminal fragments of Colicin E1, *Proc. Natl. Acad. Sci. USA.* **80**:3706–3710.

Cohen, S. N., 1976, Transposable genetic elements and plasmid evolution, *Nature* **263**:731–738.

Cole, S. T., Chen-Schmeisser, U., Hindennach, I., and Henning, U., 1983, Apparent bacteriophage binding region of an *Escherichia coli* K-12 outer membrane protein, *J. Bacteriol.* **153**:581–587.

Cose, C. D., Rinehart, K. L., Moore, M. L., and Cook, J. C., Jr., 1981, Pyochelin: Novel structure of an iron-chelating growth promoter for *Pseudomonas aeruginosa*, *Proc. Natl. Acad. Sci. USA* **78**:4256–4260.

Coulton, J. W., and Braun, V., 1979, Protein II* influences ferrichrome–iron transport in *Escherichia coli* K-12, *J. Gen. Microbiol.* **110**:211–220.

Csonka, L. N., 1981, Proline overproduction results in enhanced osmotolerance in *Salmonella typhimurium*, *Mol. Gen. Genet.* **182**:82–86.

Dallas, W. S., and Falkow, S., 1980, Amino-acid sequence homology between cholera toxin and *Escherichia coli* heat-labile toxin, *Nature* **288**:499–501.

Danner, D. B., Deich, R. A., Sisco, K. L., and Smith, H. O., 1980, An eleven-base-pair sequence determines the specificity of DNA uptake in Haemophilus transformation, *Gene* **11**:311–318.

Dassa, E., Frelat, G., and Boquet, P. L., 1978, Protein analysis of outer membrane prepared from *Escherichia coli* K-12 by different procedures, *Biochem. Biophys. Res. Commun.* **81**:616–622.

Date, T., 1983, Demonstration by a novel genetic technique that leader peptidase is an essential enzyme in *Escherichia coli*, *J. Bacteriol.* **154**:76–83.

Date, T., and Wickner, W., 1981, Isolation of the *Escherichia coli* leader peptidase gene and effects of leader peptidase overproduction in vivo, *Proc. Natl. Acad. Sci. USA* **78**:6106–6110.

de Geus, P., van Die, I., Bergmans, H., Tommassen, J., and de Haas, G., 1983, Molecular cloning of *pldA* the structural gene for outer membrane phospholipase of *Escherichia coli* K-12, *Mol. Gen. Genet.* **190**:150–155.

Diaz, J. L., and Heckels, J. E., 1982, Antigenic variation of outer membrane Protein II in colonial variants of *Neisseria gonorrhoeae* P9, *J. Gen. Microbiol.* **128**:585–591.

DiRienzo, J. M., Nakamura, K., and Inouye, M., 1978, The outer membrane proteins of gram-negative bacteria: Biosynthesis, assembly, and functions, *Ann. Rev. Biochem.* **47**:481–532.

Douglas, J. T., Lee, M. D., and Nikaido, H., 1981, Protein I of *Neisseria gonorrhoeae* outer membrane is a porin, *FEMS Mikrobiol. Lett.* **12**:305–309.

Drager, D. L., James, J. F., Brooks, G. F., and Sweet, R. L., 1980, Comparison of virulence markers of peritoneal and fallopian tube isolates with endocervical *Neisseria gonorrhoeae* isolates from women with acute salpingitis, *Infect. Immun.* **27**:882–888.

Duckworth, M., Jackson, D., Zak, K., and Heckels, J. E., 1983, Structural variations in pili expressed during gonococcal infection, *J. Gen. Microbiol.* **129**:1593–1596.

Duke, J., and Guiney, D. G., Jr., 1983, The role of lipopolysaccharide structure in the recipient cell during plasmid-mediated bacterial conjugation, *Plasmid* **9**:222–226.

Ehring, R., Beyreuther, K., Wright, J. K., and Overath, P., 1980, In vitro and in vivo products of *Escherichia coli* lactose permease gene are identical, *Nature* **283**:537–540.

Emr, S. D., and Bassford, P. J., Jr., 1982, Localization and processing of outer membrane and periplasmic proteins in *Escherichia coli* strains harboring export-specific suppressor mutations, *J. Biol. Chem.* **257**:5852–5860.

Emr, S. D., and Silhavy, T. J., 1980, Mutations affecting localization of an *Escherichia coli* outer membrane protein, the bacteriophage lambda receptor, *J. Mol. Biol.* **141**:63–90.

Emr, S. D., and Silhavy, T. J., 1983, Importance of secondary structure in the signal sequence for protein secretion, *Proc. Natl. Acad. Sci. USA* **80**:4599–4603.

Emr, S. D., Schwartz, M., and Silhavy, T. J., 1978, Mutations altering the cellular localization of the phage lambda receptor, an *Escherichia coli* outer membrane protein, *Proc. Natl. Acad. Sci. USA* **75**:5802–5806.

Emr, S. D., Hedgpeth, J., Clément, J. M., Silhavy, T. J., and Hofnung, M., 1980, Sequence analysis of mutations that prevent export of lambda receptor, an *Escherichia coli* outer membrane protein, *Nature* **285**:82–85.

Emr, S. D., Hanley-Way, S., and Silhavy, T. J., 1981, Suppressor mutations that restore export of a protein with a defective signal sequence, *Cell* **23**:79–88.

Epstein, W., and Davies, M., 1970, Potassium-dependent mutants of *Escherichia coli* K-12, *J. Bacteriol.* **101**:836–843.

Epstein, W., and Kim, B. S., 1971, Potassium transport loci in *Escherichia coli* K-12, *J. Bacteriol.* **108**:639–644.

Epstein, W., and Laimins, L. A., 1980, Potassium transport in *Escherichia coli:* Diverse systems with common control by osmotic forces, *Trends Biochem. Sci.* **5**:21–23.

Epstein, W., and Schultz, S. G., 1965, Cation transport in *Escherichia coli* V: Regulation of cation content, *J. Gen. Physiol.* **49**:221–234.

Epstein, W., Whitelaw, V., and Hesse, J., 1978, A K^+-transport ATPase in *Escherichia coli, J. Biol. Chem.* **253**:6666–6668.

Favero, M. S., Carson, L. A., Bond, W. W., and Petersen, N. J., 1971, *Pseudomonas aeruginosa:* Growth in distilled water from hospitals, *Science* **173**:836–838.

Fecker, L., and Braun, V., 1983, Cloning and expression of the *fhu* genes involved in iron(III)-hydroxamate uptake of *Escherichia coli, J. Bacteriol.* **156**:1301–1314.

Ferenci, T., Schwentorat M., Ullrich, S., and Volmart, J., 1980, Lambda receptor in the outer membrane of *Escherichia coli* as a binding protein for maltodextrins and starch polysaccharides, *J. Bacteriol.* **142**:521–526.

Fiss, E. H., Stanley-Samuelson, P., and Neilands, J. B., 1982, Properties and proteolysis of ferric enterobactin outer membrane receptor in *Escherichia coli* K-12, *Biochemistry* **21**:4517–4522.

Fornari, C. S., and Kaplan, S., 1982, Genetic transformation of *Rhodopseudomonas sphaeroides* by plasmid DNA, *J. Bacteriol.* **152**:89–97.

Fraley, R., Straubinger, R. M., Rule, G., Springer, E. L., and Papahadjopoulos, D., 1981, Liposome-mediated delivery of deoxyribonucleic acid to cells: Enhanced efficiency of delivery related to lipid composition and incubation conditions, *Biochemistry* **20**:6978–6987.

Freudl, R., and Cole, S. T., 1983, Cloning and molecular characterization of the *omp A* gene from *Salmonella typhimurium, Eur. J. Biochem.* **134**:497–502.

Furukawa, H., and Mizushima, S., 1982, Roles of cell surface components of *Escherichia coli* K-12 in bacteriophage T4 infection: Interaction of tail core with phospholipids, *J. Bacteriol.* **150**:916–924.

Furukawa, H., Kuroiwa, T., and Mizushima, S., 1983, DNA injection during bacteriophage T4 infection of *Escherichia coli, J. Bacteriol.* **154**:938–945.

Gaastra, W., and DeGraaf, F. K., 1982, Host-specific fimbrial adhesins of noninvasive enterotoxigenic *Escherichia coli* strains, *Microbiol. Rev.* **46**:129–161.

Gabay, J., Benson, S., and Schwartz, M., 1983, Genetic mapping of antigenic determinants on a membrane protein, *J. Biol. Chem.* **258**:2410–2414.

Garwin, J. L., and Beckwith, J., 1982, Genetic approaches for studying protein localization, in: *Membranes and Transport,* Vol. 1 (A. Martonosi, ed.), Plenum Press, New York, pp. 315–321.

Goldberg, D. E., Rumley, M. K., and Kennedy, E. P., 1981, Biosynthesis of membrane-derived oligosaccharides: A periplasmic phosphoglyceroltransferase, *Proc. Natl. Acad. Sci. USA* **78**:5513–5517.

Goodgal. S. H., 1982, DNA uptake in Haemophilus transformation, *Ann Rev. Genet.* **16**:169–192.

Graves, J. F., Biswas, G. D., and Sparling, P. F., 1982, Sequence-specific DNA uptake in transformation of *Neisseria gonorrhoeae, J. Bacteriol.* **152**:1071–1077.

Greco, F., Blake, M. S., Gotschlich, E. C., and Mauro, A., 1980, Major outer membrane protein of *Neisseria gonorrhoeae* forms channels in lipid bilayer membranes, *Fed. Proc.* **39**:1813.

Grinius, L., 1980, Nucleic acid transport driven by ion gradient across cell membrane, *FEBS Lett.* **113**:1–10.

Gruss, P., Greiner, J., and Martin, H. H., 1975, Amino-acid composition of the covalent rigid-layer lipoprotein in cell walls of *Proteus mirabilis, Eur. J. Biochem.* **57**:411–414.

Gubish, E. R., Jr., Chen, K. C. S., and Buchanan, T. M., 1982, Attachment of gonococcal pili to lectin-resistant clones of Chinese hamster ovary cells, *Infect. Immun.* **37**:189–194.

Gunther, T., and Dorn, F., 1966, Über den K-Transport bei der K-Mangelmutante *Escherichia coli* B 525, *Z. Naturforschg.* **216**:1082–1088.

Hackett, J., and Reeves, P., 1983, Primary structure of the *tol C* gene that codes for an outer membrane protein of *Escherichia coli* K-12, *Nucleic Acid Res.* **11**:6487–6495.

Hackett, J., Misra, R., and Reeves, P., 1983, The Tol C protein of *Escherichia coli* K-12 is synthesized in a precursor form, *FEBS Lett.* **156**:307–310.

Hale, T. L., Sansonetti, P. J., Schad, P. A., Austin, S., and Formal, S. B., 1983, Characterization of virulence plasmids and plasmic-associated outer membrane proteins in *Shigella flexneri, Shigella sonnei,* and *Escherichia coli, Infect. Immun.* **40**:340–350.

Hall, M. N., and Silhavy, T. J., 1979, Transcriptional regulation of *Escherichia coli* K-12 major outer membrane protein Ib, *J. Bacteriol.* **140**:342–350.

Hall, M. N., and Silhavy, T. J., 1981, The *omp B* locus and the regulation of the major outer membrane porin proteins of *Escherichia coli* K-12, *J. Mol. Biol.* **146**:23–43.

Hall, M. N., and Silhavy, T. J., 1981, Genetic analysis of the major outer membrane proteins of *Escherichia coli, Ann. Rev. Genet.* **15**:91–142.

Hall, M. N., and Schwartz, M., 1982, Reconsidering the early steps of protein secretion, *Ann. Microbiol. Inst. Pasteur* **133A**:123–127.

Hall, M. N., Schwartz, M., and Silhavy, T. J., 1982, Sequence information within the *lam B* gene is required for proper routing of the bacteriophage lambda receptor protein to the outer membrane of *Escherichia coli* K-12, *J. Mol. Biol.* **156**:93–112.

Hall, M. N., Gabay, J., and Schwartz, M., 1983, Evidence for a coupling of synthesis and export of an outer membrane protein in *Escherichia coli, EMBO J.* **2**:15–19.

Hanahan, D., 1983, Studies on transformation of *Escherichia coli* with plasmids, *J. Mol. Biol.* **166**:557–580.

Hancock, R. E. W., and Braun, V., 1976, The Colicin I receptor of *Escherichia coli* K-12 has a role in enterochelin mediated iron transport, *FEBS Lett.* **65**:208–210.

Hancock, R. E. W., Poole, K., and Benz, R., 1982, Outer membrane protein P of *Pseudomonas aeruginosa:* Regulation by phosphate deficiency and formation of small anion-specific channels in lipid bilayer membranes, *J. Bacteriol.* **150**:730–738.

Hantke, K., 1976, Phage T6–Colicin K receptor and nucleoside transport in *Escherichia coli, FEBS Lett.* **70**:109–112.

Hantke, K., 1983, Identification of an iron uptake system specific for coprogen and rhodotorulic acid in *Escherichia coli* K-12, *Mol. Gen. Genet.* **191**:301–306.

Hantke, K., and Braun, V., 1973, Covalent binding of lipid to protein: Diglyceride and amide-linked fatty acid to the N-terminal end of the murein-lipoprotein of the *Escherichia coli* outer membrane, *Eur. J. Biochem.* **34**:284–296.

Hantke, K., and Braun, V., 1978, Functional interaction of the ton A/ton B receptor system in *Escherichia coli, J. Bacteriol.* **135**:190–197.

Harold, F. M., 1977, Membranes and energy transduction in bacteria, *Curr. Topics Bioenerg.* **6**:83–149.

Havekes, L., Hoekstra, W., and Kempen, H., 1977, Relation between F, R1, R100, and R144 *Escherichia coli* K-12 donor strains in mating, *Mol. Gen. Genet.* **155**:185–189.

Heckels, J. E., 1981, Structural comparison of *Neisseria gonorrhoeae* outer membrane proteins, *J. Bacteriol.* **145**:736–742.

Hedgpeth, J., Clément, J. M., Marchal, C., Perrin, D., and Hofnung, M., 1980, DNA sequence encoding the NH_2-terminal peptide involved in transport of lambda receptor, an *Escherichia coli* secretory protein, *Proc. Natl. Acad. Sci. USA* **77**:2621–2625.

Heller, K., and Braun, V., 1979, Accelerated adsorption of bacteriophage T5 to *Escherichia coli* F, resulting from reversible tail fiber-lipopolysaccharide binding, *J. Bacteriol.* **139**:32–38.

Heller, K., and Braun, V., 1982, Polymannose O-antigens of *Escherichia coli*, the binding sites for the reversible adsorption of bacteriophage T5$^+$ via the L-shaped tail fibers, *J. Virol.* **41**:222–227.

Henderson, L. E., Krutzsch, H. C., and Oroszlan, S., 1983, Myristyl-amino-terminal acylation of murine retovirus proteins: An unusual posttranslational protein modification, *Proc. Natl. Acad. Sci. USA* **80**:339–343.

Hengge, R., and Boos, W., 1983, Maltose and lactose transport in *Escherichia coli:* Examples of two different types of concentrative transport systems, *Biochim. Biophys. Acta* **737**:443–478.

Henning, U., and Jann, K., 1979, Two-component nature of bacteriophage T4 receptor activity in *Escherichia coli* K-12, *J. Bacteriol.* **137**:664–666.

Hermadson, M. A., Chen, K. C. S., and Buchanan, T. M., 1978, Neisseria pili proteins: Amino-terminal amino-acid sequences and identification of an unusual amino-acid, *Biochemistry* **17**:442–445.

Higgins, C. F., and Ames, G. F. L., 1981, Two periplasmic transport proteins which interact with a common membrane receptor show extensive homology: Complete nucleotide sequences, *Proc. Natl. Acad. Sci. USA* **78**:6038–6042.

Higgins, C. F., Haag, P. D., Nikaido, H., Ardeshir, F., Garcia, G., and Ames, G. F. L., 1982, Complete nucleotide sequence and identification of membrane components of the histidine transport operon of *Salmonella typhimurium, Nature* **298**:723–727.

Huang, Y-X., Ching, G., and Inouye, M., 1983, Comparison of the lipoprotein gene among the Enterobacteriaceae: DNA sequence of *Morganella morganii* lipoprotein gene and its expression in *Escherichia coli, J. Biol. Chem.* **258**:8139–8145.

Huang, Y-X., Ching, G., Yamagata, H., and Inouye, M., 1982, Existence of several homologous sequences in the *Escherichia coli* chromosome to the gene for the major outer membrane lipoprotein, *FEBS Lett.* **137**:168–170.

Hussain, M., Ichihara, S., and Mizushima, S., 1982a, Mechanism of signal peptide cleavage in the biosynthesis of the major lipoprotein of the *Escherichia coli* outer membrane, *J. Biol. Chem.* **257**:5177–5182.

Hussain, M., Ozawa, Y., Ichihara, S., and Mizushima, S., 1982b, Signal peptide digestion in *Escherichia coli:* Effect of protease inhibitors on hydrolysis of the cleaved signal peptide of the major outer membrane lipoprotein, *Eur. J. Biochem.* **129**:233–239.

Ichihara, S., and Mizushima, S., 1978, Characterization of major outer membrane proteins O-8 and O-9 of *Escherichia coli* K-12: Evidence that structural genes of the two proteins are different, *J. Biochem.* (Tokyo) **83**:1095–1100.

Ichihara, S., Hussain, M., and Mizushima, S., 1981, Characterization of new membrane lipoproteins and their precursors of *Escherichia coli*, *J. Biol. Chem.* **256**:3125–3129.

Inokuchi, K., Mutoh, N., Matsuyama, S., and Mizushima, S., 1982, Primary structure of the *omp F* gene that codes for a major outer membrane protein of *Escherichia coli* K-12, *Nucleic Acid Res.* **10**:6957–6968.

Inouye, H., and Beckwith, J., 1977, Synthesis and processing of an *Escherichia coli* alkaline phosphatase precursor in vitro, *Proc. Natl. Acad. Sci. USA* **74**:1440–1444.

Inouye, H., Barnes, W., and Beckwith, J., 1982, Signal sequence of alkaline phosphatase of *Escherichia coli*, *J. Bacteriol.* **149**:434–439.

Inouye, M., and Halegoua, S., 1980, Secretion and membrane localization of proteins in *Escherichia coli*, *CRC Crit. Rev. Biochem.* **7**:339–371.

Inouye, S., Soberon, X., Franceschini, T., Nakamura, K., Itakura, K., and Inouye, M., 1982, Role of positive charge on the amino-terminal region of the signal sequence peptide in protein secretion across the membrane, *Proc. Natl. Acad. Sci. USA* **79**:3438–3441.

Ito, K., and Beckwith, J. R., 1981, Role of the mature protein sequence of maltose-binding protein in its secretion across the *Escherichia coli* cytoplasmic membrane, *Cell* **25**:143–150.

Ito, K., Bassford, P. J., Jr., and Beckwith, J., 1981, Protein localization in *Escherichia coli*: Is there a common step in the secretion of periplasmic and outer membrane proteins, *Cell* **24**:707–717.

Ito, K., Wittekind, W., Nomura, M., Miura, A., Shiba, K., Yura, T., and Nashimoto, H., 1983, A temperature-sensitive mutant of *Escherichia coli* exhibiting slow processing of exported proteins, *Cell* **32**:789–797.

Jackson, B., and Kennedy, E. P., 1983, The biosynthesis of membrane-derived oligosaccharides, *J. Biol. Chem.* **258**:2394–2398.

Jakes, K. S., and Model, P., 1979, Mechanism of export of Colicin E1 and Colicin E3, *J. Bacteriol.* **138**:770–778.

James, J. F., and Swanson, J., 1978, Studies on gonococcus infection XIII: Occurrence of color/opacity variants in clinical cultures, *Infect. Immun.* **19**:332–340.

Jann, K., and Jann, B., 1983, The K-antigens in *Escherichia coli*, *Progr. Allergy* **33**:53–79.

Jaurin, B., and Grundstrom, T., 1981, *amp C* cephalosporinase of *Escherichia coli* K-12 has a different evolutionary origin from that of β-lactamases of the penicillinase type, *Proc. Natl. Acad. Sci. USA* **78**:4897–4901.

Johnston, K. H., 1980, Antigenic diversity of the serotype antigen complex of *Neisseria gonorrhoeae*: Analysis by an indirect enzyme-linked immunoassay, *Infect. Immun.* **28**:101–110.

Josefsson, L. G., and Randall, L. L., 1981a, Different exported proteins in *Escherichia coli* show differences in the temporal mode of processing in vivo, *Cell* **25**:151–157.

Josefsson, L. G., and Randall, L. L., 1981b, Processing in vivo of precursor maltose-binding protein in *Escherichia coli* occurs postranslationally as well as cotranslationally, *J. Biol. Chem.* **256**:2504–2507.

Kadner, R. J., and Bassford, P. J., Jr., 1978, The role of the outer membrane in active transport, in: *Bacterial Transport* (B. P. Rosen, ed.), Marcel Dekker, New York, pp. 413–462.

Kahn, M. E., Maul, G., and Goodgal, S., 1982, Possible mechanisms for donor DNA binding and transport in Haemophilus, *Proc. Natl. Acad. Sci. USA* **79**:6370–6374.

Kalasauskaitė, E. V., Kadišaitė, D. L., Daugelavičius, R. J., Grinius, L. L., and Jasaitis, A. A, 1983, Studies on energy supply for genetic processes: Requirement for membrane potential in *Escherichia coli* infection by phage T4, *Eur. J. Biochem.* **130**:123–130.

Kanegasaki, S., Kikuchi, T., and Yoshikawa, M., 1978, Inhibition of adsorption of male-specific phage to male cells by isolated lipopolysaccharide, in: *Pili* (D. E. Bradley, E. Raizen, P. Fives-Taylor, and J. T. Ou, eds.), International Conference on Pili, Washington, D.C., pp. 291–300.

Katz, E., Losing, D., Inouye, S., and Inouye, M., 1978, Lipoprotein from *Proteus mirabilis, J. Bacteriol.* **134**:674–676.
Kawaji, H., Mizuno, T., and Mizushima, S., 1979, Influence of molecular size and osmolarity of sugars and dextrans on the synthesis of outer membrane proteins O-8 and O-9 of *Escherichia coli* K-12, *J. Bacteriol.* **140**:843–847.
Kennedy, E. P., 1982, Osmotic regulation and the biosynthesis of membrane-derived oligosaccharides, *Proc. Natl. Acad. Sci. USA* **79**:1092–1095.
Kennedy, E. P., Rumley, M. K., Schulman, H., and van Golde, L. M. G., 1976, Identification of sn-glycero-1-phosphate and phosphoethanolamine residues linked to the membrane-derived oligosaccharides of *Escherichia coli, J. Biol. Chem.* **251**:4208–4213.
Kennedy, N., Beutin, L., Achtman, M., Skurray, R., Rahmsdorf, U., and Herrlich, P., 1977, Conjugation proteins encoded by the F sex factor, *Nature* **270**:580–585.
Konisky, J., 1982, Colicins and other bacteriocins with established modes of action, *Ann. Rev. Microbiol.* **36**:125–144.
Kornberg, A., 1980, Stages in the viral life cycle, in: *DNA Replication* (A. C. Bartlett, P. Brewer, and R. McNally, eds.) W. H. Freeman and Company, San Francisco, pp. 476–478.
Koshland, D., 1982, A genetic analysis of β-lactamase, PhD Thesis, Massachusetts Institute of Technology, Cambridge, Mass.
Koshland, D., Sauer, R. T., and Botstein, D., 1982, Diverse effects of mutations in the signal sequence on the secretion of β-lactamase in *Salmonella typhimurium, Cell* **30**:903–914.
Krieger-Brauer, H. J., and Braun, V., 1980, Functions related to the receptor protein specified by the *tsx* gene of *Escherichia coli, Arch. Microbiol.* **14**:233–242.
Kumamoto, C. A., and Beckwith, J., 1983, Mutations in a new gene *sec B* cause defective protein localization in *Escherichia coli, J. Bacteriol.* **154**:253–260.
Labedan, B., and Goldberg, E. B., 1982, DNA transport across bacterial membranes, in: *Membranes and Transport,* Vol 2 (A. N. Martonosi, ed.) Plenum Press, New York, pp. 133–138.
Lai, J-S., Sarvas, M., Brammar, W. J., Neugebauer, K., and Wu, H. C., 1981, *Bacillus licheniformis* penicillinase synthesized in *Escherichia coli* contains covalently linked fatty acid and glyceride, *Proc. Natl. Acad. Sci. USA* **87**:3506–3510.
Laimins, L. A., Rhoads, D. B., Altendorf, K., and Epstein, W., 1978, Identification of the structural protein of an ATP-driven potassium transport system in *Escherichia coli, Proc. Natl. Acad. Sci. USA* **75**:3216–3219.
Laimins, L. A., Rhoads, D. B., and Epstein, W., 1981, Osmotic control of *kdp* operon expression in *Escherichia coli, Proc. Natl. Acad. Sci. USA* **78**:464–468.
Lambden, P. R., 1982, Biochemical comparison of pili from variants of *Neisseria gonorrhoeae* P9, *J. Gen. Microbiol.* **128**:2105–2111.
Lambden, P. R., Heckels, J. E., James, L. T., and Watt, P. J., 1979, Variation in surface protein composition associated with virulence properties in opacity types of *Neisseria gonorrhoeae, J. Gen. Microbiol.* **114**:305–312.
Lambden, P. R., Robertson, J. N., and Watt, P. J., 1980, Biological properties of two distinct pilus types produced by isogenic variants of *Neisseria gonorrhoeae* P9, *J. Bacteriol.* **141**:393–396.
Landick, R., and Oxender, D. L., 1982, Bacterial periplasmic binding proteins, in: *Membranes and Transport,* Vol. 2 (A. Martonosi, ed.), Plenum Press, New York, pp. 81–88.
Lanni, Y. T., 1960, Invasion by bacteriophage T5 II: Dissociation of calcium-independent and calcium-dependent processes, *Virology* **10**:514–529.
Lanni, Y. T., 1968, First-step-transfer deoxyribonucleic acid of bacteriophage T5, *Bacteriol. Rev.* **1968**:227–242.
Le Rudulier, D., and Bouillard, L., 1983, Glycine betaine, an osmotic effector in *Klebsiella pneu-*

moniae and other members of the Enterobacteriaceae, *Appl. Environm. Microbiol.* **46**:152–159.
LeRudulier, D., and Valentine, R. C., 1982, Genetic engineering in agriculture: Osmoregulation, *Trends Biochem. Sci* **7**:431–433.
Lindberg, A. A., 1977, Bacterial surface carbohydrates and bacteriophage adsorption, in: *Surface Carbohydrates of the Procaryotic Cell* (I. W. Sutherland, ed.), Academic Press, New York. pp. 289–356.
Lin, J. J. C., Kanazawa, H., and Wu, H., 1980a, Assembly of outer membrane lipoprotein in an *Escherichia coli* mutant with a single amino-acid replacement within the signal sequence of prolipoprotein, *J. Bacteriol.* **141**:550–557.
Lin, J. J. C., Giam, C. Z., and Wu, H., 1980b, Assembly of the outer membrane lipoprotein in *Escherichia coli, J. Biol. Chem.* **255**:807–811.
Low, K. B., and Porter, D. D., 1978, Modes of gene transfer and recombination in bacteria, *Ann. Rev. Genet.* **12**:249–287.
Lubin, M., and Kessel, D., 1960, Preliminary mapping of the genetic locus for potassium transport in *Escherichia coli, Biochem. Biophys. Res. Commun.* **2**:249–255.
Luckey, M., and Nikaido, H., 1980, Specificity of diffusion channels produced by lambda phage receptor protein of *Escherichia coli, Proc. Natl. Acad. Sci. USA* **77**:167–171.
Lüderitz, O., Tanamoto, K., Galanos, C., Westphal, O., Zähringer, O., Rietschel, E. T., Kusumoto, S., and Shiba, T., 1983, Structural principles of lipopolysaccharides and biological properties of synthetic partial structures, in: *Amer. Chem. Symposium Series,* Kansas City **231**:3–20.
Lugtenberg, B., and van Alphen, L., 1983, Molecular architecture and functioning of the outer membrane of *Escherichia coli* and other gram-negative bacteria, *Biochim. Biophys. Acta* **737**:51–115.
MacGregor, C. H., Bishop, C. W., and Blech, J. E., 1979, Localization of proteolytic activity in the outer membrane of *Escherichia coli, J. Bacteriol.* **137**:574–583.
Maggae, A. I., and Schlesinger, M. J., 1982, Fatty acid acylation of eucaryotic cell membrane proteins, *Biochim. Biophys. Acta* **694**:279–289.
Mäkelä, P. H., and Mayer, H., 1976, Enterobacterial common antigen, *Bacteriol. Rev.* **40**:591–632.
Maltouf, A. F., and Labedan, B., 1983, Host cell metabolic energy is not required for injection of bacteriophage T5 DNA, *J. Bacteriol.* **153**:124–133.
Mandel, M., and Higa, A., 1970, Calcium-dependent bacteriophage DNA infection, *J. Mol. Biol.* **53**:159–162.
Manning. P. A., Beutin, L., and Achtman, M., 1980, Outer membrane of *Escherichia coli:* Properties of the F sex factor Tra T protein which is involved in surface exclusion, *J. Bacteriol.* **142**:285–294.
Manoil, C., 1983, A genetic approach to defining the sites of interaction of a membrane protein with different external agents, *J. Mol. Biol.* **169**:507–519.
Manoil, C., and Rosenbusch, J. P., 1982, Conjugation-deficient mutants of *Escherichia coli* distinguish classes of functions of the outer membrane Omp A protein, *Mol. Gen. Genet.* **187**:148–156.
Mathis, L. S., and Socca, J. J., 1982, *Haemophilus influenzae* and *Neisseria gonorrhoeae* recognize different specificity determinants in the DNA uptake step of genetic transformation, *J. Gen. Microbiol.* **128**:1159–1161.
Maurer, B., Müller, A., Keller-Schierlein, W., and Zähner, H., 1968, Stoffwechselprodukte von Mikroorganismen, 61. Mitteilung, Ferribactin, ein Siderochrom aus *Pseudomonas fluorescens* Migula *Arch. Mikrobiol.* **60**:326–339.
Mayer, L. W., 1982, Rate of in vitro changes of gonococcal colony opacity phenotypes, *Infect. Immun.* **37**:481–485.

Measures, J. C., 1975, Role of amino-acids in osmoregulation of non-halophilic bacteria, *Nature* **257**:398–400.
Meury, J., and Kepes, A., 1981, The regulation of potassium fluxes for the adjustment and maintenance of potassium levels in *Escherichia coli, Eur. J. Biochem.* **119**:165–170.
Meyer, J. M., and Archibald, M. A., 1980, The siderchromes of nonfluorescent pseudomonads: Production of nocardamine by *Pseudomonas stutzeri, J. Gen. Microbiol.* **118**:125–129.
Meyer, T. F., Mlawer, N., and So, M., 1982, Pilus expression in *Neisseria gonorrhoeae* involves chromosomal rearrangement, *Cell* **30**:45–52.
Michaelis, S., and Beckwith, J., 1982, Mechanism of incorporation of cell envelope proteins in *Escherichia coli, Ann. Rev. Microbiol.* **36**:435–465.
Michaelis, S., Guarente, L., and Beckwith, J., 1983a, In vitro construction and characterization of *pho A- lac Z* gene fusions in *Escherichia coli, J. Bacteriol.* **154**:356–365.
Michaelis, S., Inouye, H., Oliver, D., and Beckwith, J., 1983b, Mutations that alter the signal sequence of alkaline phosphatase in *Escherichia coli, J. Bacteriol.* **154**:366–374.
Mickelsen, P. A., and Sparling, P. F., 1981, Ability of *Neisseria gonorrhoeae, Neisseria meningitidis,* and commensal Neisseria species to obtain iron from transferrin and iron compounds, *Infect. Immun.* **33**:555–564.
Mickelsen, P. A., Blackman, E., and Sparling, P. F., 1982, Ability of *Neisseria gonorrhoeae, Neisseria meningitidis* and commensal Neisseria species to obtain iron from lactoferrin, *Infect. Immun.* **35**:915–920.
Mizuno, T., 1981, Structure of the peptidoglycan-associated lipoprotein (PAL) of the *Proteus mirabilis* outer membrane. II. Sequence of the amino-terminal part of the peptidoglycan-associated lipoprotein, *J. Biochem.* **89**:1059–1066.
Mizuno, T., and Kageyama, M., 1979, Isolation and characterization of a major outer membrane protein of *Pseudomonas aeruginosa:* Evidence for the occurrence of a lipoprotein, *J. Biochem.* **85**:115–122.
Mizuno, T., Wurtzel, E. T., and Inouye, M., 1982, Osmoregulation of gene expression, *J. Biol. Chem.* **257**:13692–13698.
Mizuno, T., Chou, M., and Inouye, M., 1983, A comparative study on the genes for three porins of the *Escherichia coli* outer membrane: DNA sequence of the osmoregulated *omp C* gene, *J. Biol. Chem.* **258**:6932–6940.
Mooi, F. R., Wijfjers, A., and deGraaf, F. K., 1983, Identification and characterization of precursors in the biosynthesis of the K88 ab fimbria of *Escherichia coli, J. Bacteriol.* **154**:41–49.
Moreno, F., Fowler, A. V., Hall, M., Silhavy, T. J., Zabin, I., and Schwartz, M., 1980, A signal sequence is not sufficient to lead β-galactosidase out of the cytoplasm, *Nature* **286**:356–359.
Morona, R., and Reeves, P., 1982, The *tol C* locus of *Escherichia coli* affects the expression of three major outer membrane proteins, *J. Bacteriol.* **150**:1016–1023.
Morona, R., Manning, P. A., and Reeves, P., 1983, Identification and characterization of the Tol C protein, an outer membrane protein from *Escherichia coli, J. Bacteriol.* **153**:693–699.
Movva, N. R., Nakamura, K., and Inouye, M., 1980, Gene structure of the Omp A protein, a major surface protein of *Escherichia coli* required for cell–cell interaction, *J. Mol. Biol.* **143**:317–328.
Movva, N. R., Nakamura, K., and Inouye, M., 1980, Amino-acid sequence of the signal peptide of Omp A protein, a major outer membrane protein of *Escherichia coli, J. Biol. Chem.* **255**:27–29.
Munro, G. F., Hercules, K., Morgan, J., and Sauerbier, W., 1972, Dependence of the putrescine content of *Escherichia coli* on the osmotic strength of the medium, *J. Biol. Chem.* **247**:1272–1280.
Mutoh, M., Furukawa, H., and Mizushima, S., 1978, Role of lipopolysaccharide and outer mem-

brane protein of *Escherichia coli* K-12 in the receptor activity for bacteriophage T4, *J. Bacteriol.* **136**:693–699.

Mutoh, N., Inokuchi, K., and Mizushima, S., 1982, Amino-acid sequence of the signal peptide of Omp F, a major outer membrane protein of *Escherichia coli, FEBS Lett.* **137**:171–174.

Nakamura, K., and Inouye, M., 1979, DNA sequence of the gene for the outer membrane lipoprotein of *Escherichia coli:* An extremely AT-rich promotor, *Cell* **18**:1109–1117.

Nakamura, K., and Inouye, M., 1980, DNA sequence of the *Serratia marcescens* lipoprotein gene, *Proc. Natl. Acad. Sci. USA* **77**:1369–1373.

Nakamura, K., Pirtle, R. M., and Inouye, M., 1979, Homology of the gene coding for the outer membrane lipoprotein within various gram-negative bacteria, *J. Bacteriol.* **137**:595–604.

Neilands, J. B., 1982, Microbial envelope proteins related to iron, *Ann. Rev. Microbiol.* **36**:285–309.

Nicas, T. J., and Hancock, R. E. W., 1983, *Pseudomonas aeruginosa* outer membrane permeability: Isolation of a porin protein F deficient mutant, *J. Bacteriol.* **153**:281–285.

Nielsen, J. B. K., and Lampen, J. O., 1982a, Membrane bound penicillinases in gram-positive bacteria, *J. Biol. Chem.* **257**:4490–4495.

Nielsen, J. B. K., and Lampen, J. O., 1982b, Glyceride-cysteine lipoproteins and secretion by gram-positive bacteria, *J. Bacteriol.* **152**:315–322.

Nielsen, J., Hausen, F. G., Hoppe, J., Friedl, P., and von Meyenburg, K., 1981, The nucleotide sequence of the *atp* genes coding for the F_o subunits a,b,c and the F_1 subunit δ of the membrane bound ATP synthase of *Escherichia coli, Mol. Gen. Genet.* **184**:33–39.

Nikaido, H., 1979, Nonspecific transport through the outer membrane, in: *Bacterial Outer Membranes* (M. Inouye, ed.), John Wiley and Sons, New York, pp. 361–407.

Nikaido, H., and Nakae, T., 1979, The outer membrane of gram-negative bacteria, in: *Advances in Microbial Physiology,* Vol. 20 (A. H. Rose, and J. G. Morris, eds.), Academic Press, London, pp. 163–250.

Nikaido, H., and Rosenberg, E. Y., 1983, Porin channels in *Escherichia coli:* Studies with liposomes reconstituted from purified proteins, *J. Bacteriol.* **153**:241–252.

Norquist, A., Davis, J., Norlander, L., and Normark, S., 1978, The effect of iron starvation on the outer membrane protein composition of *Neisseria gonorrhoeae, FEMS Microbiol. Lett.* **4**:71–75.

Ohkawa, J., Shiga, S., and Kageyama, M., 1980, Effect of iron concentration in the growth medium on the sensitivity of *Pseudomonas aeruginosa* to pyocin S2, *J. Biochem.* **87**:323–331.

Ohki, M., 1979, The cell-cycle dependent synthesis of envelope proteins in *Escherichia coli,* in: *Bacterial Outer Membranes* (M. Inouye, ed.), John Wiley and Sons, New York, pp. 293–315.

Oliver, D. B., and Beckwith, J., 1981, *Escherichia coli* mutant pleiotropically defective in the export of secreted proteins, *Cell* **25**:765–772.

Oliver, D. B., and Beckwith, J., 1982, Regulation of a membrane component required for protein secretion in *Escherichia coli, Cell* **30**:311–319.

Oliver, D. B., Kumamoto, C., Quinlan, M., and Beckwith, J., 1982, Pleiotropic mutants affecting the secretory apparatus of *Escherichia coli, Ann. Microbiol. Inst. Pasteur* **133A**:105–110.

Ørskov, S. L., 1948, Experiments on active and passive permeability of *Bacillus coli communis, Acta Path. Microbiol. Scand.* **25**:277–283.

Ørskov, I., Ørskov, F., Jann, B., and Jann, K., 1977, Serology, chemistry and genetics of O and K antigens of *Escherichia coli, Bacteriol. Rev.* **41**:667–710.

Osborn, M. J., and Wu, H. C. P., 1980, Proteins of the outer membrane of gram-negative bacteria, *Ann. Rev. Microbiol.* **34**:369–422.

Ou, J. T., 1973, Effect of Zn^{2+} on conjugation: increase in ability of F^- cells to form mating pairs, *J. Bacteriol.* **115**:648–654.
Ou, J. T., 1982, Cellular communication between bacteria, in: The Institute for Cancer Research Twenty-Seventh Scientific Report (September 1981–September 1982), Fox Chase Cancer Center, Philadelphia, pp. 75–76.
Ou, J. T., and Anderson, T. F., 1972, Effect of Zn^{2+} on bacterial conjugation: Inhibition of mating pair formation, *J. Bacteriol.* **111**:177–185.
Ou, J. T., and Yura, T., 1982, *Escherichia coli* K-12 F^- mutants that form mating aggregates but form transconjugants with low frequencies, *Mol Gen. Genet.* **187**:202–208.
Oudega, B., Stegehuis, F., van Tiel-Menkveld, G. J., and deGraaf, F. K., 1982, Protein H encoded by plasmid CloDF 13 is involved in excretion of Cloacin DF 13, *J. Bacteriol.* **150**:1115–1121.
Overbeeke, N., and Lugtenberg, B., 1980, Expression of outer membrane protein E of *Escherichia coli* K-12 by phosphate limitation, *FEBS Lett.* **112**:229–232.
Overbeeke, N., van Scharrenburg, G., and Lugtenberg, B., 1980, Antigenic relationship between pore proteins of *Escherichia coli* K-12, *Eur. J. Biochem.* **110**:247–254.
Overbeeke, N., Bergmans, H., van Mansfeld, F., and Lugtenberg, B., 1983, Complete nucleotide sequence of *pho E*, the structural gene for the phosphate limitation inducible outer membrane pore protein of *Escherichia coli* K-12, *J. Mol. Biol.* **163**:513–532.
Oxender, D. L., Anderson, J. J., Daniels, C. J., Landick, R., Gunsalus, R. P., Zurawski, G., and Yanofsky, C., 1980, Amino-terminal sequence and processing of the precursor of the leucine-specific binding protein and evidence for conformational differences between the precursor and the mature form, *Proc. Natl. Acad. Sci. USA* **77**:2005–2009.
Ozawa, Y., and Mizushima, S., 1983, Regulation of outer membrane porin protein synthesis in *Escherichia coli* K-12: *omp F* regulates the expression of *omp C*, *J. Bacteriol.* **154**:669–675.
Paakkanen, J., Gotschlich, E. C., and Mäkelä, P. H., 1979, Protein K: A new major outer membrane protein found in encapsulated *Escherichia coli*, *J. Bacteriol.* **139**:835–841.
Pacaud, M., 1982, Purification and characterization of two novel proteolytic enzymes in membranes of *Escherichia coli:* protease IV and protease V, *J. Biol. Chem.* **257**:4333–4339.
Philson, S. B., and Llinas, M., 1982, Siderochromes from *Pseudomonas fluorescens*, *J. Biol. Chem.* **257**:8081–8085.
Plastow, G. S., and Holland, J. B., 1979, Identification of the *Escherichia coli* inner membrane peptide specified by a lambda *ton B* transducing bacteriophage, *Biochem. Biophys. Res. Commun.* **90**:1007–1014.
Postle, K., and Good, R. F., 1983, DNA sequence of the *Escherichia coli ton B* gene, *Proc. Natl. Acad. Sci. USA* **80**:5235–5239.
Pratt, J. M., Holland, J. B., and Spratt, B. G., 1981, Precursor forms of penicillin-binding proteins 5 and 6 of *Escherichia coli* cytoplasmic membrane, *Nature* **293**:307–310.
Pugsley, A. P., and Rosenbusch, J. P., 1981, Release of Colicin E2 from *Escherichia coli*, *J. Bacteriol.* **147**:186–192.
Raetz, C. R. H., and Newman, K. F., 1979, Diglyceride kinase mutants of *Escherichia coli:* Inner membrane association of 1,2-diglyceride and its relation to synthesis of membrane-derived oligosaccharides, *J. Bacteriol.* **137**:860–868.
Randall, L. L., 1983, Translocation of domains of nascent periplasmic proteins across the cytoplasmic membrane is independent of elongation, *Cell* **33**:231–240.
Randall, L. L., and Hardy, S. J., 1982, Export of protein in bacteria: Dogma and data, in: *Modern Cell Biology* (B. Satir, ed.), Alan R. Liss, Inc., New York.
Rashtchian, A., Crooks, J. H., and Levy, S. B., 1983, *tra J* independence in expression of *tra T* on F, *J. Bacteriol.* **154**:1009–1012.

Régnier, P., 1981, Identification of protease IV of *Escherichia coli:* An outer membrane bound enzyme, *Biochem. Biophys. Res. Commun.* **99**:844–854.

Reynolds, A. E., Felton, J., and Wright, A., 1981, Insertion of DNA activates the cryptic *bgl* operon in *Escherichia coli* K-12, *Nature* **293**:625–629.

Rhoads, D. B., and Epstein, W., 1977, Energy coupling to net K^+ transport in *Escherichia coli* K-12, *J. Biol. Chem.* **252**:1394–1401.

Rhoads, D. B., and Epstein, W., 1978, Cation transport in *Escherichia coli:* IX. Regulation of K transport, *J. Gen. Physiol.* **72**:283–295.

Rhoads, D. B., Waters, F. B., and Epstein, W., 1976, Cation transport in *Escherichia coli:* VIII. Potassium transport mutants, *J. Gen. Physiol.* **67**:325–341.

Rhoads, D. B., Laimins, L., and Epstein, W., 1978, Functional organization of the *kdp* genes of *Escherichia coli* K-12, *J. Bacteriol.* **135**:445–452.

Rietschel, E. T., Wollenweber, H. W., Sidorczyk, Z., Zähringer, U., and Lüderitz, O., 1983, Analysis of the primary structure of lipid A, in: *Amer. Chem. Soc. Symposion Series*, Kansas City, **231**:195–218.

Robertson, J. N., Vincent, P., and Ward, M. E., 1977, The preparation and properties of gonococcal pili, *J. Gen. Microbiol.* **102**:169–177.

Roessner, C. A., Struck, D. K., and Ihler, G. M., 1983, Injection of DNA into liposomes by bacteriophage lambda, *J. Biol. Chem.* **258**:643–648.

Roller, S. D., and Anagnostopoulos, G. D., 1982, Accumulation of carbohydrate by *Escherichia coli* B/r/1 during growth at low water activity, *J. Appl. Bacteriol.* **52**:425–435.

Rotering, H., and Raetz, C. R. H., 1983, Appearance of monoglyceride and triglyceride in the cell envelope of *Escherichia coli* mutants defective in diglyceride kinase, *J. Biol. Chem.* **258**:8068–8073.

Rosenbusch, J. P., 1974, Characterization of the major envelope protein from *Escherichia coli:* Regular arrangement on the peptidoglycan and unusual sodium dodecyl sulfate binding, *J. Biol. Chem.* **249**:8019–8029.

Ryter, A. H., Shuman, H., and Schwartz, M., 1975, Integration of the receptor for bacteriophage lambda in the outer membrane of *Escherichia coli:* Coupling with cell division, *J. Bacteriol.* **122**:295–301.

Sambucetti, L., Eoyang, L., and Silverman, P. M., 1982, Cellular control of conjugation in *Escherichia coli* K-12: Effect of chromosomal *cpx* mutations on F-plasmid gene expression, *J. Mol. Biol.* **161**:13–31.

Santos, E., and Kaback, H. R., 1981, Involvement of the proton electrochemical gradient in genetic transformation in *Escherichia coli*, *Biochem. Biophys. Res. Commun.* **99**:1153–1160.

Sarma, V., and Reeves, P., 1977, Genetic locus *(omp B)* affecting a major outer membrane protein in *Escherichia coli* K-12, *J. Bacteriol.* **132**:23–27.

Sato, T., and Yura, T., 1979, Chromosomal location and expression of the structural gene for major outer membrane protein Ia of *Escherichia coli* K-12 and of the homologous genes of *Salmonella typhimurium*, *J. Bacteriol.* **139**:468–477.

Scandella, C. J., and Kornberg, A., 1971, A membrane bound phospholipase A1 purified from *Escherichia coli*, *Biochemistry* **10**:4447–4456.

Schaefler, S., 1967, Inducible system for the utilization of β-glucosides in *Escherichia coli*, *J. Bacteriol* **93**:254–263.

Schindler, H., and Rosenbusch, J. P., 1978, Matrix protein from *Escherichia coli* outer membrane forms voltage-controlled channels in lipid bilayers, *Proc. Natl. Acad. Sci. USA* **75**:3751–3755.

Schindler, H., and Rosenbusch, J. P., 1981, Matrix protein in planar membranes: Clusters of channels in a native environment and their functional reassembly, *Proc. Natl. Acad. Sci. USA* **78**:2302–2306.

Schleie, P. O., 1969, Plasmolysis of *Escherichia coli* B/r with sucrose, *J. Bacteriol.* **98**:335–340.
Schneider, J. E., and Kennedy, E. P., 1978, A novel phosphodiesterase from *Aspergillus niger* and its application to the study of membrane-derived oligosaccharides and other glycerol containing biopolymers, *J. Biol. Chem.* **253**:7738–7744.
Schneider, J. E., Reinhold, V., Rumley, M. K., and Kennedy, E. P., 1979, Structural studies of the membrane-derived oligosaccharides of *Escherichia coli, J. Biol. Chem.* **254**:10135–10138.
Schobert, B., 1977, Is there an osmotic regulatory mechanism in algae and higher plants, *J. Theor. Biol.* **68**:17–26.
Schobert, B., and Tschesche, H., 1978, Unusual solution properties of proline and its interaction with protein, *Biochem. Biophys. Acta* **541**:270–277.
Schoolnik, G. K., Tai, J. Y., and Gotschlich, E. C., 1983, A pilus vaccine for the prevention of gonorrhea, *Progr. Allergy* **33**:314–331.
Schulman, H., and Kennedy, E. P., 1977, Relation of turnover of membrane phospholipids to synthesis of membrane-derived oligosaccharides of *Escherichia coli, J. Biol. Chem.* **252**:4250–4255.
Schulman, H., and Kennedy, E. P., 1979, Localization of membrane-derived oligosaccharides in the outer envelope of *Escherichia coli* and their occurrence in other gram-negative bacteria, *J. Bacteriol.* **137**:686–688.
Schwartz, M., Roa, M., and Debarbouille, M., 1981, Mutations that affect *lam B* gene expression at a posttranscriptional level, *Proc. Natl. Acad. Sci. USA* **78**:2937–2941.
Scocca, J. J., Poland, R. L., and Zoon, K. C., 1974, Specificity in deoxyribonucleic acid uptake by transformable *Haemophilus influenzae, J. Bacteriol.* **118**:369–373.
Scott, N. W., and Harwood, C. R., 1980, Studies on the influence of the cAMP system on major outer membrane proteins of *Escherichia coli* K-12, *FEMS Microbiol. Lett.* **9**:95–98.
Silhavy, T. J., Shuman, H. A., Beckwith, J., and Schwartz, M., 1977, Use of gene fusions to study outer membrane protein localization in *Escherichia coli, Proc. Natl. Acad. Sci. USA* **74**:5411–5415.
Silhavy, T. J., Benson, S. A., and Emr, S. D., 1983, Mechanisms of protein localization, *Microbiol. Rev.* **47**:313–344.
Silver, P., and Wickner, W., 1983, Genetic mapping of the *Escherichia coli* leader (signal) peptidase gene *(lep):* A new approach for determining the map position of a cloned gene, *J. Bacteriol.* **154**:569–572.
Simonson, C., Bremer, D., and DeVoe, J. W., 1982, Expression of a high-affinity mechanism for acquisition of transferrin iron by *Neisseria meningitidis, Infect. Immun.* **36**:107–113.
Sisco, K. L., and Smith, H. O., 1979, Sequence-specific DNA uptake in Haemophilus transformation, *Proc. Natl. Acad. Sci. USA* **76**:972–976.
Smit, J., and Nikaido, H., 1978, Outer membrane of gram-negative bacteria: XVIII. Electron microscope studies on porin insertion sites and growth of cell surface of *Salmonella typhimurium, J. Bacteriol.* **135**:687–702.
Smith, W. P., 1980, Contranslational secretion of diphtheria toxin and alkaline phosphatase in vitro: Involvement of membrane proteins, *J. Bacteriol.* **141**:1142–1147.
Smith, W. P., Tai, P. C., and Davis, B. D., 1981, *Bacillus licheniformis* penicillinase: Cleavage and attachment of lipid during cotranslational secretion, *Proc. Natl. Acad. Sci. USA* **78**:3501–3505.
So, M., and McCarthy, B. J., 1980, Nucleotide sequence of the bacterial transposon Tn1681 encoding a heat-stable (ST) toxin and its identification in enterotoxigenic *Escherichia coli* strains, *Proc. Natl. Acad. Sci. USA* **77**:4011–4015.
Sokol, P. A., and Woods, D. E., 1983, Demonstration of an iron-siderophore-binding protein in the outer membrane of *Pseudomonas aeruginosa, Infect. Immun.* **40**:665–669.
Sonntag, I., Schwarz, H., Hirota, Y., and Henning, U., 1978, Cell envelope and shape of *Esch-*

erichia coli: Multiple mutants missing the outer membrane lipoprotein and other major outer membrane proteins, *J. Bacteriol.* **136**:280–285.

Sowa, B. A., Moore, D., and Ippen-Ihler, K., 1983, Physiology of F-pilin synthesis and utilization, *J. Bacteriol.* **153**:962–968.

Spicer, E. K., Kavanaugh, W. M., Dallas, W. S., Falkow, S., Königsberg, W. H., and Schafer, D. E., 1981, Sequence homologies between subunits of *Escherichia coli* and *Vibrio cholerae* enterotoxins, *Proc. Natl. Acad. Sci. USA* **78**:50–54.

Stock, J. B., Rauch, B., and Roseman, S., 1977, Periplasmic space in *Salmonella typhimurium* and *Escherichia coli, J. Biol. Chem.* **252**:7850–7861.

Stocker, B. A. D., Nurminen, M., and Mäkelä, P. H., 1979, Mutants defective in the 33K outer membrane protein of *Salmonella typhimurium, J. Bacteriol.* **139**:376–383.

Suttcliffe, J. G., 1978, Nucelotide sequence of the ampicillin resistance gene of *Escherichia coli* plasmid pBR 322, *Proc. Natl. Acad. Sci. USA* **75**:3737–3741.

Suzuki, H., Nishimura, Y., Yamada, S., Nishimura, A., Yamada, M., and Hirota, Y., 1978, Murein—lipoprotein of *Escherichia coli:* A protein involved in the stabilization of bacterial cell envelope, *Mol. Gen. Genet.* **167**:1–9.

Swanson, J., 1979, Studies on gonococcal infection XVIII: ^{125}J-labeled peptide mapping of the major protein of the gonococcal cell wall outer membrane, *Infect. Immun.* **23**:799–810.

Takayama, K., Qureshi, N., Mascagni, P., Nashed, M. A., Anderson, L., and Raetz, C. R. H., 1983, Fatty acyl derivatives of glucosamine-1-phosphate in *Escherichia coli* and their relation to lipid A: Complete structure of a diacyl GlcN-1-P found in a phosphatidylglycerol-deficient mutant, *J. Biol. Chem.* **258**:7379–7385.

Taketo, A., 1975, Sensitivity of *Escherichia coli* to viral nucleic acid, X: Ba^{2+}-induced competence for tranfecting DNA, *Z. Naturforsch.* **30c**:520–522.

Taketo, A., 1977, Sensitivity of *Escherichia coli* to viral nucleic acid, XII: Ca^{2+}- or Ba^{2+}-facilitated transfection of cell envelope mutants, *Z. Naturforsch.* **32c**:429–433.

Taketo, A., 1978, Effect of mutations in surface lipopolysaccharide or outer membrane protein of *Escherichia coli* C on transfecting competence for microvirid phage DNA, *J. Gen. Appl. Microbiol.* **24**:51–58.

Taketo, A., and Kuno, S., 1974, Sensitivity of *Escherichia coli* to viral nucleic acid, VII: Further studies on Ca^{2+}-induced competence, *J. Biochem.* **75**:59–67.

Taylor, P. W., 1983, Bactericidal and bacteriolytic activity of serum against gram-negative bacteria, *Microbiol. Rev.* **47**:46–83.

Teintze, M., Hossain, M. B., Barnes, C. L., Leong, J., and van der Helm, P., 1981, Structure of ferric pseudobactin, a siderophore from a plant growth promoting Pseudomonas, *Biochemistry* **20**:6446–6457.

Tempest, D. W., Meers, J. C., and Brown, C. M., 1970, Influence of environment on the content and composition of microbial free amino-acid pools, *J. Gen. Microbiol.* **64**:171.

Tidmarsh, G. F., and Rosenberg, L. T., 1981, Acquisition of iron from transferrin by *Salmonella paratyphi* B, *Current Microbiol.* **6**:217–220.

Tokunaga, M., Loranger, J. M., Wolfe, P. B., and Wu, H. C., 1982a, Prolipoprotein signal peptidase in *Escherichia coli* is distinct from the M13 procoat protein signal peptidase, *J. Biol. Chem.* **257**:9922–9925.

Tokunaga, M., Tokunaga, H., and Wu, H. C., 1982b, Posttranslational modification and processing of *Escherichia coli* prolipoprotein in vitro, *Proc. Natl. Acad. Sci. USA* **79**:2255–2259.

Tommassen, J., and Lugtenberg, B., 1982, *pho*-regulon of *Escherichia coli* K-12: A minireview, *Ann. Microbiol.* **133A**:243–249.

Troy, F. A., II., 1979, The biochemistry and biosynthesis of selected bacterial capsular polymers, *Ann. Rev. Microbiol.* **33**:519–560.

van Alphen, L., and Lugtenberg, B., 1977, Influence of osmolarity of the growth medium on the outer membrane protein pattern of *Escherichia coli, J. Bacteriol.* **131**:623–630.
van Alphen, W., Lugtenberg, B., van Boxtel, R., Hack, A. M., Verhoef, C., and Havekes, L., 1979, *mer A* is the structural gene for outer membrane protein C of *Escherichia coli, Mol. Gen. Genet.* **169**:147–155.
van Alphen, L., van Kempen-DeTroye, F., and Zanen, H. C., 1983, Characterization of cell envelope proteins and lipopolysaccharides of *Escherichia coli* isolates from patients with neonatal meningitis, *FEMS Microbiol. Lett.* **16**:261–267.
van Die, I. M., Bergmans, H. E. N., and Hoekstra, W. P. M., 1983a, Transformation in *Escherichia coli:* Studies on the role of the heat shock in induction of competence, *J. Gen. Microbiol.* **129**:663–670.
van Die, I. M., van Oosterhout, A., Bergmans, H., and Hoekstra, W. P. M., 1983b, The influence of phase transition of membrane lipids on uptake of plasmid DNA in *Escherichia coli* transformation, *FEMS Microbiol. Lett.* **18**:127–130.
van Golde, L. M. G., Schulman, H., and Kennedy, E. P., 1973, Metabolism of membrane phospholipids and its relation to a novel class of oligosaccharides in *Escherichia coli, Proc. Natl. Acad. Sci. USA* **70**:1368–1372.
van Wilson, G., and Hogg, R. W., 1980, The NH$_2$-terminal sequence of a precursor form of the arabinose binding protein, *J. Biol. Chem.* **255**:6745–6750.
Varenne, S., Cavard, D., and Lazdunski, C., 1981, Biosynthesis and export of Colicin A in *Citrobacter freundii* CA31, *Eur. J. Biochem.* **116**:615–620.
Vlasuk, G. P., Inouye, S., Ito, H., Itakura, K., and Inouye, M., 1983, Effects of the complete removal of basic amino-acid residues from the signal peptide on secretion of lipoprotein in *Escherichia coli, J. Biol. Chem.* **258**:7141–7148.
von Heijne, G., and Blomberg, C., 1979, Trans-membrane translocation of proteins: The direct transfer model, *Eur. J. Biochem.* **97**:175–181.
Wagner, W., Vogel, M., and Goebel, W., 1983, Transport of haemolysins across the outer membrane of *Escherichia coli* requires two functions, *J. Bacteriol.* **154**:200–210.
Walter, P., and Blobel, G., 1980, Purification of a membrane-associated protein complex required for protein translocation across the endoplasmic reticulum, *Proc. Natl. Acad. Sci. USA* **77**:7112–7116.
Wandersman, C., and Schwartz, M., 1982, Mutations that alter the transport of the Lam B protein in *Escherichia coli, J. Bacteriol.* **151**:15–21.
Wanderman, C., Mureno, F., and Schwartz, M., 1980, Pleiotropic mutations rendering *Escherichia coli* K-12 resistant to phage TP1, *J. Bacteriol.* **143**:1374–1383.
Wanner, B. L., Sarthy, A., and Beckwith, J., 1979, *Escherichia coli* pleiotropic mutant that reduces amounts of several periplasmic and outer membrane proteins, *J. Bacteriol.* **143**:229–239.
Watt, P. J., and Ward, M. E., 1980, Adherence of *Neisseria gonorrhoeae* and other Neisseria species to mammalian cells, in: *Bacterial Adherence Receptors and Recognition,* Series B, Vol. 6 (E. H. Beachey, ed.), Chapman and Hall, London and New York, pp. 251–288.
Waxman, D. J., and Strominger, J. L., 1981a, Limited proteolysis of the penicillin-sensitive D-alanine carboxypeptidase purifed from *Bacillus subtilis* membranes, *J. Biol. Chem.* **256**:2059–2066.
Waxman, D. J., and Strominger, J. L., 1981b, Primary structure of the COOH-terminal membranous segment of a penicillin-sensitive enzyme purified from two bacilli, *J. Biol. Chem.* **256**:2067–2077.
Weckesser, J., Drews, G., and Mayer, H., 1979, Lipopolysaccharides of photosynthetic procaryotes, *Ann. Rev. Microbiol.* **33**:215–239.

Wickner, W., 1979, The assembly of proteins into biological membranes: The membrane trigger hypothesis, *Ann. Rev. Biochem.* **48**:23–45.

Wickner, W., 1980, Assembly of proteins into membranes, *Science* **210**:861–868.

Willetts, N., and Skurray, R., 1980, The conjugation system of F-like plasmids, *Ann Rev. Genet.* **14**:41–76.

Wu, H. C., Lai, J-S, Hayashi, S., and Giam, C-Z., 1982, Biogenesis of membrane lipoproteins in *Escherichia coli, Biophys. J.* **37**:307–315.

Wurtzel, E. T., Chou, M. Y., and Inouye, M., 1982, Osmoregulation of gene expression, *J. Biol. Chem.* **257**:13685–13691.

Yamada, H., Nagami, T., and Mizushima, S., 1981, Arrangement of bacteriophage lambda receptor protein (Lam B) in the cell surface of *Escherichia coli:* A reconstitution study, *J. Bacteriol.* **147**:660–669.

Yamada, M., Miki, T., and Nakazawa, A., 1982a, Translocation of Colicin E1 through the cytoplasmic membrane of *Escherichia coli, FEBS Lett.* **150**:465–468.

Yamada, M., Ebina, Y., Migata, T., Nakazawa, T., and Nakazawa, A., 1982b, Nucleotide sequence of the structural gene for Colicin E1 and predicted structure of the protein, *Proc. Natl. Acad. Sci. USA* **79**:2827–2831.

Yamagata, H., Nakamura, K., and Inouye, M., 1981, Comparison of the lipoprotein gene among the Enterobacteriaceae: DNA sequence of *Erwinia amylovora* lipoprotein gene, *J. Biol. Chem.* **256**:2194–2198.

Yamagata, H., Daishima, K., and Mizushima, S., 1983, Cloning and expression of a gene coding for the prolipoprotein signal peptidase of *Escherichia coli, FEBS Lett.* **158**:301–304.

Yamamoto, T., and Jokata, T., 1983, Sequence of heat-labile enterotoxin of *Escherichia coli* pathogenic for humans, *J. Bacteriol.* **155**:728–733.

Yancey, R. J., and Finkelstein, R. A., 1981, Siderophore production by pathogenic Neisseria spp., *Infect. Immun.* **32**:600–608.

Yoshimura, F., and Nikaido, H., 1982, Permeability of *Pseudomonas aeruginosa* outer membrane to hydrophilic solutes, *J. Bacteriol.* **152**:636–642.

Zambryski, P., Goodman, H. M., Montagu, M. V., and Schell, J., 1983, Agrobacterium tumor induction, in: *Mobile Genetic Elements* (J. A. Shapiro, ed.), Academic Press, New York, pp. 505–535.

Zwisinsky, C., Date, T., and Wickner, W., 1981, Leader peptidase is found in both inner and outer membranes of *Escherichia coli, J. Biol. Chem.* **256**:3593–3597.

Chapter 4

Biochemistry of the Sarcolemma

A. M. Kidwai
Biomembrane Laboratory
Industrial Toxicology Research Centre
Mahatma Gandhi Marg
Lucknow 226001, India

1. INTRODUCTION

In animal cells plasma membrane constitutes the cell boundary and acts as the permeability barrier for controlling the transport of water and solutes into and out of the cell. The plasma membrane also receives information and plays important roles in cell–cell interaction and in cell growth and development.

With the advent of subcellular fractionation techniques it became possible to study individual components of the cell in their isolated state. This helped biochemists to assign functions to various constituents of the cellular components; mitochondria, nuclei, lysosomes, and microsomes were studied in detail. By density gradient centrifugation techniques it is now possible to subfractionate the microsomes into plasma membrane- and endoplasmic reticulum-derived vesicles.

In muscles the cell surface structure is called the sarcolemma; it consists of three distinct layers. The outermost layer is composed of collagen fibrils, the middle layer is called basement membrane and is composed of mainly carbohydrates and collagenous proteins, and the innermost layer is the plasma membrane. However, the plasma membrane alone is frequently referred to as the sarcolemma and it is this membrane that is now the focus of attention for physiologists and biochemists alike. The plasma membrane is rich in enzymes and thus various functions have been assigned to this component of the muscle cell.

The present review deals with the isolation and properties of the plasma membranes of three types of muscle, namely, cardiac, smooth, and skeletal muscle. It is the purpose of this review to focus attention on recent isolation techniques and on various properties of the muscle cell surface currently investigated in various laboratories.

2. ISOLATION TECHNIQUES

There are two approaches to the isolation of sarcolemma or of plasma membrane. Muscle cells may be separated by gentle homogenization, followed by extraction of intracellular material by salts, or they may be disrupted by homogenization of the muscle and the resulting fragments fractionated by density gradient centrifugation.

Homogenizers most commonly used for disrupting muscle tissue are the Potter-Elvehjem, the Virtis, and the Polytron. The first two are effective with a wide variety of muscle tissues, but smooth muscle cells are enmeshed in collagen fibrils and can only be processed effectively by the use of the Polytron homogenizer at controlled speed and time (Kidwai, 1975).

2.1. Isolation of Smooth Muscle Plasma Membrane

Smooth muscle plasma membrane has been isolated from uterus, mesenteric arteries, veins, vas deferens, and trachealis (Kidwai *et al.,* 1974; Matlib *et al.,* 1979; Kwan *et al.,* 1979; Kwan *et al.,* 1981; Kwan *et al.,* 1983; Grover *et al.,* 1980).

All these methods are based on the same principle of drastic homogenization followed by density gradient centrifugation of the microsomal vesicles to separate plasma membrane and endoplasmic reticulum components. However, each tissue has to be handled slightly differently and the reader is referred to the original articles for such detail. A description of the fractionation of uterine smooth muscle is given as an example.

2.1.1. Fractionation of Rat Uterine Smooth Muscle

2.1.1.1. Method of Kidwai (1974). Myometrium from (12 rats) uterine horns were separated from endometrium, cut into small pieces, and homogenized using Polytron PT 20 for 4 sec at maximum speed. The homogenate was filtered through gauze cloth. Unbroken tissue was rehomogenized and filtered, and the final homogenate was made 20% in 0.25 M sucrose. The homogenate was centrifuged at 105,000 $\times$ g for 30 min, and the sediment suspended in a small volume of 0.25 M sucrose and layered on top of a continuous or a discontinuous gradient. The discontinuous gradient was made by layering 30%, 33%, and 40% sucrose solutions. Centrifugation was at 100,000 $\times$ g for 2 hr. At the end of the run three bands were collected that represented the plasma membrane (at 0.25 M/30% interface), endoplasmic reticulum, and mitochondria (Kidwai, 1974).

2.1.1.2. Modification Introduced by Matlib et al. (1979). Nineteen female Wistar rats weighing 160–200 g were injected with 500 µg of diethylstilbestrol in 0.1 ml of sesame oil for 2 days to bring them to a hormonal state. The animals were sacrificed on the third day by a blow on the head. The uterine horns were removed and put in Krebs solution of the following composition: KCl 4.6 mM, MgSO$_4$ 1.16 mM, CaCl$_2$ 2.5 mM, NaCl 115.5 mM, NaHCO$_3$ 21.9 mM, and glucose 11.1 mM. The uterine horns were then trimmed of fat and loosely bound connective tissue endometrium and most of the circular muscle was stripped off. This entire process can be carried out on filter paper moistened with Krebs–Ringer and maintained at 4°C on a cold plate. About 3 g of myometria was incubated at 37°C for 30 min in the presence of 95% O$_2$ and 5% CO$_2$ to revive the metabolism. Myometria was transferred to a petri dish containing 250 mM sucrose and 20 mM morpholino propane sulfonic acid (MOPS) and minced. The minced material was divided into three portions and each portion was homogenized three times for 10 sec with 5-sec intervals using Polytron PT 20 at about 15,000 rpm speed.

The homogenate was centrifuged at 1000 × g for 10 min to remove nuclei, contractile proteins, whole cells, and unbroken tissues. The supernatant was then centrifuged at 10,000 × g for 15 min. The resulting pellet was suspended in 7 ml of sucrose–MOPS solution and centrifuged at 113,000 × g for 30 min. The resulting microsomal pellet was further fractionated on a density gradient, either a linear gradient using 20–45% sucrose in a 10-ml centrifuge tube or a discontinuous gradient prepared by layering 4 ml of 50% sucrose (w/w), 2 ml of 35% sucrose, and 3 ml of 30% sucrose. One milliliter of microsomal fraction was layered on the top of the gradient and centrifuged at 112,000 × g for 2 hr in a swing-out rotor. Three distinct and two diffuse bands were seen at the end of the run and removed by pasteur pipette. Each band was diluted with cold distilled water to attain a sucrose concentration of 0.5% and centrifuged at 100,000 × g for 30 min.

The fractions near the top of the gradient were plasma membrane vesicles and as the density increased the endoplasmic reticulum fractions were obtained (Matlib et al., 1979).

2.2. Isolation of Skeletal Muscle Plasma Membrane

Early preparations of sarcolemma from frog, chicken, and rat were characterized chemically but were found not to have retained any significant enzyme activities. Retention of enzyme activities in plasma membranes isolated from rat skeletal muscle was first achieved by Peter (1970) and later by Kidwai et al. (1973). Two isolation techniques are now in common use, one featuring homogenization and centrifugal fractionation and the other using lithium bromide to extract intracellular proteins, leaving behind sarcolemmal tubes.

2.2.1. Density Gradient Centrifugation Method

2.2.1.1. Preparation of Material. Muscles from the hind legs of two rats were removed and dissected free of connective tissue and nerves. They were chopped into small pieces, weighed and suspended in 0.25 M sucrose (5 g muscle in 20 ml sucrose solution).

2.2.1.2. Homogenization. Five-gram batches of finely chopped muscle were homogenized for 15 sec in 0.25 M sucrose using a Polytron PT 20 at near maximum speed. The chopped muscle in sucrose was kept at 0–5°C in an ice bath while homogenizing. Care was taken not to homogenize too long; otherwise mitochondrial damage occurred and the resulting fragments of mitochondria contaminated other subcellular fractions.

For smaller amounts of tissue a PT 10 generator can be used, and for large amounts generator PT 35 would be suitable.

2.2.1.3. Filtration. This was an essential step in the preparation of plasma membrane from skeletal muscle. This step helped to remove contractile proteins, connective tissue, cell debris, and so on. Use of several filters of different mesh sizes helped to remove the various sizes of particles and allowed the filtration to proceed smoothly. Alternatively, the homogenate could be filtered through a series of filters set up separately, but this would be time-consuming (Kidwai, 1974).

The filtrate was clear pinkish. Slight turbidity did not interfere with the density gradient centrifugation.

The filtrate was centrifuged at 104,000 × g for 30 min. The sediment was resuspended in a small volume of 0.25 M sucrose.

All the preceding steps were performed in the cold (0–4°C).

2.2.1.4. Density Gradient Centrifugation. A discontinuous sucrose gradient can be prepared as described by Klip and Walker (1983) using 2.5 ml each of 1.3, 1.1, 0.9, and 0.7 M sucrose in buffer (5 mM CaCl$_2$, 5 mM Tris–HCl, pH 8.0, 0.5 mM dithiothreitol (DTT), and 0.1 mM phenylmethylsulfonyl flouride (PMSF)). The sediment obtained after the first centrifugation was suspended in sucrose or sucrose buffer, layered on the top of a discontinuous sucrose gradient and centrifuged for 3.5 hr at 113,000 × g. The turbid bands at the interphase of each of the four layers were collected by pasteur pipette, diluted 10 times with distilled water to reduce the density of sucrose and centrifuged at 100,000 × g for 1 hr, and resuspended in suitable buffer.

2.2.2. Toluene–Lithium Bromide Method

2.2.2.1. Preparation of the Material. Adult frogs *(Rana tigrina)* were stunned and decapitated and muscles were removed from hind limbs and kept in ice cold calcium chloride (50 mM, pH 7.0). pH was adjusted with 100 mM sodium bicarbonate. The connective tissue and nerves were dissected out. About 50 g muscle was minced and homogenized in a Scintronic homogenizer for 1 min at a speed setting of 240. The homogenate was filtered through a filter as described by Kidwai (1974); any other filter can be used to remove connective tissue and unbroken muscle lumps; this unbroken tissue was rehomogenized and filtered and the pooled muscle cell fragment suspension was centrifuged at 70 × g for 5 min in cold. The sediment was resuspended in ice cold 0.25 M sucrose (pH 7.0, adjusted with Tris). The optical density of this suspension was adjusted to 0.25 at 600 mμ in a colorimeter; this suspension consisted of cell fragments of various sizes.

2.2.2.2. Toluene Treatment. Toluene is used to increase the permeability of muscle cell surface prior to lithium bromide extraction. It was added to the suspension of cell fragments (5 ml/100 ml suspension) and gently mixed by tilting for exactly 3 min. It was allowed to settle in the cold for 5 min. The toluene layer was decanted and the remaining suspension was centrifuged at 79 × g for 5 min (Zaidi *et al.*, 1981).

2.2.2.3. Lithium Bromide Treatment. The sediment obtained after toluene treatment was suspended in 150 ml of 0.4 M lithium bromide, pH 8.4, adjusted with 1 M Tris. The suspension was stirred at minimum speed for 1.5 hr on a magnetic stirrer in cold. At the end the material was centrifuged at 900 × g for 5 min and the sediment resuspended in lithium bromide solution. The process was repeated until (after 5 hr) little further protein was extracted in the lithium bromide solution. The final sediment was suspended in 0.25 M sucrose and the pH adjusted to 7.0 by Tris. The material was washed twice by centrifugation and resuspended in 0.25 M sucrose (Zaidi *et al.*, 1981).

2.3. Isolation of Cardiac Muscle Plasma Membrane

Earlier methods for isolation of cardiac muscle plasma membrane involved removal of contractile proteins by high ionic strength solutions followed by differential centrifugation. Recently developed methods avoid high ionic strength solutions and rely mainly on density centrifugation (Kidwai *et al.*, 1971b; Stewart *et al.*, 1979).

These methods have the advantage of yielding all the subcellular fractions in one step, thus enabling one to carry out comparative studies on various subcellular components of the same tissue.

2.3.1. Homogenization

Approximately 5 g of left ventricular myocardial tissue was homogenized in 20 ml of 0.25 M sucrose in a Virtis Model 45 omnimixer for 1–2 sec at full speed. The homogenate was centrifuged in a Model PRI international portable refrigerated centrifuge at 25 × g (300 rpm) for 1 min. The supernatant fluid was decanted and saved. The residue was suspended in 20 ml of 0.25 M sucrose, rehomogenized for 1–2 sec at maximum speed, and centrifuged at 25 × g (300 rpm). This process of residue homogenization and slow centrifugation was repeated six times (Stewart et al., 1979).

The combined supernatant fluid was then filtered through two layers of 10/grade cheesecloth. The filtrate was centrifuged at 176,600 × g for 20 min in a Beckman L265N ultracentrifuge using the 65 rotor.

The pellet from this last centrifugation step was suspended in 27 ml of 0.25 M sucrose. This constituted the crude membrane suspension.

2.3.2. Sucrose Gradient

Discontinuous sucrose gradient tubes were prepared on the day of use in 13 ml tubes using three sucrose concentrations (4 ml of 1.5 M sucrose, 3 ml of 1.22 M sucrose, and 1.5 ml of 1.1 M sucrose (Stewart et al., 1979).

2.3.3. Centrifugation

Four and one-half millimeters of the crude membrane suspension was layered on each of six prepared sucrose gradient tubes. The gradients were centrifuged at 111,700 × g for 1.5 hr in a Beckman L2-65B ultracentrifuge using a SW40 rotor.

A band of sarcolemma (SL) separated at the interface of the gradient and the membrane suspension, a sarcoplasmic reticulum (SR) band separated just beneath the sarcolemma layer, and a mitochondrial band appeared at the interface between the 1.22 M sucrose and 1.5 M sucrose. Contractile proteins, nuclei, unhomogenized tissue, and cell debris accumulated at the bottom of the gradient. The subcellular fractions were collected by pasteur pipette, diluted ten times with distilled water, and centrifuged at 100,000 × g for 30 min. Each sediment was suspended in an appropriate buffer for use.

3. CHARACTERIZATION OF THE SARCOLEMMA

Sarcolemma isolated by various techniques can be characterized on the basis of morphological, chemical, and enzymatic characteristics. In addition to the sarcolemmal properties one should also look for the absence of other subcellular organelles and components.

3.1. Morphology

The membranes on homogenization form vesicles of different sizes which can be viewed by electron microscopy. It is not possible to assign the origin of these vesicles unequivocally on the basis of morphology but the thickness of the membrane can give an indication of its origin. In rat muscle the plasma membrane image was found by Kidwai *et al.* (1973) to measure 95 Å overall and sarcoplasmic reticulum 56 Å.

The sarcolemma isolated by the technique of Zaidi *et al.* (1981) maintained its tubelike appearance and was therefore identifiable by light microscopy.

3.2. Enzyme Markers

Membrane-bound enzymes are used as markers for the identification of various membranes after isolation.

Sarcolemma is known to possess $Na^+ + K^+$-ATPase, *p*-nitrophenylphosphatase (K^+-activated), 5′-nucleotidase, adenylate cyclase, phosphodiesterase, and leucylnaphthylamidase (Kidwai *et al.*, 1971a, 1973; Moffet *et al.*, 1976; Jones *et al.*, 1980; Zaidi *et al.*, 1981). The specific activities of the preceding enzymes were increased in the purified sarcolemma.

The enzymes, succinic dehydrogenase, cytochrome *c* oxidase, and monoamine oxidases are sometimes used as markers of mitochondrial membranes to check the extent of contamination in sarcolemmal preparations (Kidwai *et al.*, 1971; 1973).

3.3. Chemical Composition

The chemical constituents are sometimes useful as markers for a particular membrane. It has been reported (Kidwai *et al.*, 1971, 1973; Zaidi *et al.*, 1981) that the cholesterol:phospholipid ratio was higher in sarcolemma than in other muscle membrane fractions. The lipid composition of the sarcolemma is different from that of other subcellular fractions; phosphatidylserine and sphingomyelin are reported to be higher (Fiehn *et al.*, 1971) in sarcolemma.

Ribonucleic acid and deoxyribonucleic acid can be used to detect contamination in a sarcolemma preparation as these substances are not sarcolemmal constituents.

3.4. Miscellaneous Markers

3.4.1. Protein Biosynthesis

Glucose 6-phosphatase activity is often used as a marker for endoplasmic reticulum, but in muscles this enzyme activity is negligible and therefore this enzyme cannot be used for assessment of contamination of the sarcolemmal preparation by endoplasmic reticulum.

In rat skeletal and smooth muscle (Kidwai et al., 1973; Matlib et al., 1979) [^{3}H]leucine incorporation has been used to identify the endoplasmic reticulum.

3.4.2. Fluorescent Markers

A fluorescent dye, 4-acetamido-4'-isothocynate stilbene-2, 2'-disulfonic acid (SITS), was used by many workers (Marinetti and Gray, 1967; Kidwai et al., 1971) as a marker for surface membranes; it was found that if the intact tissue was incubated with this dye it was bound to the surface membrane and did not penetrate the cell interior.

Substances that have specific affinities for the cell surface constituents can be used as markers for sarcolemma. Thus, wheat germ agglutinin and oxytocin have been employed to identify the plasma membrane of rat smooth muscle (Matlib et al., 1979).

4. BIOCHEMICAL PROPERTIES

4.1. Enzymatic Makeup

Preparations of sarcolemma from skeletal, smooth, and heart muscle have been found to contain $Na^+ + K^+$-APTase, 5'-nucleotidase, Mg^{2+}-ATPase, $Ca^{2+} + Mg^{2+}$-ATPase, K^+-activated p-nitrophenyl phosphatase (PNPPase), adenylcyclase, phosphodiesterase, and leucyl-β-naphthylamidase.

4.1.1. Sodium-Potassium-ATPase

$Na^+ + K^+$-ATPase is known to be present in most surface membranes and is involved in the transport of sodium and potassium across the plasma membrane. In skeletal muscle plasma membrane from rat and frog, specific activities of ouabain-sensitive ATPase were 15 and 11 μmoles P_i/mg protein/hr, respectively (Kidwai et al., 1973; Zaidi et al., 1981). In smooth muscle Mg^{2+}-ATPase was reported but $Na^+ + K^+$ also activated the enzyme and this activation was sensitive to ouabain. The specific activity of $Na^+ + K^+$-ATPase in smooth muscle plasma membrane was reported by Kidwai (1977) to be 42 μmoles P_i/mg protein/hr. In heart plasma membrane it was reported to be in the range of 13.0–25.4 μmoles P_i/mg protein/hr (Kidwai et al., 1971; Jones et al., 1980; Stewart, 1979).

4.1.2. 5′-Nucleotidase

5′-Nucleotidase is known to be concentrated in sarcolemmal preparations from cardiac, smooth, and skeletal muscles (Kidwai, 1974; Matlib et al., 1979).

4.1.3. Miscellaneous Enzymes

Several additional enzyme activities have been identified in sarcolemmal preparations. These include cAMP-dependent protein kinase (Manalan and Jones, 1982), adenylate cyclase (Besch et al., 1977; Moffet et al., 1976) $Ca^{2+} + Mg^{2+}$-ATPase (Marcos and Drummond, 1980), L-leucyl-β-naphthylamidase, p-nitrophenylphosphatase, acid phosphatase, alkaline phosphatase, Mg^{2+}-ATPase, and phosphodiesterase (Kidwai et al., 1971; Zaidi et al., 1981).

4.2. Glucose Transport

The glucose transport system in rat skeletal muscle plasma membrane has been studied by Klip and Walker (1983). Plasma membrane vesicles isolated by a modification of the method of Kidwai et al. (1973) were capable of a D-glucose uptake that was sensitive to cytochalasin B (CB). Binding to the glucose carrier was inhibited by 2,4-dinitrofluorobenzene and N-ethylamaleimide. Antibodies raised against the glucose carrier of human red cell membranes cross-reacted with a polypeptide of muscle membranes that might be a glucose carrier molecule. This polypeptide had a molecular weight of 45,000–50,000 on SDS polyacrylamide gel electrophoresis.

4.3. Calcium Transport

Calcium is important in excitation–contraction coupling in all muscles. Increase of calcium in the cytosol activates myofilaments, whereas for relaxation excess calcium must be removed. A calcium ATPase is located in the endoplasmic reticulum and recently another calcium-pumping mechanism was shown to be present in plasma membrane. These maintain calcium homeostasis in the cells. Calcium can be pumped out of the muscle cells by Ca^{2+}-ATPase also via Na^+–Ca^{2+} exchange (Lamers and Stinis, 1981; Wibo et al., 1981.)

4.4. Receptors

The biochemical approach to studies on receptors for hormones, neurotransmitters, and drugs has been advanced in recent years as more and more techniques have been developed to isolate sarcolemma from muscles. Studies on cardiac β-adrenergic receptors were undertaken by Stiles *et al.* (1983); receptors for oxytocin and for prostaglandins have been identified in rat and in human myometrium by Crankshaw *et al.* (1978, 1979).

5. SPECIAL STUDIES

5.1. Membrane Orientation

A knowledge of the orientation of isolated sarcolemmal vesicles is important in the interpretation of many of the properties of sarcolemmal preparations.

Grover *et al.* (1980) and Besch *et al.* (1976) studied the intactness and orientation of sarcolemmal vesicles prepared from rat myometrium and from canine cardiac muscle, respectively. Rat myometrium plasma membrane was prepared by the method of Matlib *et al.* (1979). It was assumed that the intact vesicles were osmotically active to sucrose and also impermeable to EGTA (ethylene glycol-bis(β-amino ethyl ether)N,N,N',N'-tetraacetic acid) and to cationized ferritin. For determining sidedness, the assumptions were that ouabain-sensitive K^+-activated PNPPase, as well as galactose oxidase and trichloroacetic acid precipitable glycolipids of plasma membrane, reside on the external cell surface.

5.1.1. Vesicle Intactness

When vesicles from uterine smooth muscle plasma membrane were exposed to sucrose concentrations higher than the concentration trapped inside them, both their volume and surface area decreased (Grover *et al.*, 1980). It was concluded that 70% of the vesicles were osmotically active to sucrose. When cationized ferritin was used to label the vesicles, very few had the large molecules of this protein trapped inside them, thus indicating that the majority of the vesicles were intact. Further evidence was obtained by Ca^{2+} uptake and EGTA release studies. It was assumed that EGTA would release Ca^{2+} bound on both the surfaces of broken vesicles but only from the outside of intact vesicles. They concluded that only 25% of the vesicles were broken. This finding was also consistent with other studies which indicated that 80% of the vesicles were intact.

5.1.2. Vesicle Sidedness

In studying the sidedness of smooth muscle plasma membrane vesicles, Grover (1980) assumed that in isotonic media the external surface of the right-side-out vesicles, and both surfaces of broken vesicles, would be available to nonpermeable external cell surface markers, whereas after hypotonic shock the inside surfaces of the inside-out vesicles would also be exposed to the reaction media. If this assumption is correct, hypotonic shock would result in an enhancement of the measurable external cell surface activities. The external cell surface markers were ouabain-sensitive K^+-activated PNPPase and galactose oxidase-catalized $Na_3B^3H^4$ labeling; both markers were enhanced 55% in their activity by hypotonic shock. In these studies the vesicles were calculated to be 20% broken, 40% inside out, and 40% right side out.

5.2. Labeling Techniques

Lactoperoxidase has been used to identify the surface proteins of intact cells by iodinating tyrosine and histidine residues of exposed proteins. It is a high-molecular-weight protein that does not dissociate into subunits and will therefore not gain access to the inner surface of a membrane vesicle unless the membrane is damaged (Krall and Koreman, 1979).

6. CONCLUSIONS

Physiologists recognized the importance of muscle cell surface membranes and studies were conducted on the role of sarcolemma in cellular function, but biochemical studies lagged behind for want of a suitable method for isolating this component of the cell. Attempts were made in the 1960s to isolate sarcolemma by the use of high salt, pH, and temperature. These preparations were not suitable for enzymatic studies as during the extraction procedure most of the enzymes were denatured.

The approach taken by Kidwai (1974) was to subfractionate the total homogenate of smooth, skeletal, or heart muscle in order to collect all the subcellular fractions in one step. However, muscles, being tough tissues, have to be drastically homogenized to disrupt the sarcolemma into small fragments in order to release the intracellular material. The resulting sarcolemmal vesicles are morphologically very similar to vesicles of different origin and they vary considerably in size. The subcellular cut identified as plasma membrane may have lost material and may not represent the total composition of the plasma membrane.

There is a need to develop a technique for isolation of the sarcolemma in which it can maintain its shape. A beginning has been made in this direction by Zaidi *et al.* (1981), who isolated the frog skeletal muscle sarcolemma by increasing the permeability of the cells by toluene treatment followed by extraction of the intracellular material by lithium bromide. This preparation was claimed to possess basement membrane as well as plasma membrane while in other methods only plasma membrane was identifiable. An approach that has great potential is one in which the muscle cell is first dispersed by the use of enzymes and chemicals, followed by extractions of intracellular material to prepare intact and pure sarcolemma.

7. REFERENCES

Besch, H. R., Jones, L. R. and Watanabe, A. M., 1976, Intact vesicles of canine cardiac sarcolemma, *Cir. Res.* **39**:586–595.

Besch, H. R., Jr., Jones, L. R., Fleming, J. W., and Watanabe, A. M., 1977, Parallel unmasking of latent adenylate cyclase and (Na^+, K^+)-ATPase activities in cardiac sarcolemmal vesicles, *J. Biol. Chem.* **252**:7905–7908.

Crankshaw, D. J., Crankshaw, J., Branda, L. A., and Daniel, E. E., 1979, Receptors of E type prostaglandins in the plasma membrane of nonpregnant human myometrium, *Arch. Biochem. Biophys.* **198**:70–77.

Crankshaw, D. J., Branda, L. A., Matlib, M. A., and Daniel, E. E., 1978, Localization of the oxytocin receptor in the plasma membrane of rat myometrium, *Eur. J. Biochem.* **86**:481–486.

Fiehn, W., Peter, J. B., Mead, J. F., Gan-Flepano, M., 1971, Lipids and fatty acids of sarcolemma, sarcoplasmic reticulum, and mictochondria from rat skeletal muscle, *J. Biol. Chem.* **246**:5617–5620.

Grover, A. K., Crankshaw, J., Garfield, R. E., and Daniel, E. E., 1980, Smooth muscle membrane vesicle orientation: A study on intactness and sidedness of rat myometrium plasma membrane vesicles, *Can. J. Physiol. Pharmacol.* **58**:1202–1211.

Grover, A. K., Kannan, M. S., Daniel, E. E., 1980, Canine trachealis membrane fractionation and characterization, *Cell Calcium* **1**:135–146.

Jones, L. R., Maddock, S. W., and Besch, Jr., H. R., 1980, Unmasking effect of alamethicin on the (Na^+,K^+)APTase, β-adrenergic receptor-coupled adenylate cyclase, and cAMP-dependent protein kinase activities of cardiac sarcolemmal vesicles, *J. Biol. Chem.* **255**:9971–9980.

Kidwai, A. M., 1974, Isolation of plasma membrane from smooth, skeletal, and heart muscle, in: *Methods in Enzymology*, Vol. 31A (S. Fleischer and L. Packer, eds.), pp. 134–144, Academic Press, New York.

Kidwai, A. M., 1975, Homogenization and fractionation techniques in smooth muscle, in: *Methods in Pharmacology*, Vol. 3 (E. E. Daniel and D. M. Paton, eds.), pp. 543–554, Plenum Press, New York.

Kidwai, A. M., 1977, Smooth muscle sarcolemmal biochemistry, in: *The Biochemistry of Smooth Muscle* (N. L. Stephens, ed.), pp. 585–594, University Park Press, Baltimore.

Kidwai, A. M., Radcliffe, M. A. and Daniel, E. E., 1971a, Isolation and characterization of plasma membrane from rat myometrium, *Biochim. Biophys. Acta* **233**:538–549.

Kidwai, A. M., Radcliffe, M. A., Duchon, G., and Daniel, E. E., 1971b, Isolation of plasma membrane from cardiac muscle, *Biochem. Biophys. Res. Commun.* **45**:901–910.

Kidwai, A. M., Radcliffe, M. A., Lee, E. Y., and Daniel, E. E., 1973, Isolation and properties of skeletal muscle plasma membrane, *Biochim. Biophys. Acta* **298**:593–607.

Klip, A., and Walker, D., 1983, The glucose transport system of muscle plasma membrane: Characterization by means of [^{3}H]cytochalasin B binding, *Arch. Biochem. Biophys.* **21**:175–187.

Krall, N. F., and Koreman, S. G., 1979, Smooth muscle sarcolemma. Purification and properties of plasma membrane from the rat uterus, *Biochim. Biophys. Acta* **556**:105–117.

Kwan, C. Y., Garfield, R., Daniel, E. E., 1979, An improved procedure for the isolation of plasma membrane from rat mesenteric arteries, *J. Mol. Cell. Cardiology* **11**:639–659.

Kwan, C. Y., Lee, R. M. K. W. and Daniel, E. E. 1981, Isolation of plasma membranes from rat mesenteric veins: A comparison of their physical and biochemical properties with arterial membranes, *Blood Vessels* **18**:171–186.

Kwan, C. Y., Lee, R. M. K. W., and Grover, A. K., Isolation and characterizaton of subcellular membrane fractions from the smooth muscle rat vas deferens, *Molec. Physiol.* **3**: 53–69, 1983.

Lamers, J. M. J. and Stinis, J. T., 1981, An electrogenic Na$^+$/Ca^{2+} antiporter in additon to the Ca^{2+} pump in cardiac sarcolemma, *Biochim. Biophys. Acta.* **640**:521–534.

Manalan, A. S. and Jones, L. R. 1982, Characterization of the intrinsic cAMP-dependent protein kinase activity and endogenous substrates in highly purified cardiac sarcolemmal vesicles, *J. Biol. Chem.* **257**:10052–10062.

Marcos, N. C. and Drummond, G. I., 1980, (Ca^{2+}+Mg^{2+})-ATPase in enriched sarcolemmas from dog heart, *Biochim. Biophys. Acta.* **598**:27–39.

Marinetti, G. V. and Gray, G. M., 1967, A fluorescent chemical marker for the liver cell plasma membrane, *Biochim. Biophys. Acta.* **135**:580–590.

Matlib, M. A., Crankshaw, J., Garfield, R. E., Crankshaw, D. J., Kwan, C. Y., Branda, L. A., and Daniel, E. E. 1979, Characterization of membrane fractions and isolation of purified plasma membranes from rat myometrium, *J. Biol. Chem.* **254**:1834–1840.

Moffet, F. J., Kidwai, A. M., and Baer, H. P., 1976, Adenyl cyclase, cyclic nucleotide phosphodiesterase and norepinephrine binding in rat heart membranes, in: *Recent Advances in Studies on Cardiac Structure and Metabolism: The Sarcolemma* (P. E. Roy and N. S. Dhalla, eds.), pp. 183–191, University Park Press, Baltimore.

Peter, J. B., 1970, A (Na$^+$+K$^+$) ATPase of sarcolemma from skeletal muscle, *Biochem. Biophys. Res. Commun.* **40**:1362–1367.

Steck, T. L., and Kant, J. A. 1974, Preparation of impermeable ghosts and inside out vesicles from human erythrocyte membranes, in: *Methods in Enzymology,* Vol. 31A (S. Fleischer and L. Packer, eds.), pp. 172–180, Academic Press, New York.

Steck, T. L., and Wallach, D. F. H., 1970, The isolation of plasma membranes, in: *Methods in Cancer Research,* Vol. 5 (H. Busch, ed.), pp. 93–152, Academic Press, New York.

Stewart, M. J. J., Read, W. O., and Steffen, R. P., 1979, Isolation of pure myocardial subcellular organelles, *Anal. Biochem.* **96**:293–297.

Stiles, G. L., Strasser, R. H., Lavin, T. N., Jones, L. R., Caron, M. G., and Lefkowitz, R. J., 1983, The cardiac β-adrenergic receptor. *J. Biol. Chem.* **258**:8443–8449.

Wibo, M., Morel, N., and Godfraind, T., 1981, Differentiation of Ca^{2+} pumps linked to plasma membrane and endoplasmic reticulum in the microsomal fraction from intestinal smooth muscle, *Biochim. Biophys. Acta.* **649**:651–660.

Zaidi, S. I. M., Pandey, R. N., Kidwai, A. M., and Krishna Murti, C. R., 1981, A rapid method for preparation of sarcolemma from frog skeletal muscle, *J. Biosci.* **3**:293–302.

Chapter 5
Membrane Fusion*

Nejat Düzgüneş
Cancer Research Institute
School of Medicine
University of California
San Francisco, California 94143

1. INTRODUCTION

Membrane fusion is an important biological process that is observed in a wide variety of intra- and intercellular events. Little is known about its molecular mechanisms and few of the factors which control it in biological systems are known. Even in exhaustively studied phenomena such as neurotransmitter release, or stimulus–secretion coupling in general, the molecular events that are prerequisites for the initiation of membrane fusion between secretory vesicles and the plasma membrane are not well understood. In this communication we shall review studies on the mechanisms of membrane fusion in biological membranes and in model membrane systems and attempt to bring out the critical elements of this process from among the diverse membrane fusion systems studied. We shall nevertheless emphasize the large variety of conditions under which membrane fusion can be induced in these systems.

The different types of membrane fusion phenomena occurring in biological systems have been reviewed in detail (Poste and Allison, 1973; Poste and Nicolson, 1978). Here we shall only discuss briefly some of these phenomena, such as exocytosis, endocytosis, and the fusion of intracellular membranes with one another. We shall be concerned especially with studies that lead to interpretations at the molecular level. In this respect we shall concentrate on model membrane systems where the molecular components of the membrane and the ionic environment are strictly controlled. The early studies on the fusion of

---

*This work is dedicated to Professor Zeliha Düzgüneş, and Professor Orhan Düzgüneş, on their 68th birthday.

model membranes (Papahadjopoulos, 1978; Papahadjopoulos *et al.,* 1979; Gingell and Ginsberg, 1978) and some aspects of the more recent work have been reviewed (Düzgüneş *et al.,* 1980; Bruni and Palatini, 1982). We shall discuss the fusion of isolated secretory vesicles and other subcellular membranes such as phagosomes and lysosomes, the Ca^{2+} requirement for exocytosis, and the electron microscopic evidence for membrane reorganization during fusion. In addition, we shall review the work on the fusion of erythrocytes by chemical agents and of myoblasts during differentiation, two systems that have been studied in detail. The fusion of viruses with cells constitutes a model system in which the proteins involved in fusion are identified and thus lends itself to studies on structure–function relationships in membrane fusion; these studies will also be discussed.

2. FUSION OF SUBCELLULAR MEMBRANES

2.1. Exocytosis

Secretory vesicles containing proteins or neurotransmitters release their contents into the extracellular space in a process called exocytosis (De Duve, 1963; Palade, 1975). Exocytosis is thought to take place via fusion of the secretory vesicle membrane with the plasma membrane of the cell. Membrane fusion during secretion has been observed in a large variety of cell types (Table I). The implications of these electron microscopic observations for the molecular mechanisms of membrane fusion will be discussed in detail in Section 3.1.

Palade (1975) has described a series of membrane events preceding exocytosis. The secretory vesicles first attach to the plasmalemma and form a pentalaminar structure that then transforms into a single trilaminar diaphragm, a process we shall designate as *semifusion.* The rupture of this diaphragm is thought to result in the discharge of the contents of the secretory granule into the lumen. We consider the pentalaminar stage as adhesion or, in the case of isolated secretory vesicles (Section 3.3) and liposomes (Sections 3.8 and 3.9), as aggregation, which may or may not result in the fusion of the two membranes into a single continuous membrane. We use the term *membrane fusion* for the stage designated as "fission," by Palade in which the contents of the secretory vesicle and the extracellular space intermix. Exocytotic fusion is an example of the "planar fusions" between protoplasmic membrane surfaces in the nomenclature of Stossel *et al.* (1978).

Many exocytotic processes are believed to be initiated by the entry of Ca^{2+} into the cell. Douglas (1974) and Rubin (1974) have summarized the evidence for the direct involvement of Ca^{2+} in stimulus-secretion coupling. For example, in chromaffin cells acetylcholine induces an increase in the permeability of the

Table I
Secretory Events Involving Membrane Fusion Demonstrated by Electron Microscopy

Cell type	Secretory events	References
Mast cells	Histamine release	Lagunoff, 1973
		Lawson et al., 1977
		Chandler and Heuser, 1980
Tetrahymena	Mucocyst secretion	Satir et al., 1973
		Satir and Oberg, 1978
Frog nerve terminal (neuromuscular junction)	Neurotransmitter release	Ceccarelli et al., 1972
		Heuser et al., 1979
Sea urchin egg	Cortical granule exocytosis	Chandler and Heuser, 1979
		Epel and Vacquier, 1978
Paramecium	Trichocyst discharge	Matt et al., 1978
		Hausmann and Allen, 1976
Limulus amebocytes	Endotoxin-induced degranulation	Ornberg and Reese, 1981
Islets of Langerhans	Insulin release	Orci et al., 1973, 1977
		Berger et al., 1975
Adrenal medulla	Chromaffin granule extrusion	Smith et al., 1973
		Diner, 1967
Phytophthora zoospores	Encystment; release of cell wall precursors	Pinto da Silva and Nogueira, 1977
Luteal cells	Progesterone secretion	Gemmel and Stacy, 1979
Acinar cells of the exocrine pancreas	Extrusion of zymogen granules	Palade, 1959
		Tanaka et al., 1980
Neurohypophyseal cells	Release of oxytocin and vasopressin	Nagasawa et al., 1970
		Theodosis et al., 1976
Neutrophils	Release of β-glucuronidase	Chandler et al., 1983
Torpedo electroplaques	Neurotransmission	Nickel and Potter, 1971
Parotid acinar cells	Granule extrusion	De Camilli et al., 1976

membrane to Ca^{2+}, whose influx causes the extrusion of catecholamines. Membrane depolarization by increasing the K^+ concentration outside also leads to the influx of Ca^{2+} and exocytosis (Douglas, 1974). Neurotransmitter release at synapses follows the entry of Ca^{2+} into the presynaptic terminal on depolarization of the membrane and consequent activation of Ca^{2+} channels (Katz, 1969; Baker, 1972). Direct microinjection of Ca^{2+} into the presynaptic cell has been shown to result in the release of neurotransmitter (Miledi, 1973). Exocytosis can be induced in mast cells by intracellular injection of Ca^{2+} (Kanno et al., 1973). At present, the site of action of Ca^{2+} in these systems is not known. It is likely that Ca^{2+} participates directly in the fusion of secretory vesicles with the plasma membrane, since Ca^{2+} is required for the fusion of a large variety of biological and model membranes.

2.2. Endocytosis

The internalization of soluble or particulate substances by pinocytosis and phagocytosis, respectively, involves formation of endocytotic vacuoles from an initially planar plasma membrane. The plasma membrane undergoes a morphological change that varies with the type of cell and the stimulus (Stockem, 1977; Silverstein *et al.,* 1977; Williams, 1981). The last stage of this process is the fusion of the invaginated plasma membrane with itself and thus involves a "tubular fusion" between external membrane surfaces of the plasmalemma (Stossel *et al.,* 1978). Little is known about the molecular events responsible for membrane fusion in endocytosis. It is generally believed that metabolic energy is required for endocytosis; however, it is not clear where this energy is expended in the process. Inhibitors of microtubule and microfilament function show varying effects in different experimental systems and in pinocytosis or phagocytosis (Allison and Davies, 1974; Oliver and Berlin, 1979; Silverstein *et al.,* 1977).

Many secretory cells have been shown to undergo fluid-phase pinocytosis following secretory activity (Table II) as predicted by Palade in his early stud-

Table II
Some Cell Types in Which Pinocytotic Uptake of Plasma Membrane Follows Secretory Activity

Cell type	References
Presynaptic cells at the neuromuscular junction	Heuser and Reese, 1973
	Ceccarelli *et al.,* 1973, 1979
Cultured spinal cord neurons	Teichberg *et al.,* 1975
Pancreatic acinar cell	Geuze and Kramer, 1974
Pancreatic β cell	Orci *et al.,* 1973
Adrenal medulla	Grynszpan-Winograd, 1971
	Nagasawa and Douglas, 1972
	Holtzman *et al.,* 1973
	Smith *et al.,* 1973
	Phillips *et al.,* 1983
	Patzak *et al.,* 1984
	Koerker *et al.,* 1974
Parotid gland	Amsterdam *et al.,* 1969
	De Camilli *et al.,* 1976
Neurohypophysis	Nagasawa *et al.,* 1971
	Theodosis *et al.,* 1976
	Nordmann *et al.,* 1974
Adenohypophysis	Pelletier, 1973
Paramecium	Hausman and Allen, 1976
Amoeba	McKanna, 1973

ies (Palade, 1959). It is of particular interest for our discussion on molecular mechanisms of membrane fusion events that in parotid acinar cells following secretion the plasma membrane contains domains of intramembranous particle-rich and particle-poor domains, the latter presumably originating from the fusion of the secretory vesicles that are poor in intramembranous particles. Further, these domains disappear via endocytosis (De Camilli et al., 1976; Meldolesi and Ceccarelli, 1981). Membrane retrieval in these cells has been shown to depend on the presence of Ca^{2+} in the incubation medium (Koike and Meldolesi, 1981). In Amoeba, the Ca^{2+} ionophore A23187 induces pinocytosis, depending on the extracellular Ca^{2+} concentration (Prusch, 1980). Stockem and Klein (1979) have demonstrated an increase in Ca^{2+}-binding sites at the plasma membrane of Ameoba induced to pinocytose. Many macromolecules, hormones, and viruses that bind to cells are taken up in coated vesicles by adsorptive endocytosis (Pearse and Bretscher, 1981). The clathrin coat is thought to act as a molecular filter that selects or excludes molecules from the coated pit and to provide a cytoplasmic frame for the invagination of the plasma membrane. Endocytosis is also thought to provide a mechanism by which a net flow of lipids and receptors is achieved in the cell surface, which in turn facilitates capping of cell surface antigens in the presence of antibodies, and cell locomotion (Bretscher, 1984). Thus, intracellular mechanisms for directing the reinsertion of internalized membrane into the plasma membrane could also direct the movement of the cell.

2.3. Membrane Flow

Intracellular membranes are not static structures. They are transformed from one morphological component to another as part of the regulation of cellular metabolism. These events are accompanied either by the fusion of preexisting membranous organelles or the budding off of new vesicular structures from the endoplasmic reticulum or Golgi apparatus. Endocytotic vesicles such as phagosomes and pinosomes form secondary lysosomes on fusion with primary lysosomes (Steinman et al., 1972; Jacques, 1975; Willingham et al., 1979; Willingham and Yamada, 1978; Amano et al., 1979, 1981). Specialized vesicles formed from the plasma membrane, internalized in coated pits and containing ligand-bound receptors, and which have been termed *receptosomes*, are presumably directed to the Golgi membrane (Pastan and Willingham, 1981). Secretory products and plasma membrane components synthesized in the endoplasmic reticulum are thought to be transported to the Golgi via transfer vesicles (Farquhar, 1978; Morré, 1980). The refined products of the Golgi apparatus (Holtzman et al., 1979; Morré, 1980; Rothman, 1981) are then transported in secretion granules or carrier vesicles that are formed from the Golgi membranes by membrane fusion. Clathrin-coated vesicles may be involved in this process (Pearse and Bretscher, 1981). These secretory vesicles then fuse with the plasma membrane, releasing their internal contents to the extracellular milieu and incorporating their membranes into the plasmalemma (Farquhar, 1978; Leblond and Bennett, 1977; Palade, 1975). The insertion and retrieval of membrane transport molecules to and from the plasma membrane by exocytosis and endocytosis may also be involved in the regulation of cell membrane transport (Lienhard, 1983).

3. STUDIES ON THE MECHANISMS OF MEMBRANE FUSION

To understand the molecular mechanisms of membrane fusion, it is important to establish the involvement of specific membrane components such as phospholipids, cholesterol, glycolipids, intrinsic and extrinsic proteins and glycoproteins, as well as of components in the cytoplasm such as Ca^{2+} and other metal ions, anions, ATP, microfilaments, microtubules, enzymes, and Ca^{2+}-binding proteins. Moreover, the interactions between these components and their reorganization during the fusion reaction must be characterized and the specific stimuli that induce fusion must be identified. These aspects of membrane fusion have been investigated in a large variety of experimental systems involving both biological and artificial membranes. In this section we describe some of these systems in detail. The results of these studies will be analyzed further in Section 4 in terms of the molecular factors that regulate the membrane fusion reaction and the molecular mechanisms involved. Particular attention will be given to our recent work on the fusion of phospholipid vesicles and its modulation by lipid head-groups and proteins, since this synthetic approach provides the most clear-cut information on the role of specific membrane components and ionic environment in membrane fusion.

The complex composition of biological membranes renders them a difficult experimental system in which to identify unequivocally the molecular factors responsible for the regulation of membrane fusion. Nevertheless, the ultimate aim in discovering the molecular basis of fusion is to understand the mechanisms of exocytosis, endocytosis, the interconversion of subcellular membranes and other biological processes involving membrane fusion. It is thus essential to characterize the requirements for and the molecular events of membrane fusion in intact tissues and cells as well as in isolated subcellular membranes. In this section we discuss the results of electron microscopic studies on secretory processes in a number of cell types and on the fusion of purified secretory vesicles, summarize recent work on the Ca^{2+} concentration requirement for exocytosis, and describe some aspects of cell–cell and virus–cell fusion that may be relevant to understanding the molecular mechanisms of membrane fusion. We then describe in detail studies on membrane fusion utilizing model membrane systems such as planar bilayers and liposomes.

3.1. Electron Microscopy of Secretory Processes

Early thin-section electron microscopic observations of secretory cells indicated that the contents of secretory granules are released into the extracellular milieu by fusion of the granule membrane with the plasma membrane (Diner, 1967; Coupland, 1965; Grynszpan-Winograd, 1971; Palade and Bruns, 1968; Ekholm *et al.*, 1962; Rohlich *et al.*, 1971; Lagunoff, 1973, Gemmell and Stacy, 1979). Freeze-etch electron microscopy can reveal large areas of membrane and thus has the advantage of displaying more fusion events at the plasma membrane, compared to thin-section electron microscopy. Orci *et al.* (1973) have used the freeze-etching technique to show that the release of insulin from isolated islets of Langerhans takes place via exocytosis. Smith *et al.* (1973) have also used freeze-etching to demonstrate the attachment of secretory vesicles to the plasma membrane and their fusion in the presence of extracellular calcium; they have found that in the absence of calcium freeze-fracture profiles of fusing vesicles are rare. Later studies with freeze-fracture electron microscopy have revealed the architecture of the membranes before and after fusion. De Camilli *et al.* (1976) have observed that during isoproterenol-induced exocytosis in parotid acinar cells the protein components of the secretory granule membrane do not randomly diffuse into the plasma membrane after membrane fusion, but form patches of low particle density, corresponding to the granule membrane, within areas of high particle density, corresponding to the plasma membrane. These authors have suggested that the lack of free lateral diffusion of these proteins is due to their association with cytoplasmic peripheral proteins. Studies by Chi *et al.* (1976) and Lawson *et al.* (1977) on secretion by stimulated mast cells have shown that the intramembranous particles diffuse laterally to the edge of the region of adhesion of the secretory vesicle membrane and the plasma membrane, and they have suggested that fusion occurs between the protein-free lipid bilayers, in accordance with the hypothesis advanced by Poste and Allison (1973) and Ahkong *et al.* (1975a). Glucose-stimulated pancreatic β cells also display the clearing of intramembranous particles at the area of contact of insulin-containing secretory granules with the plasma membrane, and these protein-free areas persist temporarily after fusion is completed (Orci *et al.*, 1977). Similar observations on the displacement of intramembranous particles during membrane fusion have been described for milk fat secretion in the mammary gland (Peixoto de Menezes and Pinto da Silva, 1978), encystment in zoospores (Pinto da Silva and Nogueira, 1977), acrosome fusion in sperm (Friend *et al.*, 1977), and myoblast fusion (Kalderon and Gilula, 1979). Neutra and Schaeffer (1977) have observed extensive areas of membrane contact, poor in membrane proteins, between secretion granules in goblet cells.

More recent studies of secretion utilizing quick-freezing freeze-fracture electron microscopy have revealed, however, that the clearing of intramembranous particles may be a result of chemical fixation and treatment with glycerol before freezing. Segregation of intramembranous particles as a result of glycerination and slow freezing had been recognized earlier (Plattner et al., 1973; McIntyre et al., 1974). Chandler and Heuser (1979) observed small disks of membrane contact and multiple sites of fusion between cortical granules and the plasma membrane and between the granule membranes in sea urchin eggs fixed with glutaraldehyde and treated with glycerol; but such morphologies were not observed in quick-frozen cells. In stimulated mast cells these authors found single pores with narrow necks between secretory granules and the plasma membrane, usually with no large areas of intermembrane contact; also no clearing of intramembranous particles was observed between granule membranes. In studies of exocytosis in *Limulus* amebocytes, Ornberg and Reese (1981) have observed the apposition of secretory granule and plasma membrane, forming pentalaminar contacts, without any changes in the distribution of intramembranous particles. They have identified cytoplasmic filaments attaching the granule membrane to the plasmalemma, and the formation of small pores at the sites of fusion that subsequently widen. Cytoplasmic connections have also been described in the adrenal medulla between chromaffin granules and the plasma membrane (Aunis et al., 1979). It is not known whether these connections are involved in fusion or whether they have to be removed from the area of contact between the two membranes. Aunis et al. (1979) have suggested that contractile proteins may be involved in the redistribution of membrane proteins during exocytosis.

Fusion of synaptic vesicles with the presynaptic plasma membrane at the neuromuscular junction has been captured with quick-freezing during the release of transmitter (Heuser et al., 1979). Freeze-fracture images of the openings of synaptic vesicles into the synaptic cleft at the active zone can be observed as early as 3 msec after electrical stimulation, and the concentration of these openings reaches a maximum at about 5 msec (Heuser and Reese, 1981). The enlargement of the opening and the collapse of the vesicle structure by its merging with the plasma membrane occur faster than the complete intermixing of the membrane proteins of the vesicle and plasma membranes.

Secretion in the ciliated protozoan *Tetrahymena* takes place via the apposition of the mucocyst membrane, which has an annulus of intramembrane particles, with a special site on the plasma membrane consisting of a 60-nm-diameter rosette of 15-nm-diameter particles (Satir et al., 1972). The rosettes are reassembled for each new fusion event and are not permanent structures within the plasma membrane (Satir, 1976, 1977). The membranes undergo semifusion to form a membrane diaphragm, similar to observations with zoospores, and the particles are spread to the edge of the diaphragm, which extends to about 200 nm (Satir et al., 1972, 1973). It is considered likely that ion flow into the mucocyst induces the solubilization of the contents, leading to an increase in osmotic pressure and membrane expansion; fusion of the mucocyst and plasma membranes may then be caused by a local rupture of the semifused membranes (Satir et al., 1973; Satir, 1974). Such osmotic factors have also been proposed for exocytosis in *Limulus* amebocytes (Ornberg and Reese, 1981) and in chromaffin cells (Pollard et al., 1980, 1982).

Fusion rosettes similar to those found in *Tetrahymena* have been observed in *Paramecium* and correlated with trichocyst discharge by means of secretory mutants. These temperature-sensitive mutants secrete or assemble rosettes when grown at the permissive temperature but not when grown at the nonpermissive temperature (Beisson et al., 1976; Satir, 1976). Studies with other mutants have indicated that the attachment of the trichocyst to the plasma membrane and subsequent exocytosis depend on the properties of both the plasma membrane and the secretory vesicle membrane (Lefort-Tran et al., 1981). Using temperature-sensitive mutants, Satir and Oberg (1978) have shown that exocytosis can be initiated by Ca^{2+} and Ca^{2+} ionophores in the absence of fusion rosettes, and have suggested that the rosettes function as Ca^{2+} channels, causing an increase in the local Ca^{2+} concentration. In *Paramecium* fusion has also been shown by means of ultrathin-section electron microscopy (Hausman and Allen, 1976) to start with a narrow zone within the rosette region.

3.2. Studies with Permeabilized Cells and Ca^{2+} Indicators

The Ca^{2+} requirement of exocytosis has been demonstrated by several experimental systems. Douglas (1968) has shown that the microinjection of Ca^{2+} into mast cells induces massive release of histamine stored in secretory membranes. Injection of Ca^{2+} into the presynaptic neurone in the squid giant synapse stimulates the release of neurotransmitter (Miledi, 1973). In the same experimental system, Llinás and Nicholson (1975) have shown that depolarization of the presynaptic membrane, which causes the release of neurotransmitter, results in an increase in the free Ca^{2+} concentration as determined by light emission from aequorin previously injected into the presynaptic terminal. Chandler and Williams (1977) have demonstrated using chlortetracyline fluorescence that Ca^{2+} is released from intracellular membranes during stimulus-secretion coupling in pancreatic acinar cells. In *Paramecium* an increase in free $[Ca^{2+}]$ induces trichocyst discharge, which involves fusion of secretory vesicles with the plasma membrane (Plattner, 1974). In this system the concentration of intracellular free $[Ca^{2+}]$ required for exocytosis has been estimated to be between 10^{-4} and 10^{-6} M (Matt et al., 1978). Determination of the actual $[Ca^{2+}]$ during exocytosis is not a trivial matter, because of the Ca^{2+} buffering system in cells and the possibility that highly localized increases in $[Ca^{2+}]$ are sufficient to induce fusion between the secretory vesicles and the plasma membrane. Therefore, even intracellular Ca^{2+} electrodes (O'Doherty et al., (1980) may not be able to provide an unequivocal answer to this problem.

Using the Ca^{2+} indicator Quin 2, Rink et al. (1982) have found that the

Ca^{2+} threshold for secretion in platelets permeabilized with the Ca^{2+} inophore ionomycin is in the micromolar range. They have also reported that thrombin, diacylglycerol, and phorbol ester stimulate secretion without changing the basal Ca^{2+} levels in the cytoplasm (Rink et al., 1982, 1983). These results are intriguing in view of the role of Ca^{2+} in a wide range of secretory phenomena.

An alternative method has been developed by Baker and co-workers, who introduced various Ca^{2+} buffers into sea urchin eggs, platelets, and adrenal medullary cells by rendering their membranes leaky to low-molecular-weight substances with short high-voltage discharges (Baker and Knight, 1978; Baker and Whitaker, 1978; Baker et al., 1980). Catecholamine secretion from isolated adrenal medullary cells has been shown to be half-maximal at about 1 μM Ca^{2+} (Baker and Knight, 1978; Figure 1, triangles). Studies on adrenal medullary cells permeabilized by digitonin have indicated a similar requirement for Ca^{2+} in catecholamine release (Wilson and Kirschner, 1983; Dunn and Holz, 1983). A similar Ca^{2+} dependence has been found for the fusion of cortical granules with the plasma membrane of the sea urchin egg in the presence of ATP (Baker and Whitaker, 1978). Baker and Knight (1978) have also reported that in the absence of ATP adrenal medullary cells become refractory to Ca^{2+}-induced exocytosis. These investigators have recently found that the Ca^{2+}-activation curve shifts to lower Ca^{2+} concentrations in the presence of the phorbol ester, tetradecanoylphorbol acetate, suggesting a role of protein kinase C in exocytosis (Knight and Baker, 1983).

These observations indicate that molecular mechanisms underlying exocytotic membrane fusion must account for the high sensitivity in secretory cells to Ca^{2+} concentrations in the micromolar range. In the following section (3.3) we shall examine the sensitivity of isolated secretory vesicles to Ca^{2+} with respect to fusion and try to obtain clues as to the molecular targets of Ca^{2+} in the membranes.

3.3. Fusion of Isolated Secretory Vesicles and Other Intracellular Membranes

One form of exocytosis called "compound exocytosis" involves the tandem fusion of secretory vesicles with one another as well as the fusion of the plasma membrane with the proximal vesicle in the set of fusing vesicles. This may be contrasted to "simple exocytosis," where a single secretory vesicle is involved (Douglas, 1974). Compound exocytosis has been observed in mast cells (Rohlich et al., 1971), parotid acinar cells (Amsterdam et al., 1969), adenohypophysis (de Virgilis et al., 1968), and cells of the exocrine and endocrine pancreas (Palade, 1959; Ekholm et al., 1962; Berger et al., 1975). The fusion of isolated secretory vesicles from a variety of cell types has been studied extensively to provide insight into the mechanism of membrane fusion events taking place in compound exocytosis, and presumably also in simple exocytosis (Gratzl et al.,

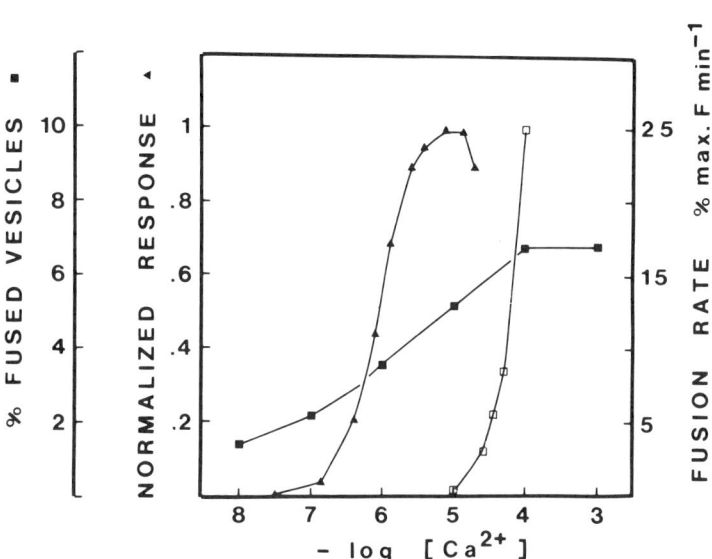

FIGURE 1. Ca^{2+} concentration dependence of exocytosis in leaky adrenal medullary cells (triangles; data from Baker et al., 1980), fusion of secretory vesicles isolated from adrenal medulla (solid squares; data from Dahl et al., 1979), and fusion of phospholipid vesicles composed of phosphatidate–phosphatidylethanolamine in the presence of synexin and Mg^{2+} (open squares; data from Hong et al., 1982a).

1980). These studies have shown that the Ca^{2+} specificity of membrane fusion can be explained by its action at the membrane level and that this action is presumably mediated through glycolipids or membrane glycoproteins (Dahl et al., 1979). Secretory vesicles isolated from rat liver (Gratzl and Dahl, 1976, 1978), islets of Langerhans (Dahl and Gratzl, 1976), adrenal medulla (Dahl et al., 1979; Ekerdt et al., 1981), and neurohypophysis (Gratzl et al., 1977) form twinned structures, as observed by freeze-fracture electron microscopy, when the Ca^{2+} concentration in the medium is raised. The fracture plane is continuous through the membranes of both vesicles; this observation has been interpreted convincingly as an indication of the fusion of the vesicles (Gratzl and Dahl, 1978). Half-maximal fusion is obtained in the range of 10^{-6} M Ca^{2+} for all these secretory vesicle types (Figure 1, filled squares). Other divalent cations, such as Ba^{2+}, Sr^{2+}, and Mg^{2+}, do not induce fusion above control levels. In fact, Mg^{2+}, when added together with Ca^{2+}, inhibits Ca^{2+}-induced fusion (Gratzl et al., 1977; Gratzl and Dahl, 1978). The fusion of secretory vesicles isolated from liver or adrenal medulla is sensitive to the enzymatic treatment of the membrane prior to the addition of Ca^{2+}. Neuraminidase, trypsin, and pronase all inhibit the extent of fusion, depending on the concentration of enzyme utilized (Gratzl and Dahl, 1978; Ekerdt et al., 1981). These results suggest that glycolipids or glycoproteins in the membrane of the vesicles are involved in the high sensitivity of the fusion reaction to Ca^{2+}. At Ca^{2+} concentrations above 10^{-3} M, a second type of fusion is observed, characterized by multiple fusions between several vesicles as opposed to the twinned vesicles found at lower Ca^{2+} concentrations (Dahl et al., 1979; Ekerdt et al., 1981). This type of fusion has a Ca^{2+} dependence similar to the fusion of sonicated liposomes prepared from extracted membrane lipids. It is interesting to note that treatment of the secretory vesicles by neuraminidase confers onto them a sensitivity to Ca^{2+} similar to that of the liposomes (Ekerdt et al., 1981).

Huber et al. (1979) and Abbs and Phillips (1980) have presented evidence that the major glycoproteins of chromaffin granules are oriented in the membrane such that the carbohydrate residues face the inside of the vesicle. Although it would appear from this finding that the substrate of neuraminidase is more likely to be gangliosides than glycoproteins, the addition of ganglioside G_{M3}, which is the major glycolipid in these membranes (Dreyfus et al., 1977), to liposomes does not confer an altered sensitivity with respect to Ca^{2+}-induced fusion (Ekerdt et al., 1981). Other workers have suggested, however, that glycoproteins are exposed on the exterior surface of the chromaffin granule membrane (Meyer and Burger, 1976). Synaptin, a membrane protein specific for the nervous system (Bock et al., 1975; Bock, 1978) is located on the cytoplasmic side of chromaffin granule membranes and of synaptosomal plasma membranes (Bock and Helle, 1977). Meyer and Burger (1979a) have also identified a protein from the plasma membrane of the adrenal medulla that binds secretory vesicles and have proposed that it anchors the chromaffin granule to the plasma membrane during exocytosis. Izumi et al. (1975, 1977) have found a "catecholamine-releasing factor" that appears to be a peripheral membrane protein and depends on the presence of Mg-ADP and have suggested that this factor may be involved in the exocytosis of the granule contents. Other proteins that have been shown to be associated with chromaffin granules include α-actinin (Jockusch et al., 1977) and actin (Burridge and Phillips, 1975; Meyer and Burger, 1979b; Gabbiani et al., 1976). Friedman et al. (1980) have suggested that depolymerization of actin inhibits stimulus-secretion coupling in adrenal medullary cells. The interaction between chromaffin granule membranes and actin is inhibited by increasing the Ca^{2+} concentration above 10^{-7} M and has been proposed to be involved in the movement of the granules to the plasma membrane (Fowler and Pollard, 1982a). Other factors, such as pH, chloride ions, and nucleotides, also affect this interaction and may control the localization of the granules in the cytoplasm via the cytoskeleton (Fowler and Pollard, 1982b). Binding sites for Ca^{2+} and Mg^{2+} have been demonstrated on the outer surface of chromaffin granule membranes (Morris and Schober, 1977; Morris et al., 1979a; Ekerdt et al., 1981). Ekerdt et al. (1981)

have found two binding sites for Ca^{2+} and have concluded that the low-affinity ($K_d = 1.2 \times 10^{-4}$ M) binding site is on the phospholipids of the membrane and the high-affinity ($K_d = 1.6 \times 10^{-6}$ M) site is on the proteins.

Chromaffin granule membranes aggregate in the presence of millimolar concentrations of Ca^{2+} or Mg^{2+} (Banks, 1966; Schober et al., 1977; Morris et al., 1979a,b; Edwards et al. 1974). Both these ions are effective to the same degree in reducing the electrophoretic mobility of the membranes (Siegel et al., 1978). Aggregation of the granules with 10 mM Ca^{2+} results in contact sites with a pentalaminar membrane structure from which the intramembranous particles are excluded (Schober et al., 1977). This process is reversible by EDTA, which also causes disaggregation of the vesicles. It is difficult to reconcile this observation with that of Ekerdt et al. (1981), who reported that these membranes undergo fusion in the presence of even 10^{-4} M Ca^{2+} and more extensive fusion at 10^{-2} M. Chromaffin granules incubated with sonicated vesicles composed of phosphatidylethanolamine–phosphatidylserine (3:1), cardiolipin, phosphatidylserine, or lipids of the inner monolayer of erythrocytes, release their contents into the medium to an extent that depends on the granule-to-phospholipid vesicle ratio as well as on the Ca^{2+} concentration in the millimolar range (Nayar et al., 1982). This effect has been attributed to the fusion of the phospholipid vesicles with the granule membrane. Incubation of the granules with a plasma membrane preparation also leads to the release of catecholamines, ATP, and dopamine β-hydoxylase from the granules in the presence of approximately 3 μM Ca^{2+} and can be considered as a simple model for exocytosis (Konings and DePotter, 1981). This release appears to depend on sialic acid-containing molecules in the membranes and suggests that the surface carbohydrates of chromaffin granules could act as substrates for lectin-like molecules on the plasma membrane (Konings and DePotter, 1982; Meyer and Burger, 1976). In this regard, soluble lectins have been shown to facilitate the Ca^{2+}-induced fusion of phospholipid vesicles containing glycolipids (Sundler and Wijkander, 1983; Düzgüneş et al., 1984b; Hoekstra et al., 1985).

A cytoplasmic protein isolated from the adrenal medulla, synexin, reduces the threshold Ca^{2+} concentration required to initiate chromaffin granule aggregation (Creutz et al., 1978). This action of synexin is Ca^{2+}-specific, with a dissociation constant of 2×10^{-4} M. Electron microscopy has shown that the vesicles form pentalaminar structures at their sites of contact, but do not fuse with one another. Synexin also binds to the cytoplasmic surface of chromaffin cell plasma membranes in a Ca^{2+}-dependent manner (Pollard et al., 1980). Other proteins that exhibit Ca^{2+}-dependent binding to chromaffin granule membranes include calmodulin (Burgoyne and Geisow, 1981; Creutz et al., 1983), chromobindins (Creutz, 1981b; Creutz et al., 1983), and calelectrins (Südhof et al., 1984).

Synaptic vesicles have also been studied in detail to understand the molecular mechanisms of exocytosis at synapses. Synaptic vesicles of high purity have been obtained from a variety of sources (Carlson et al., 1978; Eichberg et al., 1964; Ohsawa et al., 1981). Ca^{2+}-induced aggregation of synaptic vesicles from *Torpedo* has been demonstrated (Haynes et al., 1979); however, vesicles from *Narcine* do not aggregate in the presence of Ca^{2+} (Deutsch and Kelly, 1981). It should be noted, however, that the former study was performed with vesicles less pure than in the latter, with respect to acetylcholine–protein ratio. It is also possible that some water-soluble and/or peripheral membrane proteins, which may facilitate aggregation, are lost during the purification procedure. Under physiological conditions synaptic vesicles are negatively charged (Vos et al., 1968; Ryan et al., 1971; Carlson et al., 1978; Wagner et al., 1978; Ohsawa et al., 1981); this charge is not removed by neuraminidase treatment (Vos et al., 1968) and is reduced by Ca^{2+} at millimolar concentrations (Ohsawa et al., 1981). Binding constants for Ca^{2+} and Mg^{2+} to synaptic vesicle membranes have been determined to be in the millimolar range (Haynes et al., 1979; Hoss et al., 1980). Wagner and Kelly (1979) have shown that the proteins of synaptic vesicle membranes from *Narcine* are distributed asymmetrically across the vesicle membrane and that an actin-like protein is on the exoplasmic side (interior) of the vesicle. If the latter finding can be extrapolated to vesicles from brain, it would imply that actin- and myosin-like proteins are not necessarily involved in exocytosis as had been proposed by Berl et al. (1973). DeLorenzo and Freedman (1977) have found a Ca^{2+}-dependent phosphorylation of brain synaptic vesicle proteins and have suggested that this process is involved in the release of neurotransmitter. Moskowitz et al. (1982) have reported the aggregation of brain synaptic vesicles induced by a Ca^{2+}-dependent phospholipase A_2. *Torpedo* synaptic vesicles release acetylcholine in the presence of soluble, presumably proteinaceous, factors from the cytoplasm and Ca^{2+}; half-maximal release is obtained at 50 μM Ca^{2+} (Michaelson et al., 1978). Release of neurotransmitter from isolated synaptic vesicles into the medium in a Ca^{2+}-dependent manner may not be a physiologically relevant phenomenon, since the contents of the vesicles should be released to the exterior of the cell after fusion with the plasma membrane. Nevertheless, it may reflect a destabilization of the membrane by the soluble factors and Ca^{2+} and thus may have some implications for the mechanism of the interaction and fusion of the vesicles with the presynaptic membrane.

David and Lazarus (1976) have described an *in vitro* system for studying insulin release in which the interaction of secretory granules with a plasma membrane preparation is induced by $2 \times 10^{-6} M$ Ca^{2+} and is augmented by ATP. The presence of glucose in addition to Ca^{2+} and ATP enhances greatly the release of insulin. Similarly, Milutinovic et al. (1977) have shown a Ca^{2+} dependent interaction between isolated zymogen granules from the pancreas and a plasma membrane fraction, with half-maximal binding obtaining at 6.5

$\times$ 10^{-6} M Ca^{2+}. The binding is abolished by the trypsinization of the plasma membrane, which suggests that membrane proteins are involved in this process. A similar requirement for membrane proteins has been shown for the fusion of Golgi membranes with plasma membranes isolated from maize root tips (Baydoun and Northcote, 1980). In a different type of preparation, isolated cortical granules already bound to the plasma membrane of sea urchin eggs, Vacquier (1975) has shown that the granules fuse with one another and discharge their contents in the presence of Ca^{2+}. This discharge is not influenced by treatments with colchicine or cytochalasin B or the presence of ATP. Vacquier (1975) has suggested that cortical granule discharge propagates as granules release Ca^{2+}, which initiates further fusion events. Whitaker and Baker (1983) have shown that cortical granules exocytose at micromolar Ca^{2+} concentrations, that Mg^{2+} decreases the Ca^{2+} sensitivity and that calmodulin may be involved in this process. They have also concluded that actin and tubulin are not likely to participate in exocytosis. Steinhardt and Alderton (1982) have shown that an antibody to calmodulin inhibits exocytosis in this system.

Other *in vitro* systems that have been developed as models for the fusion of intracellular membranes include isolated phagolysomes, lysosomes, Golgi-derived vesicles, and coated vesicles. Isolated phagolysosomes undergo fusion in a medium consisting of a cytoplasmic homogenate and Tris buffer (Oates and Touster, 1976). These vacuoles can fuse at multiple sites where effective intermembrane contact is established in a time- and temperature-dependent manner. It is not clear at this time which cytoplasmic factor controls this fusion process. Further studies with this system have indicated that fusion is strongly inhibited by millimolar concentrations of KF and ten-fold higher concentrations of KCl, whereas KCN and dinitrophenol are not as effective (Oates and Touster, 1978). Lysosomes isolated from liver have been shown to fuse without external requirements in the homogenization medium (Raz and Goldman, 1974). Vesicles derived from the Golgi fraction of liver cells also undergo fusion: in this system the Ca^{2+} requirement has been demonstrated clearly (Gratzl and Dahl, 1976; Quinn and Judah, 1978; Judah and Quinn, 1978). Fusion of clathrin-coated vesicles with lysosomes has been demonstrated by means of a fluorescence assay measuring the hydrolsis by lysosomal esterase of carboxydiacetylfluorescein encapsulated in the coated vesicles (Altstiel and Branton, 1983). Fusion requires the removal of the coat protein from the latter and free Ca^{2+} in the medium and is not dependent on proteins of the vesicle surface. A sharp increase in the rate of fusion occurs between 10^{-6} M and 10^{-5} M Ca^{2+} in the presence of 2.5 mM Mg^{2+}. Altstiel and Branton (1983) have suggested that Ca^{2+} may be mediating fusion between the phospholipid bilayers of the lysosomes and uncoated vesicles, although the question of whether lysosomal membrane proteins are involved in this fusion process has not yet been investigated. They and Unanue *et al.* (1981) have observed that coated vesicles stripped of their coat protein fuse with each other, much as uncoated vesicles *in vivo* fuse to form endosomes (Anderson *et al.,* 1977; Brown *et al.,* 1983). The *in vivo* factors that regulate the uncoating of the vesicles, their targeting to one another or to lysosomes, and their fusion is largely unknown. Patzer *et al.* (1982) have found that cytoplasmic factors and ATP regulate the uncoating process.

3.4. Fusion of Erythrocytes Induced by Chemicals

In this section and the following two we consider two examples of cell–cell fusion, the fusion of myoblasts to form myotubes and the fusion of erythocytes by various chemical fusogens and viruses. These systems have provided considerable insight into the molecular events accompanying or preceding membrane fusion. Cell–cell fusion represents the "planar fusion" between external membrane leaflets described by Stossel *et al.* (1978).

Erythrocyte membranes fuse with one another in the presence of various lipid-soluble or water-soluble substances termed *fusogens*. These include lauric, myristic, and oleic acids, glycerylmonooleate, retinol, α-tocopherol, and lysolecithin among the lipid-soluble substances and poly(ethylene glycol), dimethylsulfoxide, and glycerol among the water-soluble ones (for reviews see Lucy, 1977; 1978). Lucy (1970) has suggested that fusion may involve the formation of micellar units of lipid and lipoprotein, whose formation would be induced by agents such as lysolecithin or retinol (Poole *et al.,* 1970; Ahkong *et al.,* 1973a). The interdigitation of these micelles in two apposed membranes could then lead to the continuity of the membranes once the conditions leading to micellization were removed and the membranes were allowed to revert to a bilayer configuration. Although lysolecithin has been shown to form micellar structures in liposomes of phosphatidylcholine, phosphatidylserine, sphingomyelin, and extracted red cell lipids, other fusogenic molecules do not give rise to such structures (Howell *et al.,* 1973), implying that micelle formation is not a prerequisite to fusogenic activity. Cullis and Hope (1978) have proposed that membrane fusion in erythrocytes and erythrocye ghosts induced by oleic acid proceeds via the formation of inverse micelles in the zone of contact. They have presented [^{31}P]NMR and freeze-fracture electron microscopic evidence for the hexagonal H_{II} phase (Luzzati *et al.,* 1968; Reiss-Husson, 1967) in erythrocyte ghosts containing about 1 mol oleic acid per mole of phospholipid. Since cell fusion can be induced by raising the temperature, as well as by treatment with glycerylmonooleate, Ahkong *et al.* (1973b) have suggested that an increase in membrane fluidity may be sufficient to induce membrane fusion, but they have also indicated that increased membrane fluidity may lead to other changes in membrane structure or to the facilitation of the activity of phospholipases.

Based on the observations that dimethylsulfoxide and glycerol, agents that cause lateral aggregation of intramembranous particles in lymphocytes and

Entameba histolytica (McIntyre *et al.*, 1974; Pinto da Silva and Martinez-Palomo, 1974), also induce the fusion of hen erythrocytes, Ahkong *et al.* (1975a) have proposed that these fusogens increase the fluidity of the lipid bilayer, resulting in protein clustering, which allows the close apposition of the lipid regions and their eventual intermixing. It should be noted that fusion is observed only after 25–30 min, when cells are treated with dimethylsulphoxide, and is a very slow process compared to the fusion events in exocytosis. Ahkong *et al.* (1978) have also demonstrated that proteinases induce cell fusion in the presence of Ca^{2+}, possibly by degrading band 3, the anion channel of erythrocyte membranes. Further support for the involvement of membrane protein aggregation and the resulting protein-free areas in cell–cell fusion has been provided by experiments utilizing other cell fusion conditions, such as the treatment of hen erythrocytes with Ca^{2+} and Ca^{2+} ionophores (Ahkong *et al.*, 1975b; Vos *et al.*, 1976) and of human erythrocytes with Ca^{2+} phosphate (Zakai *et al.*, 1976, 1977; Baker and Kalra, 1979; Majumdar and Baker, 1980; Hoekstra *et al.*, 1983). The fusion of human erythrocytes by uranyl acetate and rare earth metals (Majumdar *et al.*, 1980) and of hen erythrocytes by Ca^{2+} at high pH has also been reported (Toister and Loyter, 1971); uranyl acetate treatment has been shown to result in the redistribution of membrane-associated particles, as in the case of the other fusogens mentioned earlier.

Poly(ethylene glycol) is an effective fusogen for cultured mammalian cells (Pontecorvo, 1975; Davidson *et al.*, 1976; Pontecorvo *et al.*, 1977), for plant protoplasts (Kao and Kichayluk, 1974; Wallin *et al.*, 1974) and for the fusion of human cells with plant protoplasts (Jones *et al.*, 1976) and hen erythrocytes with yeast protoplasts (Ahkong *et al.*, 1975c). Knutton (1979) has studied the mechanism of poly(ethylene glycol)-induced fusion of human erythrocytes by freeze-fracture electron microscopy. In the regions of adhesion between cells, intramembranous particles aggregate into large patches. Removal of most of the polyethylene glycol leads to membrane fusion at particle-free areas of about 0.1 μm diameter; the fusion sites constitute cytoplasmic bridges between the cells, which eventually swell to form a spherical fusion product. Wojcieszyn *et al.* (1983) have found that no water-soluble proteins are transferred from the interior of erythrocytes to cultured mammalian cells until the poly(ethylene glycol) is removed.

Poly(ethylene glycol) induces the entry of Ca^{2+} into the cells, as do other chemical fusogens, although Ca^{2+} entry by itself is not sufficient to induce fusion (Blow *et al.*, 1978, 1979). It induces a decrease in the surface potential of phospholipids (Maggio *et al.*, 1976), an increase in their gel to liquid–crystalline phase transition temperature (Tilcock and Fisher, 1979; Herrmann *et al.*, 1983), dehydration of the polar head group of phospholipids (Arnold *et al.*, 1982), and a decrease in polarity of the aqueous phase (Herrmann *et al.*, 1983). Blow *et al.* (1978) have proposed that poly(ethylene glycol) induces the fusion of fibroblasts by altering the water of hydration of the phospholipids or the physical state of the water adjacent to the cell surface. Poly(ethylene glycol) induces the aggregation of small unilamellar phosphatidylcholine vesicles at low concentrations (2.5% w/w) and their fusion at high concentrations (35%), (Boni et al., 1981a) and the formation of structural defects in multilamellar phosphatidylcholine vesicles as detected by freeze-fracture electron microscopy and [^{31}P]NMR (Boni et al., 1981b). Honda et al. (1981) have found that [^{31}P] commercial poly(ethylene glycol) has two components: the polymer itself induces cell aggregation or the close apposition of bilayers (Wojcieszyn et al., 1983), and certain impurities cause the perturbation of the bilayer leading to membrane fusion. In contrast, Smith et al. (1982) have shown that purified poly(ethylene glycol) from four other commercial sources induces cell fusion under appropriate conditions.

When human erythrocytes are treated with Ca^{2+} and the Ca^{2+} ionophore A23187, microvesicles bud off from their membranes. The diacylglycerol content of the cell membrane increases as a result of this treatment. (Allan and Michell, 1975), and the microvesicles are enriched in this lipid (Allan et al., 1976). Since diacylglycerol is a fusogenic lipid (Ahkong et al., 1973a,b), its production in the inner monolayer of erythrocytes could lead to the fusion of the cytoplasmic surfaces with each other, after the contraction of the area of the monolayer and formation of an outward bulge in the membrane (Allan et al., 1976). Microvesiculation is an example of the "tubular fusion" between protoplasmic membrane surfaces in the nomenclature of Stossel et al. (1978). Consistent with the possible role of diacylglycerol in membrane fusion involved in outward vesiculation, treatment of cells with phospholipase C, which produces diacylglycerol or ceramide in the outer monolayer, lead to inward vesiculation (Allan et al., 1975; Allan and Michell, 1977). Allan et al. (1976) have found no cell–cell fusion or internal vesiculation with human erythrocytes treated with Ca^{2+} and Ca^{2+} ionophore, suggesting that diacylglycerol is produced in the inner monolayer and that equilibration of the fusogen between both monolayers on raising the temperature or prolonged incubation could lead to cell fusion as observed by Ahkong et al. (1975b) for hen erythrocytes. It should be noted that the latter investigators have used dextran to aggregate the cells. Based on the preceding observations, Allan and Michell (1979) have proposed a molecular model for the control of membrane fusion in secretory processes, in which a Ca^{2+}-activated phospholipase C produces diacylglycerol from polyphosphoinositides, which in turn causes membrane fusion.

3.5. Fusion of Myoblasts

Cell–cell fusion in the case of myoblasts is a biologically controlled phenomenon that occurs during the development of the embryo, resulting in the formation of myotubes. Fusion proceeds via stages of acquisition of fusion competence by myoblasts, cell–cell recognition and contact, apposition of plasma membranes to a distance of a few nanometers, and rearrangement of lipids and proteins in the membrane at a restricted site (Knudsen and Horwitz, 1977; Bischoff, 1978).

Myoblast fusion requires the presence in the medium of Ca^{2+} concentrations in the millimolar range (Shainberg et al., 1969; van der Bosch et al., 1973; Cox and Gunter, 1973; Morris et al., 1976). The entry of Ca^{2+} into the cell appears to be necessary for fusion, since a net Ca^{2+} influx occurs prior to fusion and a Ca^{2+} channel blocker inhibits fusion (David et al., 1981). Addition of the Ca^{2+} ionophore A23187 into the medium causes an earlier onset of fusion, although the final extent of fusion is not altered (David et al., 1981). Schudt and Pette (1975) have found, however, that at lower ionophore concentrations the rate of fusion is not altered. Plasma membranes isolated from myoblasts undergo fusion in a Ca^{2+}-specific fashion at physiological divalent cation concentrations (Schudt et al., 1976), but at higher concentrations Mg^{2+} and Sr^{2+} also induce fusion (Dahl et al., 1978). In contrast, high concentrations of Mg^{2+} are inhibitory to the fusion of myoblasts (Schudt et al., 1973).

Myoblast fusion is inhibited by incubation of the cells with lysolecithin (Reporter and Norris, 1973), dipalmitoylphosphatidylcholine (van der Bosch et al., 1973; Sandra, 1980), cholesterol (van der Bosch et al., 1973), fibronectin (Podleski et al., 1979), lectins (Den et al., 1975; Sandra et al., 1977), and phospholipase (Nameroff et al., 1973; Schudt and Pette, 1976). Specific inhibitors such as 5-bromodeoxyuridine (Stockdale et al., 1964), cycloheximide (Kalderon et al., 1977; Yaffe and Dym, 1972), colchicine (Bischoff and Holtzer, 1968), cytochalasin B (Sanger et al., 1971; Croop and Holtzer, 1975), and inhibitors of metalloproteases (Couch and Strittmatter, 1983) inhibit myotube formation. It is important to distinguish which step in the overall fusion process is affected by these agents. Knudsen and Horwitz (1978) have found that inhibitors of protein synthesis, trypsin, or glutaraldehyde treatment inhibit the aggregation of myoblasts mediated by Ca^{2+}, whereas inhibitors of cytoskeletal function or enrichment of the fatty acyl chains in elaidate block the irreversible adhesion or membrane union step. Inhibition of cholesterol synthesis also inhibits the aggregation activity without altering the sterol–phospholipid ratio in the cells, suggesting that cholesterol may be required locally for this activity, such as in the maintenance or insertion of recognition molecules into the plasma membrane (Cornell et al., 1980). Thus, Horwitz and collaborators have proposed that recognition or aggregation is mediated by proteins and membrane union involves membrane lipid (Knudsen and Horwitz, 1978; Horwitz et al., 1978, 1979). Prives and Shinitzky (1977) and Elson and Yguerabide (1979) have shown that the microviscosity of myoblast membranes decreases rapidly before the onset of fusion. Herman and Fernandez (1978) have also demonstrated that the fluidity of the membrane increases before fusion and that areas of contact between cells have a higher fluidity than other regions of the cell membrane. These changes in fluidity are probably not due to gross alterations in the lipid composition of the membrane since the lipid composition of plasma membranes from myoblasts, fusing cells, and myotubes are found to be similar (Kent et al., 1974). In accordance with the suggestion of the involvement of lipids in muscle cell fusion, a larger percentage of phosphatidylserine and phosphatidylethanolamine is located in the outer monolayer of myoblasts (Horwitz et al., 1982; Sessions and Horwitz, 1981, 1983) compared to other mammalian cell membranes that do not fuse under physiological conditions (Op den Kamp, 1979; Sandra and Pagano, 1978). Membranes composed of these aminophospholipids are known to undergo fusion in the presence of Ca^{2+} (Düzgüneş et al., 1980, 1985a). Sessions and Horwitz (1983) have pointed out, however, that the preferential localization of the aminophospholipids in the external leaflet of myoblasts does not itself confer fusion competence to the cells, since this configuration of the phospholipids is observed beyond the fusion-susceptible period of the myoblasts and in cells treated with 5-bromodeoxyuridine, which do not fuse. Further evidence for the possible involvement of phospholipids in myoblast fusion has been obtained by electron microscopy (Kalderon and Gilula, 1979; Kalderon, 1980). Adjacent cells first form a single bilayer, which appears to be free of intramembranous particles in freeze-fracture replicas and which eventually destabilizes to allow for cytoplasmic continuity between the cells. Unilamellar, particle-free vesicles have been observed adjacent to the particle-free plasma membrane domains; these vesicles are absent in cells that are not fusion competent, that is, cells treated with 5-bromodeoxyuridine, cycloheximide, or phospholipase C. Kalderon and Gilula (1979) have proposed that these vesicles adhere to the inner monolayer of the plasma membrane, induce the removal of proteins from the region of contact, leading to the destabilization of the plasma membrane, and fuse with the latter in these particle-free regions. These fusogenic vesicles may insert certain phospholipids into the plasma membrane, which may be necessary for the fusion reaction.

A variety of proteins are involved in the adhesion and fusion of myoblasts. Inhibition of fusion by plant lectins (Den et al., 1975; Sandra et al., 1977) and by tunicamycin, an inhibitor of glycosylation (Gilfix and Sanwal, 1980; Olden et al., 1981), suggests that (mannosylated) glycoproteins are required for fusion. Studies with myoblast cell lines resistant to lectins have indicated that sialic acid or galactose on surface glycoproteins is not obligatory for myotube formation (Gilfix and Sanwal, 1982) but that mannosylated glycoproteins are involved in this process (Parfett et al., 1981, 1983). Receptors for concanavalin A are dispersed during myoblast fusion and clustered before and after fusion activity or in fusion-inhibited cells (Sandra et al., 1977; Fernandez and Herman, 1982). The distribution of fibronectin over the cell surface also changes (Chen, 1977; Furcht et al., 1978; Gardner and Fambrough, 1983). Phosphoproteins of the plasma membrane change during myotube formation (Sénéchal et al., 1982), a 70-kD membrane protein is synthesized at the onset of myoblast fusion (Cates and Holland, 1978), and various plasma membrane proteins are transformed into lower-molecular-weight species (Moss *et al.,* 1978; Pauw and David, 1979). Schubert and LaCorbiere (1980, 1982) have identified glycoprotein complexes, called adherons, which mediate Ca^{2+}-dependent adhesion between myoblasts. Strittmatter *et al.* (1982) and Couch and Strittmatter (1983) have shown that a metalloendoprotease is required for myoblast fusion and have suggested that the enzyme may participate in posttranslational modification of plasma membrane proteins to generate a fusogenic peptide. Gartner and Podleski (1975) and Nowak *et al.,* 1976, have suggested that a membrane-bound lectin mediates fusion of myoblasts of an established rat cell line, L_6 or chick myoblasts. However, the observations that incubation of chick myoblasts (Den *et al.,* 1976) or L_8 myoblasts (Kaufman and Lawless, 1980) with the haptenic saccharide specific for the lectin does not inhibit fusion and that most of the lectin is intracellular do not support a specific role for this lectin in membrane fusion. Schudt and Pette (1976) have also found no inhibitory effects of various monosaccharides. In contrast, Podleski and Greenberg (1980) have reported that a lectin, termed *electrolectin,* appears to be on the surface of the L_6 cells or primary muscle cultures, and MacBride and Przybylski (1980) have demonstrated that purified lectin inhibited myotube formation when precautions were taken to prevent its inactivation in the medium. Thus, the role of electrolectin in myoblast fusion is still controversial (Barondes, 1981; Kaufman, 1982).

3.6. Virus–Cell Fusion

Enveloped animal viruses enter their host cell by fusing with the plasma membrane or with the membrane of the endocytotic vesicle in which they are internalized (Choppin and Compans, 1975; Bächi *et al.,* 1977; Poste and Pasternak, 1978; Helenius *et al.,* 1980; Marsh, 1984), and also induce cell–cell fusion (Poste, 1972; Hosaka and Shimizu, 1977; Poste and Pasternak, 1978; Papahadjopoulos *et al.,* 1979; Knutton and Pasternak, 1979). The fusion activity of these viruses is mediated through certain proteins of the viral envelope (Poste and Pasternak, 1978; Scheid *et al.,* 1980; White *et al.,* 1983) and thus provides a useful model system in which to identify the role of proteins in membrane fusion. Interaction of viruses with cells has also provided insights into the biosynthesis and transport of glycoproteins in the cells, the organization of plasma membrane phospholipid (Patzer *et al.,* 1979), and the nature of the endocytotic pathway (Helenius and Marsh, 1982).

Sendai virus (hemagglutinating virus of Japan) is a member of the paramyxovirus family, and its fusion with cell membranes and liposomes has been studied extensively. Fusion with cell membranes appears to be accompanied by the redistribution of intramembranous particles in some cases (Bächi *et al.,* 1977; Asano and Sekiguchi, 1978; Kim and Okada, 1981) but not in others

(Knutton, 1977b; Pinto da Silva et al., 1980). Kim and Okada (1981) have pointed out that the redistribution may be a transient event occurring at the early stages of fusion. Fusion is accompanied by the transfer of phospholipids (Kuroda et al., 1980; Lyles and Landsberger, 1979) and glycoproteins (Knutton, 1977b; Poste et al., 1980) into the cell membrane. Inhibition of the mobility of membrane proteins by lectins or antibodies appears to inhibit virus-induced hemolysis (Bächi et al., 1978) and fusion (Sekiguchi and Asano, 1978), presumably because of the inability of proteins to be displaced from the region of intermembrane contact. Removal of the extrinsic membrane protein spectrin from the erythrocyte membrane as a result of the phosphoprotein phosphatase activity of Sendai virus has been suggested as a possible mechanism by which lateral clustering of membrane proteins becomes possible (Loyter and Lalazar, 1980). Knutton (1977b, 1979b) has found that incubation of the virus with cells causes a drastic change in the structure of the virus, producing invaginations of the viral envelope as smooth linear ridges that may be the sites of fusion with the cell membrane. Other investigators, however, have not observed these changes (Bächi et al., 1973, 1977; Shimizu et al., 1976; Pinto da Silva et al., 1980). Small cytoplasmic bridges are formed between erythrocytes treated with Sendai virus (Knutton, 1979c; Pinto da Silva et al., 1980), and cell swelling induced by the virus appears to provide the driving force for the incorporation of the viral envelope into the cell membrane and for expanding the cells to form a spherical fusion product (Knutton, 1977a; 1979c; Sekiguchi et al., 1981). Cell swelling results from the increase in permeability of the cell membrane to small molecules caused by the virus in the absence of Ca^{2+} (Pasternak and Micklem, 1973; Wyke et al., 1980; Fuchs et al., 1980; Poste and Pasternak, 1978). The nonhemolytic version of Sendai virus (Homma et al., 1976) fuses with the cell membrane and forms a bridge between cells, but does not induce permeabilization of the membrane, cell swelling, or hemolysis (Knutton, 1979c).

Sendai virus can also fuse with liposomes containing gangliosides (Haywood, 1974; Haywood and Boyer, 1981, 1982) or glycophorin (Oku et al., 1982) as receptors and induce an increase in the permeability of the membrane with a temperature coefficient similar to virus-induced hemolysis (Oku et al., 1982). No Ca^{2+} is required for fusion in this system. Sendai virus is believed to bind to sialic acid residues on glycoproteins of the cell surface (Bächi et al., 1977; however, glycolipids have also been proposed as Sendai virus receptors on host cells (Markwell et al., 1981; Huang, 1983a). Haywood and Boyer (1982) have concluded that the viral proteins and not host membrane proteins determine the optimal conditions for fusion.

The Sendai virus envelope has two glycoproteins, designated HN and F. HN is associated with hemagglutinating and neuraminidase activities (Homma and Ohuchi, 1973; Scheid and Choppin, 1974). The F protein is cleaved from a precursor called F_o, which is not fusogenic, and has a hydrophobic N-terminal sequence (Gething *et al.*, 1978). Cleavage is accompanied by a conformational change of the protein, indicated by circular dichroism, and an increase in its exposed hydrophobic surface, shown by detergent binding (Hsu *et al.*, 1981). Studies on mutants lacking HN (Portner *et al.*, 1975) or requiring different proteases for the activation of F-protein (Scheid and Choppin, 1976) have indicated that HN is required for cell attachment and F is necessary for fusion and infectivity. Reconstituted vesicles containing these glycoproteins (virosomes) have been shown to induce cell fusion (Hosaka *et al.*, 1974; Volsky and Loyter, 1978; Ozawa and Asano, 1981), deliver diphtheria toxin into cells (Uchida *et al.*, 1979; Nakanishi *et al.*, 1982), and insert the virus proteins or the co-reconstituted anion channel from erythrocytes into the cell membrane (Poste *et al.*, 1980; Volsky *et al.*, 1979). A critical ratio of F to HN is required for optimal virosome–cell fusion, suggesting that interactions between the two proteins may be necessary (Nakanishi *et al.*, 1982). Certain phospholipids (Hosaka and Shimizu, 1972; Poste *et al.*, 1980; Ozawa and Asano, 1981) and cholesterol (Ozawa and Asano, 1981) appear to be necessary for the fusion activity of the virosomes. Host cell membrane cholesterol has also been reported to faciliate fusion induced by Sendai virus (Hope *et al.*, 1977). Virosomes containing only the F-protein show hemolytic activity when a lectin is used to attach them to the cell surface as a substitute for the lacking HN-protein, and membranes containing only HN have hemagglutinating and neuraminidase activities (Hsu *et al.*, 1979). The observation by Miura *et al.* (1982) that a monoclonal antibody against HN blocks viral envelope–cell fusion, cell–cell fusion, and hemolysis without impairing the hemagglutinating and neuraminidase activity of HN suggests that this protein may have functions other than those presently ascribed to it. Protease activity has also been observed in Sendai virus envelopes and has been proposed to be involved in the local hydrolysis of target membrane glycoproteins, allowing the F-protein to interact with the membrane phospholipids (Israel *et al.*, 1983).

Semliki Forest virus, influenza virus, and vesicular stomatitis virus, which are members of the togavirus, orthomyxovirus, and rhabdovirus families, respectively, differ from Sendai virus in that they fuse with host cell membranes only in a mildly acidic pH environment. White *et al.* (1981) have shown that these viruses induce the fusion of BHK-21 cells at characteristic pH, Semliki Forest virus and vesicular stomatitis virus around pH 6, and influenza virus around pH 5. Hemolysis and fusion of erythrocytes by influenza and Semliki Forest virus are also pH dependent (Maeda and Ohnishi, 1980; Väänänen and Kääriäinen, 1980, Lenard and Miller, 1981). Fusion of the viruses with cell membranes can be induced by reducing the pH of the medium (White *et al.*, 1980; Helenius *et al.*, 1980a; Matlin *et al.*, 1981, 1982). The physiological pathway of infection of these viruses is by adsorptive endocytosis in coated vesicles and fusion with lysosomal or prelysosomal vacuole membranes due to the acidification of the lumen of these subcellular compartments (Fan and Sefton, 1978; Marsh and Helenius, 1980; Helenius et al., 1980b; Miller and Lenard, 1981; Matlin et al., 1981, 1982; Marsh et al., 1983a; Marsh, 1984).

Fusion of Semliki Forest virus with liposomes takes place when the pH is reduced to 6 or lower, requires the presence of cholesterol in the target membrane, is inhibited by the proteolytic digestion of the spike glycoproteins of the virus, and does not depend on divalent cations (Helenius et al., 1980b; White and Helenius, 1980). Fusion of the virus with phosphatidylcholine–cholesterol liposomes is limited, however. Influenza virus fuses with liposomes at a lower pH, consistent with its threshold for cell fusion, but fusion is not dependent on the presence of cholesterol in the liposome membrane, on the phase state of the membrane, or on its phospholipid composition except that the presence of phosphatidylethanolamine enhances fusion (White et al., 1982a; Maeda et al., 1981). Neutral glycolipids with terminal galactose residues, phosphatidylcholine, and sphingomyelin have been proposed to interact with influenza virus after initial binding to sialic acid-containing receptors on the cell surface and to be involved in the fusion process (Huang, 1983b).

Binding of Sindbis virus, a togavirus that fuses erythrocytes at low pH (Väänänen and Kääriäinen, 1980), to liposomes is enhanced by the presence of cholesterol and phosphatidylethanolamine (Mooney et al., 1975). Vesicular stomatitis virus binds specifically to phosphatidylserine liposomes, which also inhibit the interaction of the virus with cells, but the binding occurs at neutral pH (Schlegel et al., 1983). These observations suggest that phosphatidylserine may be a virus binding site on cell membranes.

The glycoproteins of these viruses have been reconstituted with phospholipids and the resulting virosomes have been shown to have fusion activity. Virosomes containing the spike glycoproteins of Semliki Forest virus cause hemolysis and fuse with the plasma membrane of BHK-21 cells at pH 6 and lower, with reduced efficiency compared to the intact virus, but do not induce cell–cell fusion (Marsh et al., 1983b). The lipid composition of the virosome does not appear to affect fusion. Pure glycoproteins, free of lipid and in oligomeric form, do not have membrane fusion activity, indicating that they have to be inserted in a phospholipid bilayer to be active. The hydrophobic peptide segment of the E_1 glycoprotein of Semliki Forest virus has been suggested to be involved in its membrane fusion activity (Garoff et al., 1980). Experiments on the fusion activity of reconstituted glycoproteins of influenza virus have indicated that the cleaved hemagglutinin (HA) is necessary for fusion with cell membranes (Huang et al., 1980a), the cleavage having been shown to enhance the infectivity of the virus (Lazarowitz and Choppin, 1975). However, these experiments were performed at neutral pH. Kawasaki et al. (1983) have shown that reconstituted HA protein induces membrane fusion only at low pH. The reasons for this discrepancy are not clear at present. From experiments utilizing the neuraminidase protein (NA) from influenza virus co-reconstituted with HA-protein as well as free neuraminidase added to virosomes containing only the HA-protein Huang *et al.* (1980b) have concluded that neuraminidase activity is essential for HA to express fusion activity. In contrast, by infecting CV-1 cells with a recombinant simian virus 40 containing the cloned gene for HA in its genome, allowing for the expression of the protein at the cell surface, cleaving the protein into HA_1 and HA_2 subunits by trypsin and inducing cell–cell fusion by lowering the pH to 5, White *et al.* (1982b) have demonstrated that HA is the only protein necessary for fusion. HA-protein has been demonstrated to undergo a conformational change at the pH optimum for fusion which may lead to hydrophobic interactions with the lipids of the target cell membrane (Skehel *et al.*, 1982). It appears that the hydrophobic *N*-terminal region of the cleaved HA protein (HA_2) is involved in membrane fusion, but the molecular mechanism of fusion is still unknown (White *et al.*, 1983). The G-protein of vesicular stomatitis virus has also been reconstituted and shown to fuse at pH below 5 with phospholipid vesicles containing phosphatidylserine (Eidelman *et al.*, 1984). A synthetic polypeptide corresponding to the 25 amino acids of the *N*-terminal of the G-protein has a pH dependence of hemolytic activity similar to that of the virus, suggesting that this portion of the G-protein may be involved in the hemolytic function of the virus (Schlegel and Wade, 1984). It will be of interest to investigate whether this peptide is fusogenic and whether it will render liposomes to which it is covalently coupled fusogenic at low pH.

Studies on the fusion activity of viruses have thus shown that viral glycoproteins are essential for the induction of membrane fusion and that in some cases the lipid composition of the target membrane and of reconstituted virosomes is an important determinant. The glycoproteins apparently act by inserting hydrophobically into the target membrane, locally perturbing the bilayer structure and allowing the two membranes to intermix at these defect points. This mechanism of fusion also seems to be involved in the fusion of pure phospholipid bilayers (Düzgüneş and Papahadjopoulos, 1983; Düzgüneş *et al.*, 1985a; Section 4.2). Furthermore, the involvement of glycoproteins in virus–cell fusion is reminiscent of their participation in Ca^{2+}-induced fusion of isolated secretory vesicles (Section 3.3).

3.7. Studies with Planar and Spherical Phospholipid Bilayers

During membrane fusion the integrity of the two interacting membranes must be disrupted momentarily at the site of contact. The lipid bilayer functions as the main structural matrix of biological membranes (Danielli and Davson, 1934; Davson and Danielli, 1952; Singer and Nicolson, 1972). Freeze-fracture electron microscopy of several types of biological membranes that undergo fusion has revealed that the zone of contact between these membranes is free of intramembranous particles and is thus primarily the lipid bilayer (Zakai *et al.*, 1977; Chi *et al.*, 1976; Burwen and Satir, 1977; Lawson *et al.*, 1977; Schober *et al.*, 1977). In contrast, rapid-freezing experiments have indicated that the area of interaction is much smaller with no clearing of particles (Chandler and Heuser, 1979, 1980); Section 3.1. It appears that the interaction between biological membranes involves both the lipid bilayer and membrane proteins. In this section and in Sections 3.8 and 3.9 we review the factors which affect the adhesion and fusion of the phospholipid bilayer component of cellular membranes. Interactions between membranes have been studied extensively by means of relatively simple analogs of biological membranes. The development of bimolecular "black" lipid films (Mueller *et al.*, 1962; Henn and Thompson, 1970; Goldup *et al.*, 1970) and liposomes (Bangham *et al.*, 1965; Bangham, 1974; Szoka and Papahadjopoulos, 1980) has facilitated the study of the role of the phospholipid bilayer in contact and fusion phenomena.

Bimolecular phospholipid membranes surrounded by an aqueous medium can be made from solutions of phospholipids in appropriate organic solvents smeared across an aperture in a solid hydrophobic support. These planar membranes can then be subjected to an asymmetric hydrostatic pressure to yield a semispherical film. Using this technique, Liberman and co-workers have found that the interaction between semispherical membranes made of brain phospholipids is facilitated by divalent cations, increase in temperature, and decrease in pH (Blioch *et al.*, 1968; Liberman and Nenashev, 1968, 1970). Electrical resistance and capacitance measurements of the region of adhesion between two semispherical bilayers indicate that a new structure is formed in this region (Badzhinyan and Chailakhyan, 1971; Badzhinyan *et al.*, 1971, 1972; Liberman and Nenashev, 1972a,b). Measurements of the distance between bilayers by optical and electrical methods also indicate that the distance decreases with increasing ionic strength and that Ca^{2+} is more effective than monovalent cations (Yelkin and Berestovskii, 1974; Badzhinyan *et al.*, 1972). Neher (1974) has demonstrated that a single bilayer could be formed at the region of adhesion of two semispherical bilayers made of dioleoyl phosphatidylcholine or its mixture with phosphatidylserine. We designate this type of interaction between two bilayers as *semifusion* and use the term *fusion* for cases where communication is established between the aqueous contents demarcated by the membranes. Melikyan *et al.* (1983) have shown that semispherical bilayers of dioleoylphosphatidylcholine–cholesterol can be fused by the application of a discrete voltage pulse after semifusion of the bilayers.

Breisblatt and Ohki (1975, 1976) have used large spherical phospholipid membranes to demonstrate temperature-induced membrane fusion. Membranes made of phosphatidylcholine or phosphatidylserine fuse at characteristic temperatures of 43°C and 38°C, respectively, forming a larger single spherical membrane. The incorporation of lysolecithin into the membrane reduces the characteristic temperature by about 10°C (Breisblatt and Ohki, 1975). The extent of temperature-induced fusion is reduced considerably by the presence of cholesterol. In the presence of Ca^{2+}, Mg^{2+}, or Mn^{2+} the temperature dependence of fusion is not as pronounced as in their absence, while the extent of fusion at lower temperatures is enhanced. The fusion temperature is also pH dependent, increasing with a decrease in pH. These results have been interpreted in terms of the importance of membrane fluidity and temperature-induced instability of the membranes (Breisblatt and Ohki, 1976). The increase in the area per lipid molecule when the temperature is increased from the gel–liquid crystalline transition temperature to the fusion temperature is similar for the different phospholipids used in these studies, suggesting that membrane expansion is related to temperature-induced membrane fusion (Chaudhury and Ohki, 1981).

Another model membrane system that bears a geometrical and functional similarity to the interaction of exocytotic vesicles with the plasma membrane is one employing a planar phospholipid bilayer and phospholipid vesicles in one of the compartments bathing the planar membrane. Pohl *et al.* (1973) have found that fluorescent pigments in liposomes (magnesium octaethylporphyrin) are transferred to the planar membrane if the lipid is egg phosphatidylcholine, but not the negatively charged phosphatidylinositol. It was not possible, however, to distinguish between the molecular exchange of the pigments during close contact of the two membranes and fusion of the vesicles with the planar membrane. The transfer of gramicidin A or amphotericin B from liposomes to planar bilayers has been interpreted as arising from their fusion (Moore, 1976). Moore (1976) has found that alkyl solvents such as hexane and decane facilitate the transfer of these conductance probes and suggested that they increase the surface tension of the liposomes, providing a driving force for fusion. It must be noted that the high concentration of gramicidin A in the vesicle membrane (3 mol %) could distort considerably the packing of the lipids, and the ionophore itself could cause fusion because of its hydrophobic nature, similar to the effect of alamethicin on sonicated liposomes (Lau and Chan, 1975). Incubation of zwitterionic phosphatidylcholine or phosphatidylethanolamine bilayers with sonciated phosphatidylserine vesicles imparts a negative charge onto the bilayer as measured by the surface potential (Cohen and Moronne, 1976, 1978; Yegorova *et al.,* 1981a), but the transfer of charge is probably not a result of membrane fusion (Cohen and Moronne, 1976). Yegorova *et al.* (1981b) have shown, however, that when negatively charged lipids are present in both the bilayer and the liposomes, Ca^{2+} above 1 mM will cause membrane fusion. Using the cholesterol-dependent nystatin conductance of bilayers as a probe, several investigators have studied the incorporation into planar bilayers of cholesterol from the liposome membrane (Nenashev et al., 1978; Grishin et al., 1979; Sokolov and Lishko, 1979; Razin and Ginsburg, 1980). There appears to be a general requirement for negatively charged phospholipids and Ca^{2+} for the transfer of cholesterol, which has been proposed to be the result of the fusion (Razin and Ginsburg, 1980; Nenashev et al., 1978) or semifusion (Nenashev et al., 1978) of the liposome membrane with the planar membrane.

A functional similarity to exocytosis has been observed in this model system by the use of a phosphatidylserine bilayer and vesicles composed of phosphatidylcholine or mixtures of phosphatidylserine–phosphatidylcholine. The introduction of threshold concentrations of Ca^{2+} into the compartment containing the vesicles induces fluctuations of the membrane conductance that have been interpreted to be the result of the incorporation of domains of phosphatidylcholine into the membrane (Düzgüneş and Ohki, 1977). The threshold Ca^{2+} concentration, $[Ca^{2+}]_t$, appears to be about 100 μM for phosphatidylcholine vesicles and about 4 mM for phosphatidylserine–phosphatidylcholine (1:1) vesicles. Similar conductance fluctuations have been observed when sarcoplasmic reticulum vesicles were incubated with planar bilayers composed of phosphatidylserine or cardiolipin in phosphatidylethanolamine, in the presence of at least 0.5 mM Ca^{2+} and an osmotic gradient across the vesicle membrane (Miller and Racker, 1976a). The interpretation of this conductance behavior of bilayers, that new conductance pathways are introduced into the planar membrane by the fusion of the vesicles (Miller and Racker, 1976a; Düzgüneş and Ohki, 1977), has been supported by the following observations: (1) small unilamellar phosphatidylserine–phosphatidylcholine (1:1) vesicles incubated in the subphase solution of phosphatidylserine monolayers effect an increase in the surface pressure of the latter at the same $[Ca^{2+}]_t$ as that required to induce conductance fluctuations of bilayers when the area per molecule in the monolayer is similar to that estimated for bilayers (Ohki and Düzgüneş, 1979). (2) These vesicles have been shown to fuse with large (approximately 0.2-μm-diameter) unilamellar phosphatidylserine vesicles at similar concentrations of Ca^{2+} (Düzgüneş and Ohki, 1981). The appearance in planar bilayers (phosphatidylserine–phosphatidylcholine, 4:1) of the voltage-dependent anion channel isolated from mitochondria and incorporated into vesicle membranes (phosphatidylserine–phosphatidylcholine, 1:4) has also been attributed to membrane fusion; this process appears to depend on an osmotic gradient across the planar membrane and is greatly enhanced by the presence of 5–10 mM Ca^{2+} (Cohen et al. 1980). Application of an osmotic gradient across the vesicle membrane has also been shown to enhance the fusion of the vesicle and planar bilayer membranes as detected by the transfer of porin from the vesicles to the planar bilayer (Cohen et al., 1982).

Two laboratories have reported the delivery of aqueous contents of vesicles across planar bilayers as a direct demonstration of fusion. Multilamellar vesicles made of a 1:4 mixture of phosphatidylserine–phosphatidylcholine and containing carboxyfluorescein have been shown to discharge their contents (i.e., the inner lamellae) into the opposite compartment bathing a phosphatidylserine–phosphatidylcholine (4:1) bilayer in the presence of at least 20 mM Ca^{2+}, utilizing a fluoresence-activated cell sorter to detect fluorescent liposomes (Zimmerberg et al., 1980a). Similarly, [^{35}S]sulfate loaded sonicated phosphatidylserine vesicles have been shown to fuse with planar phosphatidylserine–phosphatidylethanolamine (3:7) bilayers in the presence of Ca^{2+} (Razin and Ginsburg, 1980).

3.8. Fusion of Pure Phospholipid Vesicles

Much of our current knowlege on the molecular mechanisms of membrane fusion has been obtained through studies on phospholipid vesicles. The interactions between phospholipid bilayers and their fusion have been studied by X-ray diffraction, NMR and EPR spectroscopy, freeze-fracture electron microscopy, differential scanning calorimetry, light scattering, and fluorescence techniques. Some of the techniques used to study the fusion of phospholipid membranes have been reviewed critically (Papahadjopoulos et al., 1979; Nir et al., 1983a). It appears that the most rigorous criterion for membrane fusion in phospholipid vesicles is the coalescence of the internal aqueous contents with concomitant stoichiometric intermixing of the membrane constituents. Several assay systems for monitoring these reactions have been developed recently. The intermixing of aqueous contents has been followed by the reactions of Ca^{2+} and arsenazo III (Dunham et al., 1977), ATP and firefly extract (Ingolia and Koshland, 1978) or luciferase (Holz and Stratford, 1979), trypsin and a fluorogenic peptide (Hoekstra et al., 1979), and Tb^{3+} and dipicolinic acid (Wilschut and Papahadjopoulos, 1979; Wilschut et al., 1980), each one of the pair of reactants having been encapsulated in different populations of vesicles. In this assay system, it is imperative to include in the external aqueous medium an inhibitor of the reaction such that any internal contents that may be released into the medium during fusion (Portis et al., 1979; Papahadjopoulos et al., 1977; Liao and Prestegard, 1979) do not react to contribute to the registry of fusion events. We have used extensively an assay based on complex formation between dipicolinic acid and Tb^{3+} whence the fluorescence intensity of Tb^{3+} is increased by four orders of magnitude. The reaction is quenched by EDTA and Ca^{2+}, which are present in the medium outside the vesicles so that only the aqueous contents intermixing in an environment protected from the quenchers by means of a continuous bilayer structure give rise to a fluorescence signal. The details of the assay have been presented by Wilschut et al. (1980, 1981, 1983) Düzgüneş et al. (1981a) and Bentz et al. (1983b). The mixing of membrane components during fusion has been monitored by differential scanning calorimetry (Papahadjopoulos *et al.*, 1974; 1976a), fluorescence energy transfer between fluorescent phospholipid molecules (Vanderwerf and Ullman, 1979; Struck *et al.*, 1981; Hoekstra, 1982a,c; Rosenberg *et al.*, 1983) or other lipid labels (Gibson and Loew, 1979; Uster and Deamer, 1981) used as tracers in the membrane, and the interaction between membrane proteins reconstituted into different populations of vesicles (Miller and Racker, 1976b; Gad *et al.*, 1979). When utilizing the latter two methodologies it is important to differentiate between mixing of the bilayers of two fusing vesicles and lipid (or protein) exchange during adhesion or through the aqueous phase.

Membranes composed of a single species of phospholipid provide a useful model for understanding the molecular factors that are involved in the interactions between membranes. It is important to point out that the usefulness of studies on the fusion of single-component phospholipid vesicles arises from the information they yield concerning the behavior of that particular phospholipid with respect to membrane fusion. Considerable evidence now suggests that lipids may exist in domains in biological membranes (Karnovsky *et al.*, 1982; Klausner *et al.*, 1980; Hui and Parsons, 1975; Bearer and Friend, 1980; 1982; Severs and Robenek, 1983). These domains may be generated by the physicochemical properties of the different lipid species in membranes (Shimshick and McConnell, 1973; Lee *et al.*, 1974; Wu and McConnell, 1975; Takeuchi and Nikaido, 1981; Recktenwald and McConnell, 1981; Stewart *et al.*, 1979; Hui, 1981; Lentz *et al.*, 1981; Lee, 1977) or by proteins (Boggs *et al.*, 1977a,b; Mayer and Nelsestuen, 1981). Thus, the fusion characteristics of a biological membrane could be dictated by the characteristics of particular phospholipid domains. Artificial membranes composed of a single phospholipid would be expected to reflect the properties of these lipid domains which affect their interactions with other membranes. However, such phospholipid vesicles may display some physicochemical properties that may limit their usefulness as a model system.

Vesicles made of pure phosphatidylserine are stable structures in 100 mM NaCl at pH 7.4. In the presence of Ca^{2+} at concentrations above 1 mM they aggregate, fuse with one another, and eventually form large structures called cochleate lipid cylinders (Papahadjopoulos *et al.*, 1974; 1975; Düzgüneş and Ohki, 1977; Portis *et al.*, 1979; Wilschut and Papahadjopoulos, 1979; Wilschut *et al.*, 1980, 1981). Small (approximately 25 nm in diameter) unilamellar phophatidylserine vesicles have been shown to increase their permeability to Na^+ or K^+ at similar concentrations of Ca^{2+} (Papahadjopoulos and Bangham, 1966; Papahadjopoulos and Ohki, 1969; 1970). The leakage of aqueous contents of the vesicles can be followed conveniently by the relief of self-quenching of carboxyfluorescein as it is diluted in the medium (Hagins and Yoshikami, 1977; Weinstein *et al.*, 1977; Portis *et al.*, 1979). Thus, the kinetics of aggregation, fusion, and release of contents may be compared. Wilschut and Papahadjopoulos (1979) have demonstrated that in small phosphatidylserine vesicles aggregation and fusion are concurrent, whereas leakage is a slower process. Wilschut et al. (1980) have estimated that about 10% of the contents are released per fusion event. The observation that these vesicles do not entrap aqueous space markers placed in the external medium following long incubations in Ca^{2+} (Ginsberg, 1978) is expected since the fusion products obtained at equilibrium are cochleate lipid cylinders (Papahadjopoulos et al., 1975), with no aqueous space between the bilayers (Papahadjopoulos et al., 1978; Portis et al., 1979). The lack of leakage of sucrose into vesicles during Mg^{2+}-induced fusion (Ginsberg, 1978) may be explained by the observation that only a transient leakage of encapsulated carboxyfluorescein occurs on addition of Mg^{2+} at 25°C, the fusion products being large and stable vesicles that do not leak their contents (Wilschut et al., 1981). The results of Ginsberg (1978) showing no encapsulation of external sucrose by the addition of EGTA to vesicles fused with Ca^{2+} are difficult to reconcile with the observation that macromolecules such as ferritin (Papahadjopoulos and Vail, 1978), RNA (Wilson et al., 1978) and eukaryotic initiation factor (O'Loughlin et al., 1981), and viruses (Wilson et al., 1979) can be encapsulated by the addition of EDTA to the cochleate lipid cylinders (Papahadjopoulos et al., 1975; Day et al., 1977). Finally, the criticism that the Ca^{2+}-phosphatidylserine vesicle system is an inappropriate model for biological membrane interaction, since the final product of the fusion is not vesicular (Ginsberg, 1978), has been answered by the studies on the relative kinetics of the intermixing of aqueous contents and their release (Wilschut and Papahadjopoulos, 1979; Wilschut et al., 1980, 1981, 1983). These studies have shown that fusion occurs via the intermixing of the internal compartments of two separate vesicles, separated from the external medium by their fused membranes. Recent studies have also shown that the intermixing of membrane components during Ca^{2+}-induced fusion of small phosphatidylserine vesicles occurs at a rate similar to that of the intermixing of aqueous contents (Hoekstra, 1982a; Wilschut et al., 1983). Furthermore, under certain ionic conditions, small phosphatidylserine vesicles fuse but do not lose their contents or collapse (Wilschut et al., 1983; Nir et al., 1983b).

Using an assay based on the interaction of EDTA with the Co^{2+} complex of calcein, resulting in the chelation of Co^{2+} and the relief of the quenching of calcein fluorescence, Kendall and MacDonald (1982) have reported that the Ca^{2+}-induced fusion of small phosphatidylserine vesicles occurs simultaneously with, or subsequent to the loss of, vesicle integrity since immediate and extensive leakage of encapsulated contents was observed. The authors have not explained the reasons for the discrepancy between their results and the results obtained by the Tb–dipicolinic acid assay (Wilschut and Papahadjopoulos, 1979) or lipid mixing assays, which show similar fusion rates (Hoekstra,

1982a; Wilschut *et al.*, 1983). The release of both carboxyfluorescein and Tb^{3+} during fusion has been shown to occur at similar rates (Hoekstra, 1982a; Wilschut *et al.*, 1983). It is possible that the Co^{2+}-calcein complex is more membrane permeable than other aqueous space markers because of the chelation of the carboxyl groups of calcein by Co^{2+}. The observation that 17% of the calcein is outside the vesicles even before Ca^{2+} addition may be an indication of this property. We have also found considerable leakage of the Co^{2+}-calcein complex from small phosphatidylserine vesicles within a few hours of column chromatography to separate unencapsulated material (N. Düzgüneş, J. Bentz, and K. Hong, unpublished data). The Co^{2+}-calcein complex may leak even faster in the presence of Ca^{2+}, which destabilizes the membrane. Since the reaction of EDTA with Co^{2+}-calcein is extremely slow compared with the reaction of dipicolinic acid with Tb^{3+} (N. Düzgüneş, J. Bentz and K. Hong, unpublished data), the former reactants would not be able to generate fluorescence within fusing vesicles before leaking into the medium. Clearly, the fluorescence reaction measuring intermixing of aqueous contents must be faster than the fusion reaction itself in order to be reliable.

Ca^{2+}-induced fusion of large (approximately 100 nm in diameter) unilamellar phosphatidylserine vesicles is essentially a nonleaky process at the early stages of the fusion reaction; release of contents occurs as a secondary phenomenon when fusion has proceeded considerably (Wilschut *et al.*, 1980; 1983). A representative comparison of fusion and release kinetics is given in Fig. 2 for the fusion of large phosphatidylserine vesicles induced by 2 or 3 mM Ca^{2+}. The threshold concentration of Ca^{2+} for inducing aggregation and fusion of large vesicles is slightly higher than that for small vesicles (approximately 2 mM vs. 1 mM), indicating that the repulsive electrostatic forces and hence the potential energy barrier for close approach of the vesicles are greater (Hogg *et al.*, 1966; Nir and Bentz, 1978). Phosphatidylserine vesicles also undergo fusion in the presence of other cations such as H^+ (Papahadjopoulos *et al.*, 1977), Ba^{2+}, Sr^{2+} (Düzgüneş *et al.*, 1984a), and La^{3+} (Hammoudah *et al.*, 1979, 1981).

Although Mg^{2+} can induce the fusion of small phosphatidylserine vesicles to a limited extent (Papahadjopoulos *et al.*, 1977; Wilschut *et al.*, 1981), it is ineffective on large vesicles (Wilschut *et al.*, 1981). The latter aggregate above 5 mM Mg^{2+} but do not fuse or release their contents. The small vesicles, on the other hand, undergo a few rounds of fusion and then stop fusing. It appears that when the strained packing arrangement of the phospholipids in the highly curved vesicle membrane is altered as the vesicles grow in size, the fusion susceptibility of the membrane decreases drastically. This fusion process is slower than the aggregation of the vesicles, in contrast to Ca^{2+}-induced fusion, which occurs at the same rate as aggregation at low vesicle concentrations, where aggregation is rate limiting (Wilschut *et al.*, 1980, 1981; Nir *et al.*, 1980). The specificity of Ca^{2+} over Mg^{2+} in mediating membrane fusion, especially in the

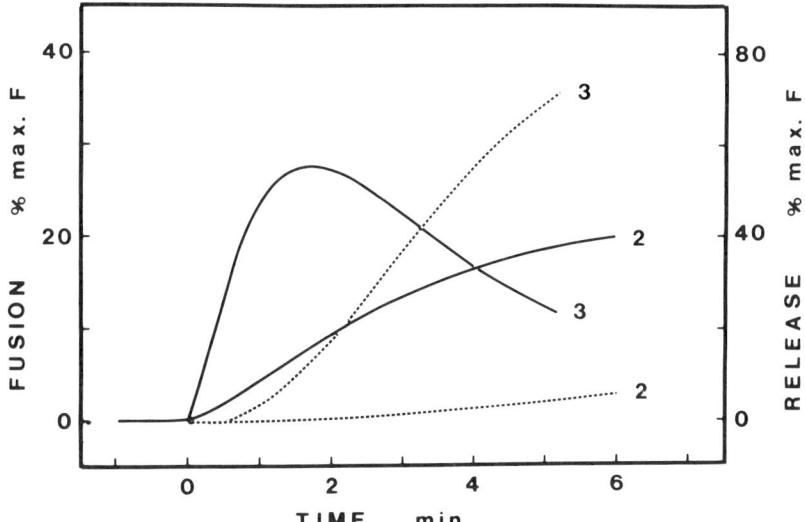

FIGURE 2. Time course of fusion (solid lines) and release of internal aqueous contents (dotted lines) of large unilamellar phosphatidylserine vesicles in the presence of Ca^{2+} at the indicated concentrations (mM). Fusion was measured by the Tb–dipicolinic acid assay and release by the enhancement of carboxyfluorescein fluorescence. 100% maximal fluorescence in the fusion experiment indicates the fluorescence that would be obtained if all the encapsulated Tb were to react with dipicolinic acid (Wilchut et al., 1980). Tb-containing and dipicolinic acid-containing vesicles were mixed in a 1:1 ratio at a final lipid concentration of 0.05 μmol/ml in 100 mM NaCl, pH 7.4. 100% maximal fluorescence in the release experiment indicates the fluorescence obtained upon lysis of the vesicles with detergent. (N. Düzgüneş, unpublished data.)

case of large phosphatidylserine vesicles, may be related to the physicochemical nature of the metal ion–PS complex (Newton et al., 1978; Portis et al., 1979; Düzgüneş and Papahadjopoulos, 1983) as well as to the isothermal phase transitions of the bilayer caused by these metal ions (Papahadjopoulos et al., 1977, 1978, 1979). This subject will be treated in detail in Section 4.4.

The rate of Ca^{2+}-induced fusion of large phosphatidylserine vesicles is enhanced up to 1000-fold in the presence of physiological levels of phosphate, and the $[Ca^{2+}]_t$ is lowered several fold (Fraley et al., 1980). Fusion is dependent on temperature, pH, ion concentration, and the composition of the calcium phosphate crystalline phase and appears to occur only when calcium phosphate precipitation is nucleated in the presence of the vesicles. Calcium phosphate also facilitates phase separation between phosphatidylserine and phosphatidylcholine in mixed membranes.

Vesicles made of other acidic phospholipids such as phosphatidylglycerol or phosphatidic acid (phosphatidate) have fusion characteristics that are different from those of phosphatidylserine, but they have not been studied in as much detail. Multilamellar phosphatidylglycerol vesicles aggregate in 5 mM Ca^{2+}, but no membrane intermixing occurs. Higher concentrations of Ca^{2+} (10 mM) are necessary to induce fusion and Mg^{2+} is much less effective than Ca^{2+} (Papahadjopoulos et al., 1976b). We have recently demonstrated the intermixing of aqueous contents during Ca^{2+}-induced fusion of large unilamellar vesicles composed of phosphatidylglycerol (Rosenberg et al., 1983). The difference in fusion susceptibility between small and large vesicles is also observed in the case of phosphatidylglycerol vesicles. Concentrations of Ca^{2+} or Mg^{2+} (5 and 20 mM, respectively), which only aggregate large vesicles, are sufficient to cause fusion in small vesicles. Cylindrical structures, similar to those observed with phosphatidylserine, are formed by phosphatidylglycerol bilayers in the presence of high concentrations of Ca^{2+} and Mg^{2+} (Tocanne et al., 1974; Verkleij et al., 1974; Ververgaert et al., 1975).

Phosphatidate vesicles have a considerably lower threshold of divalent cations that induce fusion compared to phosphatidylserine and phosphatidylglycerol vesicles. Both Ca^{2+} and Mg^{2+} cause fusion in small phosphatidate vesicles at 0.2 mM and pH 7.4 (Papahadjopoulos et al., 1976b). In this system Mg^{2+} is remarkably more effective than in phosphatidylserine or phosphatidylglycerol vesicles. The fusion of large phosphatidate vesicle has been studied recently by Sundler and Papahadjopoulos (1981), using the Tb–dipicolinic acid fusion assay. The threshold concentrations of divalent ions depend on the pH of the medium, ranging between 0.03–0.1 mM for Ca^{2+} and 0.07–0.15 mM for Mg^{2+}, the lower values obtaining at pH 8.5 and the higher ones at 6.0. At the latter pH, a 1-h incubation of small phosphatidate vesicles in the presence of Mg^{2+} produces a hexagonal H_{II} arrangement of the lipids, in contrast to the case with Ca^{2+}, which forms lamellar structures (Papahadjopoulos et al., 1976b).

Fusion of small vesicles composed of pure cardiolipin by Ca^{2+} and Mg^{2+} has been reported by Vail and Stollery (1979). These vesicles go through a series of structural transitions in the presence of Ca^{2+} as shown by freeze-fracture electron microscopy, with particles decorating the surface of fused vesicles and later transforming into the H_{II} phase.

Among the acidic phospholipids investigated, phosphatidylinositol appears to be the only one that is completely resistant to fusion. Large unilamellar vesicles composed of phosphatidylinositol aggregate but do not fuse or release their contents in the presence of Ca^{2+} or Mg^{2+}, even at high concentrations (Sundler and Papahadjopoulos, 1981). When phosphatidylethanolamine is present in the membrane in addition to phosphatidylinositol, however, the vesicles do undergo fusion at threshold concentrations of divalent cation that decrease with decreasing phosphatidylinositol content (Sundler et al., 1981). This finding has allowed for the examination of the role of phosphatidylethanolamine in Ca^{2+}-induced membrane fusion, since it is the phosphatidylethanolamine and not the acidic component that renders the vesicles susceptible to fusion (see Section 4.3).

Stollery and Vail (1977) have shown that small unilamellar vesicles prepared from phosphatidylethanolamine at high pH and low ionic strength (a prerequisite for the formation of stable vesicles composed of this phospholipid) fuse in the presence of 2 mM Ca^{2+} or 3 mM Mg^{2+}. Removal of the divalent cations by dialysis against buffer with no EDTA results in the formation of a vesicle population with a different size distribution, indicating that the divalent cations are not bound strongly to phosphatidylethanolamine. The aggregation of phosphatidylethanolamine vesicles induced by lowering the pH from 9 to 7 has led Kolber and Haynes (1979) to propose that phosphatidylethanolamine forms interbilayer hydrogen bonds that mediate intermembrane contact.

The first studies on the fusion of liposomes employed small phosphatidylcholine vesicles (Taupin and McConnell, 1972; Metcalfe et al., 1972). Small vesicles composed of phosphatidylcholine with saturated acyl chains have been shown to fuse when incubated for long periods of time below the gel-to-liquid crystalline phase transition, T_c, (Suurkuusk et al., 1976; Schullery et al., 1980; Larrabee, 1979; Lichtenberg et al., 1981; Schmidt et al., 1981; Wong et al., 1982; Wong and Thompson, 1982). Vesicles composed of dipalmitoylphosphatidylcholine grow in size to form unilamellar vesicles of about 70 nm in diameter (Schullery et al., 1980). Earlier studies with dimyristoylphosphatidylcholine had indicated, however, that appreciable fusion occurs only in the phase transition region and in the presence of myristic acid (Kantor and Prestegard, 1975, 1978) or gangliosides (Martin and MacDonald, 1976). These vesicles also undergo fusion when the temperature is cycled through the T_c (van Dijck et al., 1978). Papahadjopoulos et al. (1976a) have shown that the growth of vesicles composed of saturated acyl chains and myristic acid is due to molecular exchange between the vesicles and that considerably more mixing occurs with sonicated vesicles than with multilamellar vesicles. These investigators have also suggested that lysolecithin, which induces cell fusion (Poole et al., 1970), causes the intermixing of membrane components via diffusion of molecules rather than by fusion of entire vesicles. Felgner et al. (1981) have reported that the incorporation of trisialoganglioside into dipalmitoylphosphatidylcholine vesicles inhibits the fusion of these vesicles below the T_c. On the other hand, the presence of lipophilic spin labels enhances the rate of fusion of dipalmitoylphosphatidylcholine vesicles below the T_c and of molecular exchange between vesicles above the T_c (Defrise-Quertain et al., 1982). Martin and MacDonald (1974) have proposed that dimyristoylphosphatidylcholine vesicles grow by a molecular exchange process that is maximal at T_c. The transformation of dipalmitoyl- and distearoylphosphatidylcholine vesicles below the T_c has been attributed to the molecular packing strains in the highly curved vesicle membrane in the gel state (Schmidt *et al.*, 1981; Lichtenberg and Schmidt, 1981). The regions between the facets of the membrane of such vesicles observed by X-ray diffraction (Blaurock and Gamble, 1979) are likely to be nucleation sites for membrane fusion. Vesicles prepared by sonication below the T_c, and which display a high permeability to ions because of structural defects in the bilayer, undergo extensive fusion at rates that increase with increasing temperature and that reach a maximum at the T_c (Lawaczeck *et al.*, 1975). Gaber and Sheridan (1982) have found that the longer the acyl chain length, the more stable (i.e., resistant to fusion) the vesicles and that the fusion rate is the same when the vesicles are incubated 10°C below the T_c or at the T_c. Vesicles composed of egg phosphatidylcholine do not aggregate or fuse at physiological temperatures or in the presence of Ca^{2+} (Papahadjopoulos *et al.*, 1974; Düzgüneş and Ohki, 1977; Roseman *et al.*, 1977) and do not fuse with vesicles made of dipalmitoylphosphatidylcholine (Korn *et al.*, 1974). However, they may be induced to fuse with fusogenic molecules such as alamethicin and the *n*-alkyl bromides (Lau and Chan, 1975; Mason *et al.*, 1980).

It should be noted here that the fusion of dipalmitoylphosphatidylcholine vesicles takes place on a time scale of hours to days even at very high lipid concentrations (5–50 m*M*), whereas the Ca^{2+}-induced fusion of phosphatidylserine vesicles occurs on a time scale of seconds at lipid concentrations two to three orders of magnitude lower.

3.9. Fusion of Mixed Phospholipid Membranes

Studies on the fusion susceptibility of liposomes composed of pure acidic phospholipids demonstrate the relationship between the physicochemical properties of these molecules and their involvement in intermembrane contact and fusion. Biological membranes contain low amounts of acidic phospholipids compared to the zwitterionic phospholipids. Although the acidic phospholipids may be organized in domains, it is important to understand how the zwitterionic phospholipids modulate membrane fusion when they are mixed with the acidic phospholipids.

Higher threshold concentrations of divalent cations are required to aggregate liposomes composed of mixtures of phosphatidylserine and phosphatidylcholine compared to pure phosphatidylserine liposomes (Düzgüneş and Ohki, 1977; Ohki and Düzgüneş, 1979; Ohki *et al.*, 1982). This threshold concentration increases as the phosphatidylcholine content of the membrane is increased (Ohki *et al.*, 1982). Likewise, the initial rate and extent of fusion induced by Ca^{2+} decreases with higher mole fractions of phosphatidylcholine in the liposome membrane (Papahadjopoulos *et al.*, 1974; Düzgüneş *et al.*, 1981a). The initial rate of fusion of small, unilamellar vesicles composed of pure phosphatidylserine is compared with that of phosphatidylserine–phosphatidylcholine

(4:1 and 2:1) vesicles as a function of Ca^{2+} concentration in Figure 3. Although the initial stage of fusion of the mixed phospholipid vesicles is leaky, the vesicles retain most of their contents. Phosphatidylserine–phosphatidylcholine (4:1) vesicles, for example, release about 20% of their contents during fusion in the presence of 3 mM Ca^{2+}. The extremely inhibitory effect of phosphatidylcholine on membrane fusion is clearly demonstrated with large unilamellar vesicles composed of an equimolar mixture of phosphatidylserine and phosphatidylcholine. These vesicles aggregate in the presence of Ca^{2+} but do

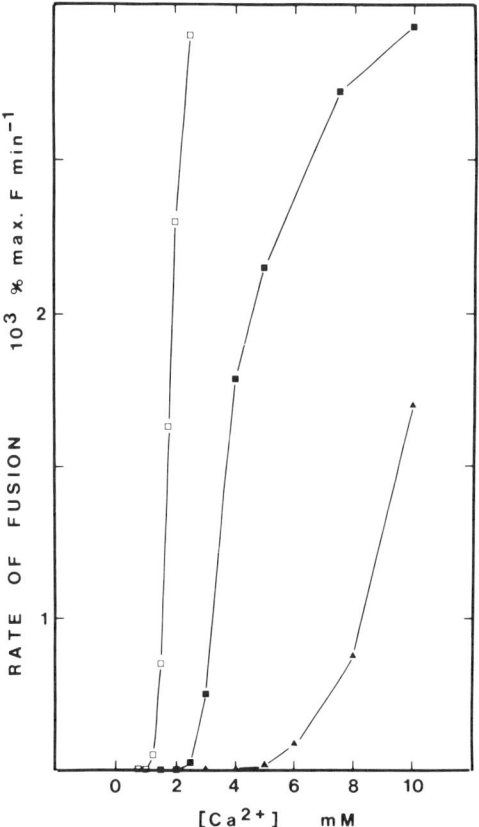

FIGURE 3. Ca^{2+} concentration dependence of the initial rate of fusion of small unilamellar vesicles composed of phosphatidylserine (open squares), and its mixtures with phosphatidylcholine at mole ratios of 4:1 (solid squares) and 2:1 (triangles). The initial rate is given as the percent maximal fluorescence per minute. The vesicle concentration was 0.05 μmol phospholipid/ml in 100 mM NaCl, pH 7.4 25°C. (N. Düzgüneş and J. Wilschut, unpublished data.)

not fuse or release their contents even at Ca^{2+} concentrations as high as 50 mM (Düzgüneş et al., 1981b; Uster and Deamer, 1981). Small vesicles of the same composition, however, do undergo limited fusion above 10 mM Ca^{2+} (N. Düzgüneş, unpublished data), indicating the relative instability of small vesicles compared to large vesicles.

In contrast to phosphatidylcholine, when phosphatidylethanolamine is mixed with phosphatidylserine, it does not impart fusion resistance to the membrane. Large liposomes composed of phosphatidylserine–phosphatidylethanolamine (1:1) undergo rapid and extensive fusion in the presence of Ca^{2+}, the threshold concentration being slightly higher and the initial rate being slower than that for pure phosphatidylserine vesicles (Düzgüneş et al., 1980, 1981b). The fusion of these vesicles is initially nonleaky (Düzgüneş et al., 1981b; Morris et al., 1983). Liposomes composed of only 25 mol % phosphatidylserine in phosphatidylethanolamine fuse in the presence of about 3 mM Ca^{2+}. Miller and Racker (1976b), Vanderwerf and Ullman (1980), and Gad et al. (1979) have made similar observations on the difference between phosphatidylcholine and phosphatidylethanolamine in modulating Ca^{2+}-induced membrane fusion. The possible physicochemical basis of the difference in fusion behavior of vesicles containing phosphatidylcholine or phosphatidylethanolamine will be discussed in Section 4.3.

The extent of Ca^{2+}-induced aggregation of phosphatidate–phosphatidylcholine (1:1) vesicles is also considerably less than that of pure phosphatidate vesicles (Lansman and Haynes, 1975). When the phosphatidylcholine mol fraction is kept above 0.5 in such membranes, Ca^{2+} induces a transformation of small, unilamellar vesicles to larger ones of well-defined size determined by the Ca^{2+} concentration and mol fraction of the lipids (Liao and Prestegard, 1979). These vesicles also retain their contents during fusion. Koter et al., (1978) have reported that small phosphatidate–phosphatidylcholine vesicles aggregate reversibly at a Ca^{2+}–phosphatidate ratio of 0.3 and fuse above a ratio of 0.5. Liao and Prestegard (1980) have found that the rate of fusion of phosphatidate–phosphatidylcholine (1:2) vesicles is faster in the presence of Ca^{2+} than in the presence of Mg^{2+} or Ba^{2+}.

Sundler et al. (1981) have investigated the kinetics of fusion of large unilamellar vesicles composed of phosphatidate and phosphatidylcholine using the Tb–dipicolinic acid assay. They found that the mixed membranes require a higher threshold concentration of Ca^{2+} for fusion compared to pure phosphatidate vesicles, that the extent of fusion and the rate and extent of release of vesicle contents are considerably lower, and that Mg^{2+} is unable to fuse the mixed membranes, although it induces aggregation at 5 mM. The latter observation, when compared with the results of Liao and Prestegard (1980), emphasizes the vesicle size dependence for the induction of fusion by Mg^{2+} in the phosphatidate–phosphatidylcholine system. When phosphatidylethanolamine is present in the vesicle membrane up to 80 mol %, the threshold concentrations of Ca^{2+} or Mg^{2+} for fusion increases only moderately compared to pure phosphatidate vesicles (Sundler *et al.*, 1981). In the presence of subthreshold concentrations of Mg^{2+}, phosphatidate–phosphatidylethanolamine (1:4) vesicles could be induced to fuse by Ca^{2+} concentrations as low as 0.1 mM.

Fusion of equimolar mixtures of cardiolipin and phosphatidylcholine in the presence of Ca^{2+} has been demonstrated by freeze-fracture electron microscopy and the intermixing of aqueous contents (Verkleij *et al.*, 1979; Rand *et al.*, 1980; Wilschut *et al.*, 1982; Düzgüneş *et al.*, 1982; Bearer *et al*, 1982). The threshold Ca^{2+} concentration for the fusion of large unilamellar vesicles is around 9 mM, and the initial stage of fusion is essentially nonleaky (Wilschut *et al.*, 1982; Figure 4); leakage of contents is a secondary phenomenon, as in the case of Ca^{2+}-induced fusion of phosphatidylserine vesicles (Wilschut *et al.*, 1980; Figure 2). Wilschut *et al.* (1983) have shown that vesicles composed of a 1:1 mixture of *Bacillus subtilis* cardiolipin (a highly saturated cardiolipin species) and dioleoylphosphatidylcholine undergo about two rounds of fusion in the presence of Ca^{2+} at 10°C with complete retention of contents. When multilamellar cardiolipin–phosphatidylcholine liposomes are incubated

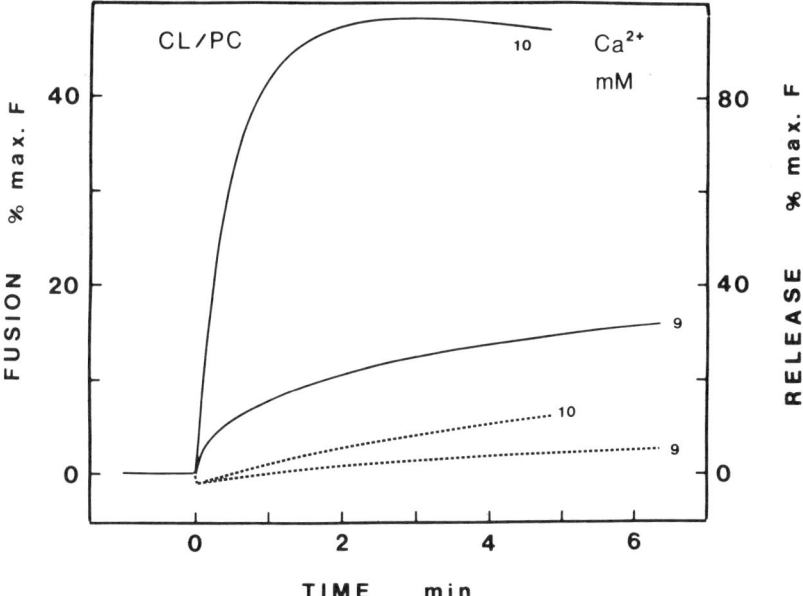

FIGURE 4. Time course of fusion of large (approximately 200 nm diameter) unilamellar vesicles composed of cardiolipin–phosphatidylcholine (1:1) (solid lines), and the release of aqueous contents (dotted lines) in the presence of Ca^{2+} (mM). (Data from Düzgüneş *et al.*, 1982.)

with Ca^{2+} at concentrations as low as 10^{-6} M, helical structures are formed that have been attributed to Ca^{2+}-mediated intermembrane binding (Lin et al., 1982). The Ca^{2+}-induced fusion of small unilamellar vesicles composed of cardiolipin–phosphatidylethanolamine, as well as mixtures of phosphatidylethanolamine with other acidic phospholipids, such as phosphatidylserine, phosphatidylglycerol, phosphatidylinositol, and phosphatidate, has been demonstrated by freeze-fracture electron microscopy (Hope et al., 1983).

4. MOLECULAR MECHANISMS OF MEMBRANE FUSION

In Section 3 we have reviewed the experimental systems used to study membrane fusion and outlined the molecular requirements for various types of membranes to undergo fusion. In this section we discuss the physicochemical basis for these observations, such as the specificity of Ca^{2+} over Mg^{2+}, and describe the molecular events that are thought to occur during the close approach and fusion of membranes and the role of certain proteins, peptides, and polyamines in modulating membrane fusion. Finally, we discuss the implications of the findings on the fusion of model and biological membranes for the fusion of subcellular membranes.

4.1. Close Approach of Membranes

The overall process of membrane fusion can be separated into two distinct stages that are coupled kinetically. The first stage is the close approach of the membranes or, in the case of membranous vesicles, aggregation of the vesicles. The second stage is the membrane fusion reaction *per se*, involving the local and momentary loss of structural integrity of the membrane and the resulting coalescence of the internal aqueous contents of the vesicles with concomitant intermixing of membrane components (Nir et al., 1982, 1983a; Bentz et al., 1983a; Düzgüneş et al., 1985a). The distance of approach of two membranes is determined by the free energy of interaction between the two surfaces. The Derjaguin–Landau–Verwey–Overbeek (DLVO) theory expresses the free energy of interaction between colloidal particles of like charge as the sum of the repulsive electrostatic or Coulombic interaction energy and the attractive van der Waals interaction energy due to fluctuating charges (Verwey and Overbeek, 1948; Parsegian, 1973; Nir, 1977; Rand, 1981). Liposomes made of acidic phospholipids or their mixtures with zwitterionic phospholipids experience mutual electrostatic repulsion in physiological salt solutions and thus cannot form stable aggregates. The screening of negative surface charges by cations reduces the repulsive barrier to close approach, and binding of the cations further reduces the barrier by decreasing the magnitude of the surface charge density. Aggregation of vesicles is expected to occur only if there is a local minimum in the free energy of interaction, and the average time that the vesicles can spend in this state depends on the depth of the minimum. The vesicles would diffuse away if the depth of the minimum were less than kT (relative to infinite separation), where k is Boltzmann's constant and T is the absolute temperature, due to thermal fluctuations. Local minima in the free energy of interaction can occur either at surface separations less than 10 Å (i.e., at the primary minimum) or at separations of 20–100 Å (the secondary minimum). To approach molecular contact, vesicles must overcome the potential energy barrier between the secondary and primary minima (Nir et al., 1983a). Small unilamellar vesicles made of pure phosphatidylserine aggregate in the presence of 1 mM Ca^{2+} or 4 mM Mg^{2+} in 100 mM NaCl at pH 7.4 (Papahadjopoulos et al., 1977; Düzgüneş and Ohki, 1977; Portis et al., 1979). This observation can be predicted by the DLVO theory when ion binding to the phospholipid is accounted for explicitly (Nir and Bentz, 1978), the difference between Ca^{2+} and Mg^{2+} being explained by the differences in their binding constants to phosphatidylserine (Newton et al., 1978; Ohki and Sauvé, 1978; Portis et al., 1979; McLaughlin et al., 1981; Ohki and Kurland, 1981). Aggregation of phosphatidylserine vesicles induced by divalent and monovalent cations has been related to the slope of the repulsive interaction energy (repulsive force) at the Debye distance of two interacting membranes (Ohki et al., 1982). Aggregation is a dynamic process where vesicles aggregate and disaggregate in the primary minimum, since the secondary minimum is too shallow for measurable aggregation (Nir and Bentz, 1978). Large unilamellar vesicles, however, may aggregate in the secondary minimum under certain ionic conditions. Nir et al. (1981) have demonstrated this mode of aggregation for large phosphatidylserine vesicles in an aqueous medium containing Na^+ and Mg^{2+}, under which conditions the small vesicles do not aggregate in the primary minimum.

Studies on the kinetics of fusion of phosphatidylserine vesicles have shown that the rate of fusion induced by Ca^{2+} in the presence of 100 mM NaCl is second-order with respect to vesicle concentration; this observation indicates that the rate-limiting step in the overall fusion process is the rate of vesicle aggregation, which is determined by the frequency of collisions between vesicles (Wilschut et al., 1980). The mass action kinetic description of the time course of fusion also indicates that aggregation is the rate-limiting step (Nir et al., 1980).

As the phospholipid bilayers approach each other at distances less then about 25 Å, they experience a repulsive force, referred to as a hydration force, caused by the tightly bound water of hydration around the phospholipid head group, which would have to be removed for the bilayers to come into contact (LeNeveu et al., 1976, 1977; Rand, 1981). This force has been observed both for zwitterionic phospholipids such as phosphatidylcholine and phosphatidylethanolamine (Lis *et al.,* 1982) and for negatively charged phospholipids such as phosphatidylglycerol (Cowley *et al.,* 1978).

The close approach of phospholipid bilayers mediates a number of changes in the biophysical properties of the membranes (Rand, 1981; Düzgüneş and Papahadjopoulos, 1983). Binding of ions to the phospholipid polar groups is altered as the bilayers come into contact; binding of Ca^{2+} to phosphatidylcholine decreases (Lis *et al.,* 1981), whereas its binding to phosphatidylserine increases (Portis *et al.,* 1979; Rehfeld *et al.,* 1981; Ekerdt and Papahadjopoulos, 1982). The molecular area occupied by phospholipids (Reiss-Husson, 1967) and the lateral compressibility of the bilayer decrease as bilayers approach to less than 25 Å (Rand, 1981). The distance between bilayers and the hydration of the head groups also affect the fluidity of the hydrocarbon chains, causing the lipid to undergo a transition from liquid crystalline to the gel phase (Ladbrooke and Chapman, 1969; Portis *et al.,* 1979; Lis *et al.,* 1982) and inducing a lateral phase separation of lipids in mixed lipid systems (Rand, 1981). In the next section we shall discuss the possible role of lateral phase separations in membrane fusion and how contact-mediated changes in membrane structure may lead to destabilization of membranes at the point of contact, leading to the intermixing of membrane components and membrane fusion.

4.2. Lateral Reorganization and Destabilization of Membranes

In Section 3.1 we summarized electron microscopic studies on membrane events occurring during exocytosis. Although many of these studies have shown the clearing of intramembranous particles from the area of adhesion of secretory vesicles to the plasma membrane, quick-freezing freeze-fracture electron microscopy has produced evidence to the contrary and has indicated that fusion occurs over a limited area, forming a pore between the extracellular environment and the interior of the secretory vesicles. It appears, therefore, that large areas of apposed, protein-free lipid bilayers are not necessary for fusion during exocytosis. It is obvious that proteins or glycoproteins that constitute a steric hindrance to the close approach of membranes must be removed from the region of adhesion before membrane fusion can occur (Poste and Allison, 1973; Maroudas, 1975; Ahkong *et al.,* 1975a). On the other hand, membrane proteins in the region of adhesion may be involved in the recognition of the two membranes or may be fusion-inducing proteins much like certain proteins of lipid-enveloped viruses (Section 3.6). Clearly, this is an area that warrants further exploration.

Here we shall discuss the lateral phase separation of different lipid species in mixed phospholipid bilayers, which may be relevant to their susceptibility to fusion, and the possible mechanisms of temporary destabilization of bilayers that must take place during membrane fusion. We focus on phospholipid bilayers because this is the system on which much of our current knowledge of the mechanisms of membrane fusion is based and because they constitute the backbone of biological membranes that must be destabilized during membrane fusion in biological systems. In Section 4.6 we shall outline the effects of proteins on membrane fusion.

Early studies on the fusion of phospholipid vesicles revealed a close correlation between the induction of fusion by divalent cations and their ability to phase-separate acidic phospholipids from the zwitterionic phospholipids in the membrane (Papahadjopoulos et al., 1974). Papahadjopoulos and co-workers have proposed that the key event in membrane fusion is a Ca^{2+}-induced lateral phase separation of acidic and zwitterionic phospholipids, and the formation of domain boundaries between solid areas of Ca^{2+} complexes of acidic phospholipids and fluid areas of zwitterionic phospholipids (Papahadjopoulos et al., 1977, 1978b; Papahadjopoulos, 1978). Thus, the fusion susceptibility of a particular membrane would be determined by the mol fraction, species, and transbilayer distribution of acidic phospholipids. Sun et al. (1979) have proposed that Ca^{2+} first induces molecular segregation of phosphatidylserine from phosphatidylcholine and that fusion is initiated between the phosphatidylserine domains in apposed membranes. Phase separation of phosphatidylserine from phosphatidylcholine is induced by Ca^{2+} and not Mg^{2+}, and the same specificity is found for the fusion of these membranes (Papahadjopoulos et al., 1974; Düzgüneş and Papahadjopoulos, 1983; Ohnishi and Ito, 1974; Ohnishi and Tokutomi, 1981). Phase separation can also be induced in phosphatidate–phosphatidylcholine membranes by Ca^{2+} (Ito and Ohnishi, 1974; Jacobson and Papahadjopoulos, 1975), and these membranes do undergo fusion in the presence of Ca^{2+} (Sundler et al., 1981). Recent studies have shown, however, that macroscopic phase separation as observed by differential scanning calorimetry is not strictly correlated with membrane fusion (Düzgüneş et al., 1984a). For example, large unilamellar vesicles composed of phosphatidylserine–dimyristoylphosphatidylethanolamine (1:1) fuse in the presence of Mg^{2+} without any observable phase separation. Furthermore, phosphatidylserine–dipalmitoylphosphatidylcholine (1:1) vesicles can fuse in the presence of Ca^{2+} within a narrow temperature range without exhibiting any massive phase separation of the lipid species. The occurrence of maximal rate of fusion within the gel–liquid crystalline phase transition range suggests, however, that temperature-induced lateral phase separation of phosphatidylserine from phosphatidylcholine before the addition of Ca^{2+} renders the membrane fusion-susceptible by creating fusogenic domains of phosphatidylserine (Düzgüneş et al., 1984a). Thus, the fusogenicity of a membrane can be controlled by the lateral distribution of the phospholipids; this observation may have significant implications for the control of membrane fusion among subcellular membranes (Düzgüneş et al., 1985a). Membrane proteins may be involved in the creation of such domains (Boggs et al., 1977a,b; Mayer and Nelsestuen, 1981; Bernard et al., 1982; Warren et al., 1975; Armitage et al., 1977). It is interesting to note that in experimental systems where the phospholipid bilayers are in close apposition micromolar concentrations of Ca^{2+} cause lateral phase separation of acidic and zwitterionic phospholipids (Tokutomi et al., 1981). Thus, concentrations of Ca^{2+} attained in the cytoplasm during stimulation for exocytosis may be sufficient for local phase separation of the lipids of the plasma membrane and secretory vesicles if intermembrane contact has been achieved by means of recognition proteins.

Hoekstra (1982b,c) has studied the relative rates of fusion and phase separation in small unilamellar vesicles containing phosphatidylserine and has found that the rate of fusion is considerably faster than the rate of phase separation. These results suggest that macroscopic phase separation may facilitate membrane fusion but is not involved in the fusion reaction itself.

Based on observations that extensive fusion occurs when small phosphatidylserine vesicles are incubated with Ca^{2+}, Mg^{2+}, or H^+ at temperatures at which an isothermal phase change from the liquid-crystalline to the gel state takes place, Papahadjopoulos et al. (1977) have proposed that this phase change induces transient destabilization of the membranes, leading to fusion. Studies on the initial kinetics of the fusion of phosphatidylserine vesicles have shown, however, that small vesicles do fuse in the presence of Mg^{2+} at temperatures above the T_c of the Mg^{2+}-phosphatidylserine complex, that is, under conditions where no isothermal phase change has been induced (Wilschut et al., 1981, 1985), whereas large vesicles are resistant to fusion except at very high temperatures ($>40\,°C$; J. Wilschut, H. Düzjüneş, and D. Papahadjopoulos, unpublished data).

Other divalent cations such as Ba^{2+} and Sr^{2+} induce fusion of large phosphatidylserine vesicles both below and above the T_c of the metal ion complex of the lipid, indicating that an isothermal phase change from the liquid crystalline to the gel phase is not necessary for fusion to occur (Düzgüneş et al., 1984a). Utilizing a mass action kinetic analysis of fusion that yields the rate constants for the aggregation step and the fusion reaction per se, Bentz et al. (1985) have found that the rate constants for fusion increase almost linearly over the temperature range spanning the broad phase transition of the metal ion–phosphatidylserine complex and concluded that the ionotropic phase transition is not the driving force for the initial fusion event. La^{3+} also induces fusion both below and above the T_c of the La^{3+}-phosphatidylserine complex, but fusion is more extensive at the T_c (Hammoudah et al., 1981). Even under conditions where an isothermal phase transition takes place at equilibrium (i.e., after a long-term incubation of the lipid with the metal ion), it is not known whether such a transition occurs within the time scale of fusion. It is possible, however, that phase transitions or phase separations occur in the limited region of contact between two membranes and thus may not be readily detectable by methods that monitor average properties of the membrane. It appears that intermembrane contact induced by Ca^{2+} between phosphatidylserine membranes leads to the dehydration of the polar groups (Papahadjopoulos *et al.*, 1978a; Portis *et al.*, 1979). Thus, dehydration at the point of contact may be the critical event in the fusion of phospholipid vesicles (Portis *et al.*, 1979; McIver, 1979; Düzgüneş *et al.*, 1981b; Hoekstra, 1982c). The possible sequence of events from the introduction of Ca^{2+} to membrane fusion may be summarized as follows (Düzgüneş and Papahadjopoulos, 1983):

1. Prior to the aggregation of the vesicles Ca^{2+} causes a partial dehydration of the polar head groups and condensation of the outer monolayer, leading to an increase in lateral compressibility of the membrane.
2. The vesicles aggregate at a threshold concentration of Ca^{2+}. If the intermembrane distance is close enough, Ca^{2+} may interact with the polar groups of both membranes and form an anhydrous interbilayer *trans* complex at the point of contact.
3. The difference in packing between phospholipids forming a *trans* complex with Ca^{2+} and those that form an intrabilayer *cis* complex could create point defects in the membrane.
4. The two apposed bilayers may intermix at these point defects that tend to expose the hydrophobic interior of the lipid bilayer.

Fusion of phosphatidylethanolamine–phosphatidylcholine membranes induced by freezing and thawing has been proposed to proceed via the formation of point defects (Hui *et al.*, 1981). Ohki (1982) has concluded that divalent cations induce the fusion of small phosphatidylserine vesicles by increasing the hydrophobicity of the membrane surface detected as an increase in surface tension, corroborating an earlier proposal by Ohki and Düzgüneş (1979). Membrane fusion induced by increasing the temperature has also been suggested to involve an increased hydrophobic interaction and the disorder of structural water at the region of contact (Breisblatt and Ohki, 1976; Chaudhury and Ohki, 1981).

Recently, a mass-action kinetic analysis of phospholipid vesicle fusion has been developed that yields the rate constants for vesicle dimerization and for the membrane fusion reaction (Nir *et al.*, 1982, 1983a; Bentz *et al.*, 1983a). The rate of fusion of small phosphatidylserine vesicles in $[Ca^{2+}]$ above threshold is about two orders of magnitude higher than that of large vesicles. Fusion of the former is expected to occur within several milliseconds of dimerization. Miller and Dahl (1982) have indeed shown that these vesicles fuse within 10 msec of Ca^{2+} addition, using fast-freezing freeze-fracture electron microscopy. Düzgüneş *et al.* (1982) and Bearer *et al.* (1982) have also demonstrated the fusion of large phosphatidylserine–phosphatidylethanolamine and cardiolipin–phosphatidylcholine vesicles within 1–2 sec of stimulation with Ca^{2+}.

It is clear that membrane fusion involves a temporary and local change in the bilayer configuration of the interacting membranes. The observation of the hexagonal H_{II} phase when vesicles composed of certain phospholipids are treated with Ca^{2+} has led to the suggestion that these vesicles undergo fusion as a result of the induction of nonbilayer intermediary structures by Ca^{2+} (Cullis and Verkleij, 1979; Hope and Cullis, 1979). An analogous proposal has been made for the chemically induced fusion of erythrocyte ghosts (Cullis and Hope, 1978). It has also been suggested that fusion proceeds through the formation of inverted micellar structures at the region of adhesion, based on observations of lipidic particles at the edges of fusing cardiolipin–phosphatidylcholine vesicles in the presence of Ca^{2+} (Verkleij *et al.*, 1979) or phosphatidylethanolamine–phosphatidylcholine–cholesterol vesicles at high temperature (Verkleij *et al.*, 1980). Rand (1981) and Lis *et al.* (1982) have proposed that phosphatidylethanolamine would preferentially accumulate over phosphatidylcholine in the region of contact, based on the energy of interaction of bilayers made of these phospholipids. Since the removal of water can result in the transformation of phosphatidylethanolamine from the bilayer to the hexagonal phase (Reiss-Husson, 1967; Rand *et al.*, 1972), dehydration in the region of contact may lead to such a transformation. Siegel (1984) has calculated that the rate of formation of inverted micellar intermediates is consistent with the rates of vesicle–vesicle fusion and concluded that these inverted micelles are likely to be involved in membrane fusion but that they may be difficult to visualize in fast-freezing experiments because of their short half-lives. The Ca^{2+}-induced fusion of vesicles composed of phosphatidylethanolamine and various negatively charged phospholipids has also been proposed to involve the formation of nonlamellar, possibly inverted, micellar structures (Hope *et al.*, 1983). These conclusions are based on electron-microscopic observations of vesicles incubated for long periods of time and in the presence of glycerol used as a cryoprotectant. When the vesicles are quick-frozen (Heuser *et al.*, 1976, 1979) within seconds of stimulation with Ca^{2+} no lipidic particles can be observed at the sites of membrane fusion, although they are observed after long-term incubations and the addition of glycerol increases the number of vesicles displaying the particles (Düzgüneş *et al.*, 1982; Bearer *et al.*, 1982). Others have made similar observations (Verkleij *et al.*, 1984). Thus, lipidic particles as defined by their morphology in freeze fracture electron microscopy are not likely to be involved as intermediate structures in membrane fusion even in membrane systems in which they are observed at equilibrium (Bearer *et al.*, 1982). It should be noted, however, that lipid conformations which transiently deviate from the stable bilayer configuration may occur at the site of fusion but may be too fast to be visualized by morphological studies or may be confined to a very small area of the zone of contact between membranes (Bearer et al., 1982; Siegel, 1984; Düzgüneş et al., 1985a). This "intermediate" could be an inverted micellar or other nonbilayer configuration, or a domain of more condensed lipid in the bilayer state. It could well be a local perturbation of the lipid bilayer structure that allows for the mixing of lipid molecules between contacting vesicles, leading to the intermixing of aqueous contents. In time these intermediates could convert to more stable structures such as lipidic particles, the hexagonal H_{II} phase or the crystalline bilayer, depending on the composition of the membrane (Bearer et al., 1982).

4.3. Phospholipid Specificity in Membrane Fusion

In Sections 3.8 and 3.9 we have reviewed the properties of various phospholipid vesicles with respect to their susceptibility to fusion under different environmental conditions. Liposomes composed of acidic phospholipids fuse in the presence of Ca^{2+} at threshold concentrations that vary from approximately 10^{-4} M for phosphatidate (Papahadjopoulos et al., 1976b; Sundler and Papahadjopoulos, 1981) through 10^{-3} M for phosphatidylserine (Papahadjopoulos et al., 1977; Portis et al., 1979; Wilschut et al., 1980; Day et al., 1977), to 10^{-2} M for phosphatidylglycerol (Papahadjopoulos et al., 1976b; Rosenberg et al., 1983). Large phosphatidylinositol vesicles are resistant to Ca^{2+}-induced fusion (Sundler and Papahadjopoulos, 1981). The $[Ca^{2+}]_t$ for vesicles composed of equimolar mixtures of phosphatidate and phosphatidylcholine is raised to 3 mM (Sundler et al., 1981). Large phosphatidylserine–phosphatidylcholine (1:1) vesicles do not fuse in the presence of Ca^{2+} (Düzgüneş et al., 1981b). When phosphatidylethanolamine is substituted for phosphatidylcholine, however, the vesicles fuse at $[Ca^{2+}]_t$ slightly above that for pure phosphatidylserine (Düzgüneş et al., 1981b).

This difference between phosphatidylcholine and phosphatidylethanolamine probably arises from the differences in their physicochemical properties. Phosphatidylethanolamine head groups form a compact lattice through electrostatic interactions and hydrogen bonding between the phosphate oxygens and ammonium nitrogens, but phosphatidylcholine head groups do not allow for such interactions, the phosphate groups being intercalated with water molecules (Hauser et al., 1981). Phosphatidylcholine has a higher affinity for water than phosphatidylethanolamine, and increased unsaturation of the hydrocarbon chains of phosphatidylcholine increases the hydration of the head group. The presence of cholesterol in the membrane increases the water absorbed by phosphatidylcholine (Jendrasiak and Hasty, 1974; Jendrasiak and Mendible, 1976). Phosphatidylethanolamine molecules pack more closely than phosphatidylcholine molecules in monolayers (Phillips and Chapman, 1968). The hydrocarbon chains of dipalmitoylphosphatidylcholine bilayers below the

T_c are tilted relative to the normal to the surface of the membrane, whereas those of dipalmitoylphosphatidylethanolamine are normal to the surface, probably because of the size and conformation of the phosphatidylcholine head group (McIntosh, 1980). The head group of phosphatidylethanolamine has a space requirement similar to that of the cross section of two crystalline hydrocarbon chains; however, when the chains expand in the liquid crystalline state there would be a tendency to form a concave surface and in certain cases the bilayer would revert to the hexagonal H_{II} phase at the appropriate temperature (Hauser *et al.*, 1981; Reiss-Husson, 1967; Rand *et al.*, 1971; Cullis and deKruijff, 1978; Hui *et al.*, 1981). On the other hand, the head group of phosphatidylcholine is more flexible and can adapt to changes in the packing of the hydrocarbon chains, forming lamellar structures under a variety of conditions (Hauser *et al.*, 1981; Luzzati and Tardieu, 1974; Deamer *et al.*, 1970; Reiss-Husson, 1967). Phosphatidylethanolamine membranes experience the repulsive hydration forces at shorter interbilayer separations compared to phosphatidylcholine membranes, allowing the former to approach to closer distances (Rand, 1981; Lis *et al.*, 1982).

The inhibitory role of phosphatidylcholine is particularly apparent in the Mg^{2+}-induced fusion of membranes containing phosphatidylserine and phosphatidylethanolamine. Although Mg^{2+} does not induce the fusion of large phosphatidylserine vesicles (Wilschut *et al.*, 1981), it does become fusogenic when phosphatidylethanolamine is also present in the membrane. However, when 10% of this lipid is replaced by phosphatidylcholine, Mg^{2+} is again ineffective (Düzgüneş *et al.*, 1981b). Concerning the possible involvement in membrane fusion of the lamellar-hexagonal H_{II} phase transition of phosphatidylethanolamine, discussed in Section 4.2, the fusion of phosphatidylserine–phosphatidylethanolamine vesicles in the presence of Mg^{2+} indicates that such a macroscopic transition is not necessary for fusion (Düzgüneş *et al.*, 1981b, 1985a), since Mg^{2+} does not induce this phase transition in such mixtures (Cullis and Verkleij, 1979; Tilcock and Cullis, 1981).

The presence of phosphatidylethanolamine in phosphatidylinositol membranes also allows these membranes to undergo fusion in the presence of Ca^{2+} or Mg^{2+}, although pure phosphatidylinositol membranes do not (Sundler *et al.*, 1981). The observation that the threshold concentration of the ions is decreased as the phosphatidylethanolamine mol fraction is increased suggests that this lipid is directly involved in the membrane fusion reaction. The temperature dependence of Ca^{2+}-induced fusion of vesicles composed of equimolar mixtures of phosphatidylinositol and egg phosphatidylethanolamine or dimyristoylphosphatidylethanolamine indicates that fusion is dependent on the gel to liquid-crystalline transition of the mixture but not on the bilayer to hexagonal H_{II} transition of the phosphatidylethanolamine species (Sundler *et al.*, 1981).

The conversion of phosphatidylinositol to phosphatidate as a consequence of secretory stimuli in several cell types, a process termed the *phosphatidylinositol response* (Hokin and Hokin, 1953; Michell, 1975; Michell et al., 1981), suggests the possibility that this conversion may be involved in the fusion of secretory vesicles with the plasma membrane, particularly since phosphatidylinositol is resistant to Ca^{2+}-induced fusion and since this phenomenon is associated with the mobilization of cellular Ca^{2+}. Phosphatidate is very susceptible to undergoing fusion at low Ca^{2+} concentrations (Sundler and Papahadjopoulos, 1981) and the presence of proteins such as synexin reduces the threshold concentrations of Ca^{2+} by about two orders of magnitude (Hong et al., 1982a,b, Section 4.6). The difference in fusion susceptibility between phosphatidate and phosphatidylinositol is most likely due to the hydrated and bulky inositol moiety of the latter phospholipid, which would be unable to form dehydrated interbilayer complexes in the presence of Ca^{2+}.

4.4. Cation Specificity in Membrane Fusion: Ion Binding and Dehydration

Divalent cations exhibit a marked specificity in causing the fusion of certain phospholipid membranes. For example, large phosphatidylserine vesicles fuse extensively in the presence of Ca^{2+} (Wilschut et al., 1980) but are resistant to fusion when Mg^{2+} is added, although the vesicles aggregate (Wilschut et al., 1981, 1984). Small phosphatidylserine vesicles, however, do undergo limited fusion with Mg^{2+}, although at higher concentrations than Ca^{2+} (Papahadjopoulos et al., 1977; Wilschut et al., 1981; Ohki, 1982, 1984). Even at saturation concentrations of divalent cation, the initial rate of fusion is an order of magnitude higher for Ca^{2+} than for Mg^{2+}. The rate of vesicle aggregation would be influenced by the residual electrostatic repulsion between vesicles even around 15 mM of the cations, since complete charge neutralization occurs at considerably higher cation concentrations (Nir et al., 1980; McLaughlin et al., 1981). The repulsion would be greater in the presence of Mg^{2+} than in that of Ca^{2+}, as Ca^{2+} binds more avidly to phosphatidylserine than Mg^{2+} (Portis et al., 1979; Newton et al., 1978; Nir et al., 1978). Studies on the kinetic order of fusion indicate that fusion in the presence of Mg^{2+} and 100 mM Na^+ is delayed with respect to aggregation, whereas for Ca^{2+} fusion occurs as fast as aggregation. Moreover, Mg^{2+}-induced fusion is initially more leaky than Ca^{2+}-induced fusion; yet the vesicles retain most of their aqueous contents, unlike the collapse of the internal aqueous space in the presence of Ca^{2+}, as fusion ceases after a few rounds when the vesicles have grown to a limiting size (Wilschut et al., 1981). It appears that the vesicles aggregate in the primary minimum of the interbilayer interaction energy versus distance curve (Nir and

Bentz, 1978; Nir et al., 1981), are destabilized because of Mg^{2+} binding, and their membranes eventually intermix at the points of destabilization when these points are aligned in the apposed bilayers (Wilschut et al., 1981).

A critical ratio of bound divalent cation per phosphatidylserine molecule is required for the destabilization and fusion of small vesicles composed of phosphatidylserine or its mixtures with phosphatidylcholine (Düzgüneş et al., 1981a). The critical amount of bound cation depends on whether the aggregation step or the fusion reaction per se is rate-limiting to the overall fusion process (Bentz et al., 1983b). Under conditions where the aggregation is rate-limiting (Na^+ or Li^+ concentrations of 100 mM in the medium), the sequence of effectiveness of the divalent cation in terms of the bulk concentrations decreases as $Ba^{2+} > Ca^{2+} > Sr^{2+} > Mg^{2+}$ (Bentz et al., 1983b; Ohki, 1982), which is the same as the sequence for aggregation (Ohki and Düzgüneş, 1979; Ohki et al., 1982). Under conditions where the rate of bilayer destabilization is rate limiting (Na^+ or Li^+ concentrations at or above 500 mM) and where the vesicles aggregate reversibly in the presence of subfusogenic concentrations of divalent cations, Ca^{2+} and Ba^{2+} appear to be effective to the same degree. When the amount of divalent cation bound per phosphatidylserine is considered, however, Ca^{2+} is more effective than Ba^{2+} (Bentz et al., 1983b). A similar observation has been made for large phosphatidylserine vesicles, using Mg^{2+} to enhance the aggregation and rendering the fusion step rate-limiting (Düzgüneş et al., 1983b; Bentz and Düzjüneş, 1985). Considerably more Sr^{2+} needs to be bound to the large vesicles than to the small ones for the induction of fusion, whereas the amount of Ca^{2+} or Ba^{2+} bound at the threshold of fusion does not appear to depend on vesicle size. Monovalent cations displace divalent cations from the membrane surface at high concentrations, increasing the bulk threshold concentration of the divalent cation necessary for fusion (Düzgüneş et al., 1981a). At high concentrations that render fusion per se rate-limiting, the sequence of effectiveness of the monovalent ions in inhibiting Ca^{2+} or Mg^{2+}-induced fusion of small phosphatidylserine vesicles is $Li^+ > Na^+ > K^+ >$ tetramethylammonium$^+$ (Nir et al., 1983b), which is the same as the sequence of their binding constants to phosphatidylserine (Eisenberg et al., 1979).

The specificity of Ca^{2+} over Mg^{2+} in membrane fusion has also been observed in large vesicles composed of phosphatidylserine, phosphatidylethanolamine, and a small amount of phosphatidylcholine, the latter component being the determinant of the resistance to fusion in the presence of Mg^{2+} (since phosphatidylserine–phosphatidylethanolamine vesicles do undergo fusion with Mg^{2+}) (Düzgüneş et al., 1981b). The enhancement of the rate and extent of fusion by the cytosolic protein synexin also exhibits an absolute specificity for Ca^{2+} over Mg^{2+} (Creutz et al., 1978; Düzgüneş et al., 1980; Hong et al., 1981, 1982a,b) and will be discussed further in Section 4.6. The physicochemical differences between the Ca^{2+} and Mg^{2+} complexes of phosphatidylserine offer an explanation for the specificity of Ca^{2+} over Mg^{2+} in membrane fusion. Ca^{2+} binds phosphatidylserine more avidly than Mg^{2+} (Newton *et al.*, 1978; Nir *et al.*, 1978; Portis *et al.*, 1979) and its water of hydration can be removed more readily (Gresh, 1980). Ca^{2+} has a coordination number of 7 or 8, with variable bond angles and lengths, whereas Mg^{2+} has a coordination number of 6, with the bonds shaped as a regular octahedron. The flexible nature of its coordination bonds makes Ca^{2+} a favorable ion for reversibly cross-linking molecules (Williams, 1974, 1976). X-ray diffraction and NMR studies (Hauser *et al.*, 1975, 1977; Papahadjopoulos *et al.*, 1978a; Portis *et al.*, 1979) indicate that the interbilayer space is dehydrated when the Ca^{2+}-phosphatidylserine complex is formed but the Mg^{2+} complex maintains its layer of hydration. The anhydrous Ca^{2+} complex is formed on interbilayer contact (Portis *et al.*, 1979) and the Ca^{2+}/Mg^{2+} specificity is likely to arise from this interaction. It has been proposed that a new binding mode of Ca^{2+}, called the *trans* complex, occurs when bilayers are brought into contact, allowing Ca^{2+} to bind to two membrane surfaces simultaneously, thus increasing abruptly the affinity of Ca^{2+} for phosphatidylserine, dehydrating the point of contact between membranes, and inducing membrane fusion (Portis *et al.*, 1979; Rehfeld *et al.*, 1981; Ekerdt and Papahadjopoulos, 1982; Düzgüneş and Papahadjopoulos, 1983).

4.5. Osmotic Effects in Membrane Fusion

We have noted that certain types of small phospholipid vesicles tend to fuse under conditions where large vesicles are resistant to fusion. The driving force for these fusion events may be the high curvature of the bilayer and molecular packing defects resulting from it (Wilschut *et al.*, 1981; Schmidt *et al.*, 1981). Vesicles that grow to a limiting size and stop fusing under isoosmotic conditions (across the vesicle bilayer) may be induced to undergo further fusion when an osmotic gradient is applied across the membrane. Miller *et al.*, (1976) have found that liposomes composed of asolectin–phosphatidylserine (6:4) and containing cytochrome oxidase in one population and mitochondrial hydrophobic protein in the other as elements of a fusion assay monitoring the intermixing of membrane components (Miller and Racker, 1976b) grow from a size of about 40 nm to 100–200 nm via fusion in the presence of Ca^{2+}. When an osmotic gradient is applied with the internal osmotic strength higher than that in the medium, the vesicles undergo a secondary fusion process, growing up to 500 nm in diameter and up to 2 μm if sequential osmotic gradients are applied. Taupin *et al.* (1975) have proposed that osmotic pressure induces pores in egg phosphatidylcholine and dipalmitoylphosphatidylcholine vesicles and that the difference in the number of pore sites between the two vesicle types accounts for the difference in their fusion behavior. Cohen et al. (1982) have shown that multilamellar phosphatidylcholine–phosphatidylserine (4:1) vesicles in the presence of 15 mM Ca^{2+} can be induced to fuse with a planar membrane composed of crude soybean phospholipids, as indicated by the transfer of porin, when the vesicles are swollen osmotically. Ohki (1984) has found that large (1-μm- and 0.6-μm-diameter) phosphatidylserine vesicles aggregated by Mg^{2+} undergo fusion if an osmotic pressure gradient exists across the membrane and that the larger vesicles require a smaller osmotic gradient. The effect of the osmotic pressure gradient has been interpreted to be due to the increase in surface energy of the membrane, similar to the effects of divalent cations and temperature in this system.

The osmotic properties of isolated chromaffin granules in the presence of Mg^{2+}-ATP and Cl^- (Oka et al., 1965; Pollard et al., 1979; Pazoles and Pollard, 1978; Creutz and Pollard, 1980) and of platelet secretory granules in the presence of Ca^{2+} (VanderMeulen and Grinstein, 1982) have led to the proposal that the osmotic swelling of secretory granules during stimulation of the cells provides the driving force for membrane fusion during exocytosis. The possible role of osmotic forces during exocytosis has been proposed for a number of secretory cells, such as *Limulus* amebocytes (Ornberg and Reese, 1981), *Tetrahymena* (Satir et al., 1972, 1973), adrenal chromaffin cells (Pollard et al., 1979, 1981; Hampton and Holz, 1983), human platelets (Pollard et al., 1977), bovine parathyroid cells (Brown et al., 1978), and pancreatic cells (Orci and Malaisse, 1980). The mechanism by which secretory granules swell osmotically may be different in different cellular systems, such as Cl^- transport in chromaffin cells and OH^- transport in platelets. In contrast, Grinstein et al. (1982) have proposed that H^+-alkali cation countertransport is the cause of granule swelling. It should be noted, however, that the osmotic properties of granules may be altered on their isolation, rendering them more susceptible to lysis than granules *in situ* (Hampton and Holz, 1983).

4.6. Modulation of Membrane Fusion by Proteins, Polypeptides, and Polyamines

The free Ca^{2+} concentration at which exocytosis is induced in a variety of cellular systems is in the micromolar range (Baker and Knight, 1984; Dunn and Holz, 1983; Wilson and Kirschner, 1983; Baker et al., 1980). The fusion of isolated secretory vesicles with each other also occurs at much lower Ca^{2+} concentrations than those required for the fusion of phospholipid vesicles (Figure 1), and this Ca^{2+} sensitivity is abolished when the vesicles are treated with proteases and neuraminidase, suggesting that glycoproteins are involved in their fusion (Dahl et al., 1979; Gratzl et al., 1980; Ekerdt et al., 1981). Ca^{2+} may act directly on these proteins, which may then promote the close approach and/or fusion of the membranes in which they are embedded. In this sense, these proteins may be similar to the viral membrane proteins that are activated by an elevated H^+ concentration (White et al., 1983).

The only Ca^{2+}-binding protein promoting membrane fusion so far characterized is a cytosolic protein originally isolated from bovine adrenal medulla and subsequently found in liver, brain, and platelets and is called synexin. This protein mediates the aggregation of isolated chromaffin granules at Ca^{2+} concentrations above 6 μM and also self-aggregates (Creutz et al., 1978; 1979). The fusion-enhancing effect of synexin was first demonstrated by the use of phospholipid vesicles. Synexin increases the initial rate of fusion of pure phosphatidylserine and mixed phosphatidylserine–phosphatidylethanolamine vesicles in the presence of Ca^{2+} and lowers the threshold Ca^{2+} concentration (Düzgüneş et al., 1980; Hong et al., 1981, Figure 5). The effect of synexin is specific for the phospholipid composition of the membrane. For example, phosphatidylserine–phosphatidylcholine vesicles aggregate at a lower Ca^{2+} concentration in the presence of synexin, but their fusion is not enhanced (Hong et al., 1981) and the Ca^{2+}-induced fusion of phosphatidylinositol–phosphatidylethanolamine vesicles is inhibited by synexin, although the vesicles aggregate (Hong et al., 1982a). Phosphatidate–phosphatidylethanolamine vesicles exhibit the most dramatic enhancement of fusion in the presence of synexin. Here the initial rate of fusion increases by three orders of magnitude and the $[Ca^{2+}]_t$ is reduced to less than 100 μM (Hong et al., 1982a). Ca^{2+} concentrations as low as 10 μM can induce fusion in the presence of Mg^{2+} concentrations that are below the threshold for fusion. Figure 6 shows the time course of fusion in the presence of 7 μg/ml synexin, 1 mM Mg^{2+}, and 20 μM Ca^{2+}; in the

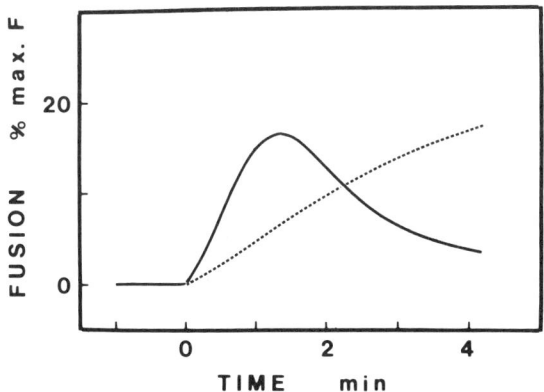

FIGURE 5. Fusion of large unilamellar phosphatidylserine vesicles in the presence (solid line) and absence (dotted line) of 10 μg/ml synexin, induced by 2 mM Ca^{2+}. Other conditions were as in Figure 2. (K. Hong and N. Düzgüneş, unpublished data.)

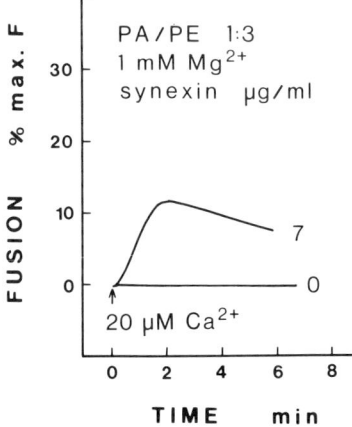

FIGURE 6. The effect of synexin on the fusion of large unilamellar phosphatidate–phosphatidylethanolamine (1:3) vesicles in the presence of 20 μM free Ca^{2+} and 1 mM Mg^{2+}. Other conditions were as in Figure 2. (K. Hong and N. Düzgüneş, unpublished data.)

absence of synexin no fusion can be detected. Thus, although synexin is not very specific in enhancing the Ca^{2+}-induced aggregation of vesicles composed of a variety of acidic phospholipids or of certain biological membranes (Hong et al., 1981, 1982a; Morris and Hughes, 1979), it is specific in enhancing Ca^{2+}-induced fusion of particular membranes. The phospholipid specificity of synexin in facilitating membrane fusion is summarized in Table III.

Synexin is most likely to interact with the phospholipid bilayer of chromaffin granules since it is also active in aggregating trypsin-treated granule membranes (Morris et al., 1982). However, unlike its effect on certain phospholipid membranes, synexin in the presence of Ca^{2+} does not mediate the fusion of granule membranes. This may be due to the phospholipid composition of the cytoplasmic monolayer of the membrane. Granules aggregated by Ca^{2+}

Table III
Modulation by Synexin of Ca^{2+}-Induced Fusion of Phospholipid Vesicles[a]

Phospholipid vesicle composition	Reduction of the threshold of aggregation	Enhancement of the rate of fusion	$[Ca^{2+}]$ threshold (mM)
Phosphatidylserine	+	+	1.0
Phosphatidylserine–phosphatidylethanolamine (1:3)	+	+	1.0
Phosphatidate–phosphatidylethanolamine (1:3)	+	+	0.1
Phosphatidylinositol–phosphatidylethanolamine (1:3)	+	Inhibited	3.0
Phosphatidylserine–phosphatidylcholine (3:1)	+	No effect	5.0

[a]Data from Hong et al., 1982a.

and synexin can be induced to fuse by the addition of *cis* unsaturated fatty acids such as arachidonic acid and oleic acid but not by *trans* unsaturated or saturated fatty acids, glycerylmonooleate, detergents, or lysolecithin (Creutz, 1981a). These observations have been made at pH 6.0 and increasing the pH to 7.2 decreases the rate of fusion. In this respect it should be noted that liposomes composed of oleic acid and phosphatidylethanolamine also fuse rapidly and extensively around pH 6.0 (Düzgüneş *et al.*, 1983a, 1985b). Fusion of the chromaffin granule membranes by the combined action of Ca^{2+}, synexin and *cis* unsaturated fatty acids does not appear to depend on the chemiosmotic properties of the granule membrane, although the internal volume of the fusion product must increase from that of the initial vesicles. The fusogenic action of arachidonic acid also suggests that the free fatty acid formed during the stimulation of platelets may be involved in membrane fusion during exocytosis in these cells (Creutz, 1981a).

Ca^{2+}-binding proteins other than synexin have been shown to be either slightly inhibitory to fusion, as with calmodulin from bovine brain, electroplax and bovine heart, and parvalbumin from rabbit muscle (Düzgüneş *et al.*, 1980; Hong *et al.*, 1981), or strongly inhibitory, as in the case of prothrombin and its proteolytic fragment 1, which retains the Ca^{2+}-binding activity of the parent protein (Hong *et al.*, 1982a, Table IV). In contrast, a Ca^{2+}-binding protein isolated from calf brain synaptic membranes (Abood *et al.*, 1976) and incorporated into planar bilayers composed of asolectin–diphytanoylphosphatidylcholine (9:1) stimulates the fusion of the latter with phosphatidylcholine–phosphatidylserine (4:1) liposomes in the presence of 10 μM Ca^{2+} and an osmotic gradient across the planar membrane (Zimmerberg *et al.*, 1980b). The fusion effects of prothrombin and fragment 1 may be the result of the inability of the exposed surface of the protein to bind to another vesicle after binding to one;

Table IV
Modulation of Ca^{2+}-Induced Membrane Fusion by Various Ca^{2+}-Binding Proteins[a]

Proteins	Effect on fusion of phospholipid vesicles
Calmodulin	Slight inhibition
Parvalbumin	Slight inhibition
Prothrombin	Inhibition
Fragment 1 of prothrombin	Inhibition
Synexin	Enhancement

Data from Hong *et al.*, 1981, 1982a.

that is, these proteins are monopolar with respect to their ability to bind to a membrane surface. Synexin, however, is bipolar in nature, as evidenced by its property of self-aggregation (Creutz et al., 1979), and would be able to bind two vesicles simultaneously (Hong et al., 1982a). The energetics of close interbilayer approach would thus be altered at points where synexin cross-links the two bilayers.

Several other proteins have been isolated that associate with membranes in a Ca^{2+}-dependent manner. Odenwald and Morris (1983) have identified a synexin-like protein that aggregates chromaffin granules but whose molecular weight and aggregation kinetics are different from synexin. Pollard and Scott (1982) have isolated from bovine liver a protein called synhibin that inhibits the action of synexin. Südhof et al. (1982) have found a protein (called calelectrin) in *Torpedo* electric organ that aggregates synaptic vesicles from the same source and also chromaffin granules in the presence of Ca^{2+}. They have subsequently identified calelectrin of different molecular weights in bovine liver, brain, and adrenal medulla (Südhof et al., 1984). Creutz (1981b) and collaborators (Creutz et al., 1983) have characterized a series of proteins (called chromobindins) that bind to immobilized chromaffin granule membranes to varying degrees, depending on the Ca^{2+} concentration; two of these proteins are synexin and calmodulin. How the preceding proteins (except the latter two) affect the fusion of phospholipid vesicles or secretory vesicles is not known.

Lectins have also been shown to facilitate the fusion of phospholipid vesicles containing glycolipids. Sundler and Wijkander (1983) have found that the Ca^{2+} threshold for fusion of phosphatidate–phosphatidylethanolamine–phosphatidylethanol-lacto-bionamide vesicles is reduced by an order of magnitude when preincubated with *Ricinus communis* agglutinin. The preagglutination of the vesicles increases the specificity of Ca^{2+} over Mg^{2+} in this system from less than 2 to greater than 10. Thus, the establishment of interbilayer contact by means of the lectin increases both the sensitivity to Ca^{2+} and the Ca^{2+}/Mg^{2+} selectivity in this system, similar to the effect of synexin on phosphatidate–phosphatidylethanolamine vesicles. Soybean agglutinin enhances the initial rate of fusion of phosphatidylserine–globoside vesicles induced by Ca^{2+}, and this effect is inhibited by the presence of the haptenic sugar in the medium. Wheat germ agglutinin, likewise, enhances the rate and reduces the threshold Ca^{2+} concentration of fusion of phosphatidate–phosphatidylethanolamine–disialoganglioside vesicles (Düzgüneş et al., 1984b). The enhancement of the Ca^{2+}-induced fusion of phosphatidate–phosphatidylethanolamine–globoside vesicles by soybean agglutinin requires the addition of Ca^{2+} after the lectin, although the rate of agglutination of the vesicles in the presence of lectin and Ca^{2+} is independent of the order of addition (Hoekstra et al., 1985). Thus, fusion appears to require the alteration of the bilayer by the lectin before Ca^{2+}

addition, as evidenced by the increase in membrane permeability during incubation of the vesicles with the lectin. The lectin concanavalin A has also been reported to initiate the fusion of small dipalmitoylphosphatidylcholine vesicles at the phase transition temperature (van der Bosch and McConnell, 1975). We have suggested that endogenous lectin-like molecules may be involved in the recognition and facilitation of the fusion of intracellular membranes (Düzgüneş et al., 1984b).

Several other proteins have been investigated with respect to their capacity to modulate fusion. Spectrin from human erythrocytes inhibits the formation of the *trans* Ca^{2+}-phosphatidylserine complex, the release of encapsulated material from phosphatidylserine vesicles and their fusion in the presence of Ca^{2+} (Portis et al., 1979; Düzgüneş et al., 1980; Figure 7). Spectrin also inhibits the fusion of dimyristoylphosphatidylglycerol vesicles induced by divalent cations (Mombers et al., 1977). The major proteolipid apoprotein from myelin (N-2 or lipophilin) causes the fusion of dimyristoylphosphatidylcholine vesicles with dipalmitoylphosphatidylcholine vesicles above the phase transition of both vesicles (Papahadjopoulos et al., 1976a). In contrast, N-2 (or glycophorin) incorporated into phosphatidylserine vesicles increases the $[Ca^{2+}]_t$ for aggregation (Ohki and Leonards, 1982). The fusion of mixed phosphatidylserine–phosphatidylcholine vesicles containing these proteins with phosphatidylserine monolayers is also inhibited. Myelin basic protein induces the fusion of vesicles composed of phosphatidylcholine (Smith, 1977), phosphatidylethanolamine (Stollery and Vail, 1977), and mixtures of phosphatidylglycerol with dipalmi-

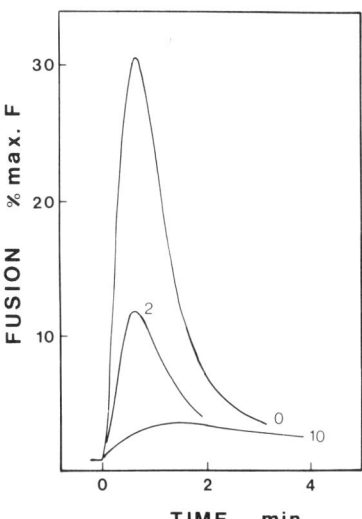

FIGURE 7. The effect of spectrin (μg) on the fusion of large unilamellar phosphatidylserine vesicles induced by 5 mM Ca^{2+}. Other conditions were as in Figure 2. (N. Düzgüneş and J. Wyatt, unpublished data.)

toylphosphatidylcholine or with lysolecithin and phosphatidylcholine (Lampe and Nelsestuen, 1982). Albumin and fragments of albumin induce the fusion of small phosphatidylcholine vesicles at pH below 4 (Schenkman *et al.*, 1981; Garcia *et al.*, 1983). The presence of sulfatide in phosphatidylcholine vesicles greatly enhances fusion induced by albumin, presumably because of the generation of sulfatide domains and membrane instability (Cestaro *et al.*, 1983). Clathrin, the major protein of coated vesicles, induces the fusion of small dioleoylphosphatidylcholine vesicles at pH 6 (Blumenthal *et al.*, 1983), although this may not be a general phenomenon for vesicles made of other types of phosphatidylcholines, since clathrin does not cause the destabilization of small or large egg phosphatidylcholine vesicles (Hong *et al.*, 1983a) in the way that it does small dioleoylphosphatidylcholine or dipalmitoylphosphatidylcholine vesicles (Steer *et al.*, 1982).

Certain polypeptides have been shown to induce the fusion of small unilamellar vesicles. The cyclic oligopeptide, alamethicin, causes the fusion of phosphatidylcholine vesicles (Lau and Chan, 1975), and the antibiotic polymyxin B induces fusion in cardiolipin–phosphatidylethanolamine–phosphatidylcholine vesicles (Gad and Eytan, 1982). Polylysine treatment of vesicles containing various acidic phospholipids leads to their fusion, dependent on the size of the polycation and the presence of phosphatidylethanolamine in the vesicles (Gad *et al.*, 1982). The lipophilic peptide of bee venom, mellitin, causes fusion of vesicles composed of a variety of phosphatidylcholines (Morgan *et al.*, 1983) or a mixture of phosphatidylethanolamine–cardiolipin–phosphatidylcholine (Eytan and Almary, 1983).

Recently, Lucy (1984) has proposed that hydrophobic peptides produced by the proteolysis of membrane or cellular proteins may induce membrane fusion in cellular systems. These polypeptides would be functionally similar to the proteins of lipid-enveloped viruses. For example, treatment of chicken erythrocytes with Ca^{2+} and ionophore, which leads to microvesicle release and nucleus-plasma membrane fusion, is associated with the breakdown of certain proteins by endogenous proteases (Thomas *et al.*, 1983); fusion of human erythrocytes by chlorpromazine and Ca^{2+} is related to the activation of serine- and cysteine proteinases (Lang *et al.*, 1984); fusion of rat myoblasts is accompanied by the Ca^{2+} activation of a neutral proteinase (Kaur and Sanwal, 1981); exocytotic events involved in synaptic transmission are associated with a metalloendoprotease activity (Baxter *et al.*, 1983).

Polyamines are very effective modulators of membrane fusion. Spermine and spermidine induce the aggregation but not fusion of large unilamellar vesicles composed of phosphatidylserine, phosphatidate, and mixtures of phosphatidate with phosphatidylcholine at physiological concentrations (Schuber *et al.*, 1983). Vesicles composed of acidic phospholipids, cholesterol, and a high mole fraction of phosphatidylethanolamine fuse in the presence of these polyamines and in the absence of divalent cations (Hong et al., 1983b; Schuber et al., 1983). Pure phosphatidylcholine vesicles do not aggregate in the presence of polyamines. Spermine drastically increases the rate of Ca^{2+}-induced fusion of phosphatidate liposomes and decreases the $[Ca^{2+}]_t$. These effects are not as large in the case of phosphatidylserine or phosphatidate–phosphatidylcholine vesicles. We have proposed that polyamines may be regulators of membrane fusion processes involved in cell growth, cell division, exocytosis, and fertilization (Schuber et al., 1983).

4.7. Implications for Subcellular Membrane Fusion

A localized hydrophobic interaction between two membranes within the area of contact, such as the insertion of a hydrophobic polypeptide on one membrane into the other or the apposition of hydrophobic point defects in molecular packing in two membranes, appears to be common to many types of membrane fusion. Such an interaction leading to membrane fusion can be achieved under a large variety of conditions. The composition, curvature, and phase state of the membrane, the ionic environment, osmotic gradients, membrane proteins and cytoplasmic proteins, as well as low-molecular-weight modulators such as polyamines and phosphate ions, all determine the fusion susceptibility of the membrane. These factors could be utilized by the cell to varying degrees to control the fusion of subcellular membranes.

Ca^{2+} appears to be intimately involved in stimulus-secretion coupling (Douglas, 1974; Rubin, 1974), and the threshold Ca^{2+} concentration for exocytosis is estimated to be in the micromolar range (Baker et al., 1980). Creutz et al. (1982), however, have argued that the Ca^{2+} concentration at the site of adhesion between secretory granules and the plasma membrane may be as high as 400 μM–1 mM. Baker and Knight (1984) have countered the proposals of locally high Ca^{2+} concentrations during fusion by the argument that the free Ca^{2+} concentration inducing exocytosis is the same even at high EGTA concentrations that would be expected to buffer the intracellular Ca^{2+} very effectively and that the Ca^{2+} threshold in permeabilized cells is compatible with the Ca^{2+} transient recorded in Quin-2 loaded cells. Factors other than Ca^{2+}, such as Mg^{2+}–ATP, are also required for exocytosis. The site of action of Ca^{2+} is not known. Whether the presence of divalent cation-stimulated ATPase activity in secretory vesicles (Trifaró et al., 1976; Breer et al., 1977) and at exocytosis sites (Plattner et al., 1977) is involved in fusion is not known (Plattner, 1978). The Ca^{2+}-dependent phosphorylation of the proteins of synaptic vesicles (DeLorenzo and Freedman, 1977), the Ca^{2+}-dependent trifluoperazine-sensitive phosphorylation of an insulinoma cell protein associated with insulin release (Schubart et al., 1980), and the reduction of the Ca^{2+} threshold for exocytosis by phorbol ester (Knight and Baker, 1983) suggest that a Ca^{2+}-

sensitive protein kinase may be involved in this process. Plattner (1978) has suggested that the phosphorylation of membrane proteins may affect their lateral distribution as well as influence the formation of lipid domains. It is possible that Ca^{2+} has several targets that are involved both in the redistribution of membrane components and in the close approach, destabilization, and fusion of the membranes.

On the other hand, the induction of exocytosis in a number of cell types without a rise in cytoplasmic free Ca^{2+} as determined by Quin-2 fluorescence (Rink et al., 1982, 1983; Korchak et al., 1984; Shoback et al., 1984) raises the possibility that other cellular factors activated by the secretagogues cause the fusion of secretory granules with the plasma membrane.

If the local Ca^{2+} concentration at exocytosis sites is high, the direct interaction of Ca^{2+} with phospholipids such as phosphatidylserine and phosphatidate in the cytoplasmic monolayer of the plasma membrane and secretory granule is likely. The enrichment of phosphatidylethanolamine and acidic phospholipids in the inner monolayer of the plasma membrane of various cells (Op den Kamp, 1979; Zwaal and Bevers, 1983; Sandra and Pagano, 1978; Fontaine et al., 1980), and the cytoplasmic Mg^{2+}, would contribute to the fusion susceptibility of the membrane when the Ca^{2+} concentration increased transiently. The phospholipid bilayer of coated vesicles is also thought to be directly involved in the fusion on the "uncoated" vesicles with lysosomes (Altstiel and Branton, 1983). The generation of phosphatidate at the expense of phosphatidylinositol during stimulation of cells by ligands (Michell, 1975; Michell et al., 1981) may also be involved in the generation of fusogenic conditions, especially in the presence of Ca^{2+}-binding proteins such as synexin (Sundler and Papahadjopoulos, 1981; Hong et al., 1982a). The threshold Ca^{2+} concentration for the fusion of phosphatidate–phosphatidylethanolamine in the presence of millimolar Mg^{2+} is reduced to about 10 μM by synexin (Hong et al., 1982a,b). This Ca^{2+} concentration is two orders of magnitude lower than that necessary for fusion in the absence of synexin but is still an order of magnitude higher than that required for exocytosis, assuming that the bulk concentration reflects the Ca^{2+} concentration at exocytosis sites. Since the function of synexin in enhancing the rate of fusion and reducing the threshold Ca^{2+} concentration is inhibited by the presence of low mol fractions of randomly distributed phosphatidylcholine in the membrane, it is expected that the fusion susceptible lipid would be laterally segregated into domains that would interact with synexin (Hong et al., 1982b; Düzgüneş et al., 1985a). Phospholipid domains have been demonstrated in certain cell membranes (Bearer and Friend, 1980, 1982; Karnovsky et al., 1982). Exocytotic sites in the plasma membrane of pancreatic β cells have been shown to be rich in cholesterol (Orci et al., 1981; Orci, 1982). Higher concentrations of anionic lipids have been detected in the sperm plasma membrane in regions where the acrosome reaction takes place (Bearer and Friend, 1982). Although it is possible that synexin, or a similar Ca^{2+}-binding protein, is the intracellular receptor for Ca^{2+} in cells that undergo Ca^{2+}-induced exocytosis (Creutz *et al.*, 1978, 1982; Hong *et al.*, 1982a,b), several other cytoplasmic and membrane factors may contribute to the lower threshold Ca^{2+} concentrations in cellular systems. Future studies on the interaction and fusion of isolated secretory vesicles with inside-out plasma membrane preparations are expected to facilitate the identification of these factors.

ACKNOWLEDGMENTS

The work of the author has been supported by the U.S. National Institutes of Health (Fellowship CA-06190; Grants GM-28117, CA-25526 and GM-26369), the March of Dimes Foundation (Grant 1-758), the Scientific Affairs Division of the North Atlantic Treaty Organization (Research Grant RG 151.81), and the American Heart Association (Grant-in-Aid, with funds contributed by the California affiliate). Jean Swallow is gratefully acknowledged for preparing the manuscript.

5. REFERENCES

Abbs, M. T. and Phillips, J. H., 1980, Organization of the proteins of the chromaffin granule membrane, *Biochim. Biophys. Acta* **595**:200–221.

Abood, L. G., Hong, J. S., Takeda, F., and Tometsko, A. M., 1976, Preparation and characterization of calcium-binding and other hydrophobic proteins from synaptic membranes, *Biochim. Biophys. Acta* **443**:414–427.

Ahkong, Q. F., Fisher, D., Tampion, W., and Lucy, J. A., 1973a, The fusion of erythrocytes by fatty acids, esters, retinol and α-tocopherol, *Biochem. J.* **136**:147–155.

Ahkong, Q. F., Cramp, F. C., Fisher, D., Howell, J. I., Tampion, W., Verrinder, M., and Lucy, J. A., 1973b, Chemically-induced and thermally-induced cell fusion: Lipid–lipid interactions, *Nature New Biol.* **242**:215–217.

Ahkong, Q. F., Fisher, D., Tampion, W., and Lucy, J. A., 1975a, Mechanisms of cell fusion, *Nature* **253**:194–195.

Ahkong, Q. F., Tampion, W., and Lucy, J. A., 1975b, Promotion of cell fusion by divalent cation ionophores, *Nature* **256**:208–209.

Ahkong, Q. F., Howell, J. I., Lucy, J. A., Safwat, F., Davey, M. R., and Cocking, E. C., 1975c, Fusion of hen erythrocytes with yeast protoplasts induced by polyethylene glycol, *Nature* **255**:66–67.

Ahkong, Q. F., Blow, A. M. J., Botham, G. M., Launder, J. M., Quirk, S. J., and Lucy, J. A., 1978, Proteinases and cell fusion, *FEBS Lett* **95**:147–152.

Allan, D., and Michell, R. H., 1975, Accumulation of 1,2-diacylglycerol in the plasma membrane may lead to echinocyte transformation of erythrocytes, *Nature* **258**:348–349.

Allan, D., and Michell, R. H., 1977, Calcium-ion dependent diacyglycerol accumulation in erythrocytes is associated with microvesiculation but not with efflux of potassium ions, *Biochem. J.*, **166**:495–499.

Allan, D., and Michell, R. H., 1979, The possible role of lipids in control of membrane fusion during secretion, *Symp. Soc. Exp. Biol.* **33**:323–336.

Allan, D., Low, M. G., Finean, J. B., and Michell, R. H., 1975, Changes in lipid metabolism and cell morphology following attack by phospholipase C *(Clostridium perfringens)* on red cells or lymphocytes, *Biochim. Biophys. Acta* **413**:308–316.

Allan, D., Billah, M. M., Finean, J. B. and Michell, R. H., 1976, Release of diacylglycerol-enriched vesicles from erythrocytes with increased intracellular [Ca^{2+}], *Nature* **261**:58–60.

Allison, A. C., and Davies, P., 1974, Mechanisms of endocytosis and exocytosis, *Symp. Soc. Exp. Biol.* **28**:419–446.

Altstiel, L., and Branton, D., 1983, Fusion of coated vesicles with lysosomes: Measurement with a fluorescence assay, *Cell* **32**:921–929.

Amano, F., Hashida, R., and Mizuno, D., 1979, Membrane fusion of phagocytic vesicles with lysosomes in polymorphonuclear leukocytes without phagocytic stimuli, *FEBS Lett.* **106**:171–175.

Amano, F., Hashida, R., and Mizuno, D., 1981, Lysosomal fusion in endocytosis and exocytosis. I. Demonstration and characterization of two fusion reactions proceeding simultaneously in non-phagocytosing Guinea pig polymorphonuclear leukocytes, *Exptl. Cell Res.* **136**:15–26.

Amsterdam, A., Ohad, I. and Schramm, M., 1969, Dynamic changes in the ultrastructure of the acinar cell of the rat parotid gland during the secretory cycle, *J. Cell Biol.* **41**:753–773.

Anderson, R. G. W., Brown, M. S. and Goldstein, J. L., 1977, Role of the coated endocytic vesicle in the uptake of receptor-bound low density lipoprotein in human fibroblasts, *Cell* **10**:351–364.

Armitage, I. M., Shapiro, D. L., Furthmayr, H., and Marchesi, V. T., 1977, ^{31}P-Nuclear magnetic resonance evidence for polyphosphoinositide associated with the hydrophobic segment of glycophorin A, *Biochemistry* **16**:1317–1320.

Arnold, K., Pratsch, L. and Gawrisch, K., 1983, Effect of poly(ethyleneglycol) on phospholipid hydration and polarity of the external phase, *Biochim. Biophys. Acta* **728**:121–128.

Asano, A., and Sekiguchi, K., 1978, Redistribution of intramembrane particles of human erythrocytes induced by HVJ (Sendai virus): A prerequisite for the virus-induced cell fusion, *J. Supramolec. Struct.* **9**:441–452.

Aunis, D., Hesketh, J. E. and Devilliers, G., 1979, Freeze-fracture study of the chromaffin cell during exocytosis: Evidence for connections between the plasma membrane and secretory granules and for movements of plasma membrane-associated particles, *Cell Tissue Res.* **197**:433–441.

Bächi, T., Aguet, M., and Howe, C., 1973, Fusion of erythrocytes by Sendai virus studied by immuno-freeze-etching, *J. Virol.* **11**:1004–1012.

Bächi, T., Deas, J. E., and Howe, C., 1977, Virus-erythrocyte membrane interactions, *Virus Infection and the Cell Surface* (G. Poste and G. L. Nicolson, eds.), pp. 83–127, Elsevier/ North-Holland Biomedical Press, Amsterdam.

Bächi, T., Eichenberger, G., and Hauri, H. P., 1978, Sendai virus hemolysis: Influence of lectins and analysis by immune fluorescence, *Virology* **85**:518–530.

Badzhinyan, S. A. and Chailakhyan, L. M., 1971, Measurement of the electrical capacitance of the region of contact of two bimolecular phospholipid membranes, *Biofizika* **16**:1141–1143.

Badzhinyan, S. A., Dunin-Barkovskii, V. L., Kovalev, S. A. and Chailakhyan, L. M., 1971, Electrical resistance of the region of adhesion of bimolecular phospholipid membranes, *Biofizika* **16**:1019–1024.

Badzhinyan, S. A., Berkinblit, M. B., Kovalev, S. A. and Chailakhyan, L. M., 1972, Investiga-

tion of the electrical structure of the region of adhesion of two bimolecular phospholipid membranes modified by TTPB, *Biofizika* **17**:428-434.
Baker, P. F., 1972, Transport and metabolism of calcium ions in nerve, *Prog. Biophys. Mol. Biol.* **24**:177-223.
Baker, P. F. and Knight, D. E., 1978, Calcium-dependent exocytosis in bovine adrenal medullary cells with leaky plasma membranes, *Nature* **276**:620-622.
Baker, P. F. and Knight, D. E., 1984, Calcium control of exocytosis in bovine adrenal medullary cells, *Trends Neurosci.* **7**:120-126.
Baker, P. F., and Whitaker, M. J., 1978, Influence of ATP and calcium on the cortical reaction in sea urchin eggs, *Nature* **276**:513-515.
Baker, P. F., Knight, D. E., and Whitaker, M. J., 1980, Calcium and the control of exocytosis, in: *Calcium Binding Proteins: Structure and Function* (F. L. Siegel, E. Carafoli, R. H. Kretsinger, D. H. MacLennan, and R. H. Wasserman, eds.), pp. 47-55, Elsevier/North-Holland, New York.
Baker, R. F., and Kalra, V. K., 1979, Chemically induced fusion of fresh human erythrocytes, *Biochem. Biophys. Res. Commun.* **86**:920-928.
Bangham, A. D., 1968, Membrane models with phospholipids, *Prog. Biophys. Mol. Biol.* **18**:29-95.
Bangham, A. D., Standish, M. M., and Watkins, J. C., 1965, Diffusion of univalent ions across lamellae of swollen phospholipids, *J. Mol. Biol.* **13**:238-252.
Banks, P., 1966, An interaction between chromaffin granules and calcium ions, *Biochem. J.* **101**:18c-20c.
Barondes, S. H., 1981, Lectins: Their multiple endogenous cellular functions, *Ann. Rev. Biochem.* **50**:207-231.
Baxter, D. A., Johnston, D., and Strittmatter, W. J., 1983, Protease inhibitors implicate metalloendoprotease in synaptic transmission at the mammalian neuromuscular junction, *Proc. Natl. Acad. Sci. USA* **80**:4174-4178.
Baydoun, E. A-H., and Northcote, D. H., 1980, Measurement and characteristics of fusion of isolated membrane fractions from maize root tips, *J. Cell Sci.* **45**:169-186.
Bearer, E. L., and Friend, D. S., 1980, Anionic lipid domains: Correlation with functional topography in a mammalian cell membrane, *Proc. Natl. Acad. Sci. USA* **77**:6601-6605.
Bearer, E. L., and Friend, D. S., 1982, Modifications of anionic-lipid domains preceding membrane fusion in guinea pig sperm, *J. Cell Biol.* **92**:604-615.
Bearer, E. L., Düzgüneş, N., Friend, D. S., and Papahadjopoulos, D., 1982, Fusion of phospholipid vesicles arrested by quick-freezing. The question of lipidic particles as intermediates in membrane fusion, *Biochim. Biophys. Acta* **693**:93-98.
Beisson, J., Lefort-Tran, M., Pouphile, M., Rossignol, M. and Satir, B., 1976, Genetic analysis of membrane differentiation in *Paramecium:* Freeze-fracture study of the trichocyst cycle in wild-type and mutant strains, *J. Cell Biol.* **69**:126-143.
Bentz, J. and Düzjüneş, H., 1985, Fusogenic capacities of divalent cations and the effect of liposome size, *Biochemistry* (in press).
Bentz, J., Nir, S., and Wilschut, J., 1983a, Mass action kinetics of vesicle aggregation and fusion, *Colloids and Surfaces* **6**:333-363.
Bentz, J., Düzgüneş, N., and Nir, S., 1983b, Kinetics of divalent cation induced fusion of phosphatidylserine vesicles: Correlation between fusogenic capacities and binding affinities, *Biochemistry* **22**:3320-3330.
Bentz, J., Düzgüneş, N., and Nir, S., 1985, Temperature dependence of divalent cation induced fusion of phosphatidyl serine liposomes: Evaluation of the kinetic rate constants, *Biochemistry* **24** 1064-1072.
Berger, W., Dahl, G., and Meissner, H. P., 1975, Structural and functional alterations in fused

membranes of secretory granules during exocytosis in pancreatic islet cells of the mouse, *Cytobiologie* **12**:119-139.
Berl, S., Puszkin, S., and Nicklas, W. J., 1973, Actomyosin-like protein in brain, *Science* **179**:441-446.
Bernard, E., Faucon, J.-F., and Dufourcq, J., 1982, Phase separations induced by mellitin in negatively-charged phospholipid bilayers as detected by fluorescence polarization and differential scanning calorimetry, *Biochim. Biophys. Acta* **688**:152-162.
Bischoff, R., 1978, Myoblast fusion, in: *Membrane Fusion* (G. Poste and G. L. Nicolson, eds.), pp. 127-179, Elsevier/North-Holland Biomedical Press, Amsterdam.
Bischoff, R., and Holtzer, H., 1968, The effect of mitotic inhibitors on myogenesis in vitro, *J. Cell Biol.* **36**:111-127.
Blaurock, A. E., and Gamble, R. C., 1979, Small phosphatidylcholine vesicles appear to be faceted below the thermal phase transition, *J. Memb. Biol.* **50**:187-204.
Blioch, Z. L., Glagoleva, I. M., Liberman, E. A., and Nenashev, V. A., 1968, A study of the mechanism of quantal transmitter release at a chemical synapse, *J. Physiol.* **199**:11-35.
Blitz, A. L., and Fine, R. E., 1974, Muscle-like contractile proteins and tubulin in synaptosomes, *Proc. Natl. Acad. Sci. USA* **71**:4472-4476.
Blow, A. M. J., Botham, G. M., Fisher, D., Goodall, A. H., Tilcock, C. P. S. and Lucy, J. A., 1978, Water and calcium ions in cell fusion induced by poly(ethylene glycol), *FEBS Lett* **94**:305-310.
Blow, A. M. J., Botham, G. M., and Lucy, J. A., 1979, Calcium ions and cell fusion: Effects of chemical fusogens on the permeability of erythrocytes to calcium and other ions, *Biochem. J.* **182**:555-563.
Blumenthal, R., Henkart, M., and Steer, C. J., 1983, Clathrin-induced pH-dependent fusion of phosphatidylcholine vesicles, *J. Biol. Chem.* **258**:3409-3415.
Bock, E., 1978, Nervous system specific proteins, *J. Neurochem.* **30**:7-14.
Bock, E., and Helle, K. B., 1977, Localization of synaptin on synaptic vesicle membranes, synaptosomal plasma membranes and chromaffin granule membranes, *FEBS Lett.* **82**:175-178.
Bock, E., Jorgensen, O. S., Dittmann, L., and Eng, L. F., 1975, Determination of brain-specific antigens in short-term cultivated rat astroglial cells and in rat synaptosomes, *J. Neurochem.* **25**:867-870.
Boni, L. T., Stewart, T. P., Alderfer, J. L., and Hui, S. W., 1981a, Lipid–polyethylene glycol interactions: I. Induction of fusion between liposomes, *J. Memb. Biol.* **62**:65-70.
Boni, L. T., Stewart, T. P., Alderfer, J. L. and Hui, S. W., 1981b, Lipid–polyethylene glycol interactions: II. Formation of defects in bilayers, *J. Memb. Biol.* **62**:71-77.
Breer, H., Morris, S. J. and Whittaker, V. P., 1977, Adenosine triphosphatase activity associated with purified cholinergic synaptic vesicles of *Torpedo marmorata, Eur, J. Biochem.* **80**:313-318.
Breisblatt, W., and Ohki, S., 1975, Fusion in phospholipid spherical membranes I. Effect of temperature and lysolecithin, *J. Memb. Biol.* **23**:385-401.
Breisblatt, W. and Ohki, S., 1976, Fusion in phospholipid spherical membranes II. Effect of cholesterol, divalent ions and pH, *J. Memb. Biol.* **29**:127-146.
Bretscher, M. S., 1984, Endocytosis: Relation to capping and locomotion, *Science* **224**:681-686.
Brown, M. S., Anderson, R. G. W. and Goldstein, J. L., 1983, Recycling receptors: The round-trip itinerary of migrant membrane proteins, *Cell* **32**:663-667.
Brown, W. M., Pazoles, C. J., Creutz, C. E., Aurbach, G. D. and Pollard, H. B., 1978, Role of anions in parathyroid hormone release from dispersed bovine parathyroid cells, *Proc. Natl. Acad. Sci. USA* **75**:876-880.
Bruni, A., and Palatini, P., 1982, Biological and pharmacological properties of phospholipids, *Prog. Med. Chem.* **19**:111-203.

Burgoyne, R. D., and Geisow, M. J., 1981, Specific binding of ^{125}I-calmodulin to and protein phosphorylation in adrenal chromaffin granule membranes, *FEBS Lett.* **131**:127–131.

Burridge, K., and Phillips, J. H., 1975, Association of actin and myosin with secretory granule membranes, *Nature* **254**:526–529.

Burwen, S. J., and Satir, B. H., 1977, A freeze-fracture study of early membrane events during mast cell secretion, *J. Cell Biol.* **73**:660–671.

Carlson, S. S., Wagner, J. A., and Kelly, R. B., 1978, Purification of synaptic vesicles from Elasmobranch electric organ and the use of biophysical criteria to demonstrate purity, *Biochemistry* **17**:1188–1199.

Cates, G. A., and Holland, P. C., 1978, Biosynthesis of plasma-membrane proteins during myogenesis of skeletal muscle *in vitro*, *Biochem. J.* **174**:873–881.

Ceccarelli, B., Hurlbut, W. P., and Mauro, A., 1973, Turnover of transmitter and synaptic vesicles at the frog neuromuscular junction, *J. Cell Biol.* **57**:499–524.

Ceccarelli, B., Grohovaz, F., and Hurlbut, W. P., 1979, Freeze-fracture studies of frog neuromuscular junction during intense release of neurotransmitter II. Effects of electrical stimulation and high potassium, *J. Cell Biol.* **81**:178–192.

Cestaro, B., Cervato, G., Barenghi, L., Pistolesi, E., and Pizzini, G., 1983, Fusion of sulfatide-containing vesicles of phosphatidylcholine, *Eur. J. Biochem.* **133**:229–233.

Chandler, D. E., and Williams, J. A., 1977, Fluorescent probe detects redistribution of cell calcium during stimulus-secretion coupling, *Nature* **268**:659–660.

Chandler, D. E., and Heuser, J., 1979, Membrane fusion during secretion. Cortical granule exocytosis in sea urchin eggs as studied by quick-freezing and freeze-fracture, *J. Cell Biol.* **83**:91–108.

Chandler, D. E. and Heuser, J., 1980, Arrest of membrane fusion events in mast cells by quick freezing, *J. Cell Biol.* **86**:666–674.

Chandler, D. E., Bennett, J. P., and Gomperts, B., 1983, Freeze-fracture studies of chemotactic peptide-induced exocytosis in neutrophils: Evidence for two patterns of secretory granule fusion, *J. Ultrastruct. Res.* **82**:221–232.

Chaudhury, M. K. and Ohki, S., 1981, Correlation between membrane expansion and temperature-induced membrane fusion, *Biochim. Biophys. Acta* **642**:365–374.

Chen, L. B., 1977, Alteration in cell surface LETS protein during myogenesis, *Cell* **10**:393–400.

Chi, E. Y., Lagunoff, D., and Koehler, J. K., 1976, Freeze-fracture study of mast cell secretion, *Proc. Natl. Acad. Sci. USA* **73**:2823–2827.

Choppin, P. W., and Compans, R. W., 1975, Reproduction of paramyxoviruses, in: *Comprehensive Virology* (H. Fraenkel-Conrat and R. Wagner, eds.), Vol. 4, pp. 95–178, Plenum Press, New York.

Cohen, F. S., Zimmerberg, J., and Finkelstein, A., 1980, Fusion of phospholipid vesicles with planar phospholipid bilayer membranes II. Incorporation of a vesicular membrane marker into the planar membrane, *J. Gen. Physiol.* **75**:251–270.

Cohen, F. S., Akabas, M. H., and Finkelstein, A., 1982, Osmotic swelling of phospholipid vesicles causes them to fuse with a planar phospholipid bilayer membrane, *Science* **217**:458–460.

Cohen, J. A., and Moronne, M. M., 1976, Interaction of charged lipid vesicles with planar bilayer lipid membranes: Detection by antibiotic membrane probes, *J. Supramol. Struct.* **5**:409–416.

Cohen, J. A., and Moronne, M. M., 1978, The monazomycin probe as detector of the interaction of charged lipid vesicles, polypeptides, proteins and divalent cations with planar bilayer lipid membranes, *Biochim. Biophys. Res. Commun.* **83**:1275–1283.

Cornell, R. B., Nissley, S. M. and Horwitz, A. F., 1980, Cholesterol availability modulates myoblast fusion, *J. Cell Biol.* **86**:820–824.

Couch, C. B. and Strittmatter, W. J., 1983, Rat myoblast fusion requires metalloendoprotease activity, *Cell* **32**:257–265.

Coupland, R. E., 1965, Electron microscopic observations on the structure of the rat adrenal medulla I. The ultrastructure and organization of chromaffin cells in the normal adrenal medulla, *J. Anat* **99**:231–254.

Cowley, A. C., Fuller, N. L., Rand, R. P., and Parsegian, V. A., 1978, Measurement of repulsive forces between charged phospholipid bilayers, *Biochemistry* **17**:3163–3168.

Cox, P. G., and Gunter, M., 1973, The effect of calcium ion concentration on myotube formation in vitro, *Exp. Cell Res.* **79**:169–178.

Creutz, C. E., 1981a, cis- unsaturated fatty acids induce the fusion of chromaffin granules aggregated by synexin, *J. Cell Biol.* **91**:247–256.

Creutz, C. E., 1981b, Secretory vesicle-cytosol interactions in exocytosis: Isolation by Ca^{2+}-dependent affinity chromatography of proteins that bind to the chromaffin granule membrane, *Biochem. Biophys. Res. Commun.* **103**:1395–1400.

Creutz, C. E., and Pollard, H. B., 1980, A biological model for the chromaffin granule: Accurate description of the kinetics of ATP and Cl^- dependent granule lysis, *Biophys. J.* **31**:255–270.

Creutz, C. E., Pazoles, C. J., and Pollard, H. B., 1978, Identification and purification of an adrenal medullary protein (synexin) that causes calcium-dependent aggregation of isolated chromaffin granules, *J. Biol. Chem.* **253**:2858–2866.

Creutz, C. E., Pazoles, C. J., and Pollard, H. B., 1979, Self-association of synexin in the presence of calcium: Correlation with synexin-induced membrane fusion and examination of the structure of synexin aggregates, *J. Biol. Chem.* **254**:553–558.

Creutz, C. E., Scott, J. H., Pazoles, C. J., and Pollard, H. B., 1982, Further characterization of the aggregation and fusion of chromaffin granules by synexin as a model for compound exocytosis, *J. Cell. Biochem.* **18**:87–97.

Creutz, C. E., Dowling, L. G., Sando, J. J., Villar-Palasi, C., Whipple, J. H., and Zaks, W. J., 1983, Characterization of the chromobindins: Soluble proteins that bind to the chromaffin granule membrane in the presence of Ca^{2+}, *J. Biol. Chem.* **258**:14664–14674.

Croop, J., and Holtzer, H., 1975, Response of myogenic and fibrogenic cells to cytochalasin B and to colcemid I. Light microscope observations, *J. Cell Biol.* **65**:271–285.

Cullis, P. R., and Hope, M. J., 1978, Effects of fusogenic agent on membrane structure of erythrocyte ghosts and the mechanism of membrane fusion, *Nature* **271**:672–674.

Cullis, P. R., and deKruijff, B., 1978, The polymorphic phase behavior of phosphatidylethanolamines of natural and synthetic origin. A ^{31}P-NMR study, *Biochim. Biophys. Acta* **513**:31–42.

Cullis, P. R. and Verkleij, A. J., 1979, Modulation of membrane structure by Ca^{2+} and dibucaine as detected by ^{31}P NMR, *Biochim. Biophys. Acta* **552**:546–551.

Dahl, G., and Gratzl, M., 1976, Calcium-induced fusion of isolated secretory vesicles from the islet of Langerhans, *Cytobiologie* **12**:344–355.

Dahl, G., Schudt, C., and Gratzl, M., 1978, Fusion of isolated myoblast plasma membranes: An approach to the mechanism, *Biochim. Biophys. Acta* **514**:105–116.

Dahl, G., Ekerdt, R., and Gratzl, M., 1979, Models for exocytotic membrane fusion, *Symp. Soc. Exp. Biol.* **33**:349–368.

Danielli, J. F., and Davson, H., 1934, A contribution to the theory of permeability of thin films, *J. Cell. Comp. Physiol.* **5**:495–508.

David, J. D., See, W. M., and Higginbotham, C.-A., 1981, Fusion of chick embryo skeletal myoblasts: Role of calcium influx preceding membrane union, *Dev. Biol.* **82**:297–307.

Davidson, R. L., O'Malley, K. A., and Wheeler, T. B., 1976, Polyethylene glycol-induced mammalian cell hybridization: effect of polyethylene glycol molecular weight and concentration, *Somat. Cell Genet.* **2**:271–280.

Davis, B., and Lazarus, N. R., 1976, An *in vitro* system for studying insulin release caused by secretory granules-plasma membrane interaction: Definition of the system, *J. Physiol.* **256**:709–729.

Davson, H., and Danielli, J. F., 1952, The permeability of natural membranes, Cambridge University Press, Cambridge.

Day, E. P., Ho, J. T., Kunze, R. K., Jr., and Sun, S. T., 1977, Dynamic light scattering study of calcium-induced fusion in phospholipid vesicles, *Biochim. Biophys. Acta* **470**:503–508.

Deamer, D. W., Leonard, R., Tardieu, A., and Branton, D., 1970, Lamellar and hexagonal lipid phases visualized by freeze-etching, *Biochim. Biophys. Acta* **219**:47–60.

De Camilli, P., Peluchetti, D., and Meldolesi, J., 1976, Dynamic changes of the luminal plasmalemma in stimulated parotid acinar cells. A freeze-fracture study, *J. Cell Biol.* **70**:59–74.

de Duve, D., 1963, Endocytosis (footnote), in *Lysosomes, Ciba Foundation Symposium Lysosomes,* (deReuck, A. V. S., and Cameron, M. P., eds.), p. 126, J. & A. Churchill, Ltd., London.

de Duve, C., and Wattiaux, R., 1966, Functions of lysosomes, *Ann. Rev. Physiol.* **28**:435–492.

Defrise-Quertain, F., Chatelain, P., Ruysschaert, J.-M. and Delmelle, M., 1982, Fusion and lipid exchange in vesicles containing lipophilic spin labels, *Biochim. Biophys. Acta* **688**:116–122.

DeLorenzo, R. J. and Freedman, S. D., 1977, Calcium-dependent phosphorylation of synaptic vesicle proteins and its possible role in mediating neurotransmitter release and vesicle function, *Biochem. Biophys. Res. Commun.* **77**:1036–1043.

Den, H., Malinzak, D. A., Keating, H. J., and Rosenberg, A., 1975, Influence of concanavalin A, wheat germ agglutinin and soybean agglutinin on the fusion of myoblasts in vitro, *J. Cell Biol.* **67**:826–834.

Den, H., Malinzak, D. A., and Rosenberg, A., 1976, Lack of evidence for the involvement of a β-D-galactosyl-specific lectin in the fusion of chick myoblasts, *Biochem. Biophys. Res. Commun.* **69**:621–627.

Deutsch, J. W., and Kelly, R. B., 1981, Lipids of synaptic vesicles: Relevance to the mechanism of membrane fusion, *Biochemistry* **20**:378–385.

de Virgilis, G., Meldolesi, J., and Clementi, F., 1968, Ultrastructure of growth hormone producing cells in rat pituitary after injection of hypothalamic extract, *Endocrinology* **83**:1278–1284.

Diner, O., 1967, L'expulsion des granules de la médullo-surrénale chez le Hamster, *C. R. Acad. Sci. Paris,* Ser. D., **265**:616–619.

Douglas, W. W., 1968, Stimulus-secretion coupling: The concept and clues from chromaffin and other cells, *Brit. J. Pharmacol* **34**:451–474.

Douglas, W. W., 1974, Involvement of calcium in exocytosis and the exocytosis-vesiculation sequence, *Biochem. Soc. Symp.* **39**:1–28.

Dunn, L. A. and Holz, R. W., 1983, Catecholamine secretion from digitonin-treated adrenal medullary chromaffin cells, *J. Biol. Chem.* **258**:4989–4993.

Dunham, P., Babiarz, P., Israel, A., Zerial, A., and Weissmann, G., 1977, Membrane fusion: Studies with a calcium sensitive dye, arsenazo III, in liposomes, *Proc. Natl. Acad. Sci. USA* **74**:1580–1584.

Düzgüneş, N., and Ohki, S., 1977, Calcium-induced interaction of phospholipid vesicles and bilayer lipid membranes, *Biochim. Biophys. Acta* **467**:301–308.

Düzgüneş, N., and Ohki, S., 1981, Fusion of small unilamellar liposomes with phospholipid planar bilayer membranes and large single-bilayer vesicles, *Biochim. Biophys. Acta* **640**:734–747.

Düzgüneş, N., and Papahadjopoulos, D., 1983, Ionotropic effects on phospholipid membranes: Calcium-magnesium specificity in binding, fluidity, and fusion, in: *Membrane Fluidity in Biology,* Vol. 2, *General Principles* (R. C. Aloia, ed.), pp. 187–216, Academic Press, New York.

Düzgüneş, N., Hong, K., and Papahadjopoulos, D., 1980, Membrane fusion: The involvement of phospholipids, proteins and calcium binding, in: *Calcium-Binding Proteins: Structure and Function* (F. L. Siegel, E. Carafoli, R. H. Kretsinger, D. H. MacLennan, and R. H. Wasserman, eds.), pp. 17–22, Elsevier/North-Holland, New York.

Düzgüneş, N., Nir, S., Wilschut, J., Bentz, J., Newton, C., Portis, A., and Papahadjopoulos, D., 1981a, Calcium- and magnesium-induced fusion of mixed phosphatidylserine/phosphatidylcholine vesicles: Effect of ion binding, *J. Memb. Biol.* **59**:115–125.

Düzgüneş, N., Wilschut, J., Fraley, R., and Papahadjopoulos, D., 1981b, Studies on the mechanism of membrane fusion: Role of head-group composition in calcium- and magnesium-induced fusion of mixed phospholipid vesicles, *Biochim. Biophys. Acta* **642**:182–195.

Düzgüneş, N., Bearer, E., and Papahadjopoulos, D., 1982, Phospholipid vesicle fusion monitored by rapid-freezing and mixing of aqueous contents, *Biophys. J.* **37**:25a.

Düzgüneş, N., Straubinger, R. M., and Papahadjopoulos, D., 1983a, pH-dependent membrane fusion, *J. Cell Biol.* **97**:178a.

Düzgüneş, N., Bentz, J., Freeman, K., Nir, S., and Papahadjopoulos, D., 1983b, Role of divalent cations in membrane fusion, *Fed. Am. Soc. Exp. Biol. Proc.* **42**:1771.

Düzgüneş, N., Paiement, J., Freeman, K. B., Lopez, N. G., Wilschut, J. and Papahadjopoulos, D., 1984a, Modulation of membrane fusion by ionotropic and thermotropic phase transitions, *Biochemistry* **23**:3486–3494.

Düzgüneş, N., Hoekstra, D., Hong, K. and Papahadjopoulos, D., 1984b, Lectins facilitate calcium-induced fusion of phospholipid vesicles containing glycosphingolipids, *FEBS Lett.* **173**:80–84.

Düzgüneş, N., Wilschut, J., and Papahadjopoulos, D., 1985a, Control of membrane fusion by divalent cations, phospholipid head-groups and proteins, in: *Physical Methods on Biological Membranes and Their Model Systems* (F. Conti, W. E. Blumberg, J. de Gier, and F. Pocchiari, eds.), pp. 193–218, Plenum Press (New York).

Düzgüneş, N., Straubinger, R. M., Baldwin, P. A., Friend, D. S., and Papahadjopoulos, D., 1985b, Proton-induced fusion of oleic acid/phosphatidylethanolamine liposomes, *Biochemistry* **24** (in press).

Edwards, W., Phillips, J. H., and Morris, S. J., 1974, Structural changes in chromaffin granules induced by divalent cations, *Biochim. Biophys. Acta* **356**:164–173.

Eichberg, J., Whittaker, V. P., and Dawson, R. M. C., 1964, Distribution of lipids in subcellular particles of Guinea pig brain, *Biochem. J.* **92**:91–100.

Eidelman, O., Schlegel, R., Tralka, T. S., and Blumenthal, R., 1984, pH-dependent fusion induced by vesicular stomatitis virus glycoprotein reconstituted into phospholipid vesicles, *J. Biol. Chem.* **259**:4622–4628.

Eisenberg, M., Gresalfi, T., Riccio, T., and McLaughlin, S., 1979, Adsorption of monovalent cations to bilayer membranes containing negative phospholipids, *Biochemistry* **18**:5213–5223.

Ekerdt, R., Dahl, G., and Gratzl, M., 1981, Membrane fusion of secretory vesicles and liposomes. Two different types of fusion, *Biochim. Biophys. Acta* **646**:10–22.

Ekerdt, R., and Papahadjopoulos, D., 1982, Intermembrane contact affects calcium binding to phospholipid vesicles, *Proc. Natl. Acad. Sci. USA* **79**:2273–2277.

Ekholm, R., Zelander, T., and Edlund, Y., 1962, The ultrastructural organization of the rat exocrine pancreas. I. Acinar cells, *J. Ultrastruc. Res.* **7**:61–72.

Elson, H. F., and Yguerabide, J., 1979, Membrane dynamics of differentiating cultured embryonic chick skeletal muscle cells by fluorescence microscopy techniques. *J. Supramol. Struct.* **12**:47–61.

Epel, D., and Vacquier, V. D., 1978, Membrane fusion events during invertebrate fertilization, in: *Membrane Fusion* (G. Poste and G. L. Nicolson, eds.), pp. 1-63, Elsevier/North-Holland Biomedical Press, Amsterdam.

Eytan, G. D., and Almary, T., 1983, Mellitin-induced fusion of acidic liposomes, *FEBS Lett.* **156**:29–32.

Fan, D. P., and Sefton, B. M., 1978, The entry into host cells of sindbis virus, vesicular stomatitis virus and Sendai virus, *Cell* **15**:985–992.

Farquhar, M. G., 1978, Traffic of products and membranes through the Golgi complex, in: *Transport of Macromolecules in Cellular Systems* (S. C. Silverstein, ed.), pp. 341–362, Dahlem Konferenzen, Berlin.

Felgner, P. L., Freire, E., Barenholz, Y., and Thompson, T. E., 1981, Asymmetric incorporation of trisialoganglioside into dipalmitoylphosphatidylcholine vesicles, *Biochemistry* **20**:2168–2172.

Fenwick, E. M., Fajdiga, P. B., Howe, N. B. S., and Lirett, B. H., 1978, Functional and morphological characterization of isolated bovine adrenal medullary cells, *J. Cell Biol.* **76**:12–30.

Fernandez, S. M., and Herman, B. A., 1982, Topography and mobility of concanavalin A receptors during myoblast fusion, in: *Muscle Development: Molecular and Cellular Control* (M. L. Pearson and H. F. Epstein, eds.), pp. 319–327, Cold Spring Harbor Laboratory, Cold Spring Harbor, New York.

Fontaine, R. N., Harris, R. A., and Schroeder, F., 1980, Aminophospholipid asymmetry in murine synaptosomal plasma membrane, *J. Neurochem.* **34**:269–277.

Fowler, V. M., and Pollard, H. B., 1982a, Chromaffin granule membrane-F-actin interactions are calcium sensitive, *Nature* **295**:336–339.

Fowler, V. M., and Pollard, H. B., 1982b, In vitro reconstitution of chromaffin granule-cytoskeleton interactions: Ionic factors influencing the association of F-actin with purified chromaffin granule membranes, *J. Cell. Biochem.* **18**:295–311.

Fraley, R., Wilschut, J., Düzgüneș, N., Smith, C., and Papahadjopoulos, D., 1980, Studies on the mechanism of membrane fusion: Role of phosphate in promoting calcium ion induced fusion of phospholipid vesicles, *Biochemistry* **19**:6021–6029.

Friedman, J. E., Lelkes, P. I., Rosenhech, K., and Oplatka, A., 1980, The possible implication of membrane associated actin in stimulus-secretion coupling in adrenal chromaffin cells, *Biochem. Biophys. Res. Commun.* **96**:1717–1723.

Friend, D. S., Orci, L., Perrelet, A., and Yanagimachi, R., 1977, Membrane particle changes attending the acrosome reaction in guinea pig spermatozoa, *J. Cell Biol.* **74**:561–577.

Fuchs, P., Gruber, E., Gitelman, J., and Kohn, A., 1980, Nature of permeability changes in membrane of HeLa cells adsorbing Sendai virus, *J. Cell. Physiol.* **103**:271–278.

Furcht, L. T., Mosher, D. F., Wendelschafer-Crabb, G., 1978, Immunocytochemical localization of fibronectin (LETS Proteins) on the surface of L6 myoblasts: Light and electron microscopic studies, *Cell* **13**:263–271.

Gabbiani, G., da Prada, M., Richards, G., and Pletscher, A., 1976, Actin association with membranes of monoamine storage organelles, *Proc. Soc. Exp. Biol. Med.* **152**:135–138.

Gaber, B. P., and Sheridan, J. P., 1982, Kinetic and thermodynamic studies of the fusion of small unilamellar phospholipid vesicles, *Biochim. Biophys. Acta* **685**:87–93.

Gad, A. E., and Eytan, G. D., 1983, Chlorophylls as probes for membrane fusion: Polymyxin B-induced fusion of liposomes, *Biochim. Biophys. Acta* **727**:170–176.

Gad, A. E., Broza, R., and Eytan, G. D., 1979, Calcium-induced fusion of proteo-liposomes and protein-free liposomes. Effect of their phosphatidylethanolamine content on the structure of fused vesicles, *Biochim. Biophys. Acta* **556**:181–195.

Gad, A. E., Silver, B. L., Eytan, G. D., 1982, Polycation-induced fusion of negatively-charged vesicles, *Biochim. Biophys. Acta* **690**:124–132.

Garcia, L. A. M., Schenkman, S., Araujo, P. S., and Chaimovich, H., 1983, Fusion of small unilamellar vesicles induced by bovine serum albumin fragments, *Brazilian J. Med. Biol. Res.* **16**:89–96.

Gardner, J. M., and Fambrough, D. M., 1983, Fibronectin expression during myogenesis, *J. Cell Biol.* **96**:474–485.

Garoff, H., Frischauf, A.-M., Simons, K., Lehrach, H., and Delius, H., 1980, Nucleotide sequence of cDNA coding for Semliki Forest virus membrane glycoproteins, *Nature* **288**:236–241.

Gartner, T. K., and Podleski, T. R., 1975, Evidence that a membrane bound lectin mediates fusion of L6 myoblasts, *Biochem. Biophys. Res. Commun.* **67**:972–978.

Gemmell, R. T., and Stacy, B. D., 1979, Granule secretion by the luteal cell of the sheep: The fate of the granule membrane, *Cell Tissue Res.* **197**:413–419.

Gething, M.-J., White, J., and Waterfield, M., 1978, Purification of the fusion protein of Sendai virus: Analysis of the NH_2-terminal sequence generated during precursor activation, *Proc. Natl. Acad. Sci. USA* **75**:2737–2740.

Geuze, J. J., and Kramer, M. F., 1974, Function of coated membranes and multivesicular bodies during membrane regulation in stimulated exocrine pancreas cells, *Cell Tissue Res.* **156**:1–20.

Gibson, G. A., and Loew, L. M., 1979, Phospholipid vesicle fusion monitored by fluorescence energy transfer, *Biochem. Biophys. Res. Commun.* **88**:135–140.

Gilfix, B. M., and Sanwal, B. D., 1980, Inhibition of myoblast fusion by tumicamycin and pantomycin, *Biochem. Biophys. Res. Commun.* **96**:1184–1191.

Gilfix, B. M., and Sanwal, B. D., 1982, Lectin-resistant myoblasts, in: *Muscle Development: Molecular and Cellular Control* (M. L. Pearson and H. F. Epstein, eds.), pp. 329–336, Cold Spring Harbor Laboratory, Cold Spring Harbor, New York.

Gingell, D., and Ginsberg, L., 1978, Problems in the physical interpretation of membrane interaction and fusion, in: *Membrane Fusion* (G. Poste and G. L. Nicolson, eds.), pp. 791–833, Elsevier/North-Holland Biomedical Press, Amsterdam.

Ginsberg, L., 1978, Does calcium cause fusion or lysis of unilamellar lipid vesicles? *Nature* **275**:758–760.

Goldup, A., Ohki, S., and Danielli, J. F., 1970, Black lipid films, *Recent Prog. Surface Sci.* **3**:193–260.

Gratzl, M., and Dahl, G., 1976, Calcium-induced fusion of Golgi-derived secretory vesicles isolated from rat liver, *FEBS Lett.* **62**:142–145.

Gratzl, M., and Dahl, G., 1978, Fusion of secretory vesicles isolated from rat liver, *J. Memb. Biol.* **40**:343–364.

Gratzl, M., Dahl, G., Russell, J. T., and Thorn, N. A., 1977, Fusion of neurohypophyseal membranes *in vitro*, *Biochim. Biophys. Acta* **470**:45–57.

Gratzl, M., Schudt, C., Ekerdt, R., and Dahl, G., 1980, Fusion of isolated biological membranes: A tool to investigate basic processes of exocytosis and cell–cell fusion, in: *Membrane Structure and Function*, Vol. 3 (E. E. Bittar, ed.), pp. 59–92, Wiley, New York.

Gresh, N., 1980, Intermolecular chelation of two serine phosphates by Ca^{2+} and Mg^{2+}: A theoretical structural investigation, *Biochim. Biophys. Acta* **597**:345–357.

Grishin, A. F., Nenashev, V. A., and Berestovskii, G. M., 1980, Interaction of liposomes with bimolecular membranes, *Biofizika* **24**:482–487.

Grinstein, S., VanderMeulen, J., and Furuya, W., 1982, Possible role of H^+-alkali cation countertransport in secretory granule swelling during exocytosis, *FEBS Lett.* **148**:1–4.

Grynszpan-Winograd (Diner), O., 1971, Morphological aspects of exocytosis in the adrenal medulla, *Phil. Trans. R. Soc. Lond. B.* **261**:291–292.

Hagins, W. A., and Yoshikami, S., 1977, Intracellular transmission of visual excitation in photoreceptors: Electrical effects of chelating agents introduced into rods by vesicle fusion, in: *Vertebrate Photoreception* (H. B. Barlow and P. Fatt, eds.), pp. 97–138, Academic Press, London.

Hammoudah, M., Nir, S., Isac, T., Kornhauser, R., Stewart, T. P., Hui, S. W., and Vaz, W. L. C., 1979, Interactions of La^{3+} with phosphatidylserine vesicles: Binding, phase transition, leakage and fusion, *Biochim. Biophys. Acta* **558**:338–343.

Hammoudah, M. M., Nir, S., Bentz, J., Mayhew, E., Stewart, T. P., Hui, S. W., and Kurland, R. J., 1981, Interactions of La^{3+} with phosphatidylserine vesicles. Binding, phase transition, leakage, ^{31}P-NMR and fusion, *Biochim. Biophys. Acta* **645**:102–114.

Hampton, R. Y., and Holz, R. W., 1983, Effects of changes in osmolality on the stability and function of cultured chromaffin cells and the possible role of osmotic forces in exocytosis, *J. Cell Biol.* **96**:1082–1088.

Hauser, H., Phillips, M. C., and Barratt, M. D., 1975, Differences in the interaction of inorganic and organic (hydrophobic) cations with phosphatidylserine membranes, *Biochim. Biophys. Acta* **413**:341–353.

Hauser, H., Finer, E. G., and Darke, A., 1977, Crystalline anhydrous Ca-phosphatidylserine bilayers, *Biochem. Biophys. Res. Commun.* **76**:267–274.

Hauser, H., Pascher, I., Pearson, R. H., and Sundell, S., 1981, Preferred conformation and molecular packing of phosphatidylethanolamine and phosphatidylcholine, *Biochim. Biophys. Acta* **650**:21–51.

Hausmann, K., and Allen, R. D., 1976, Membrane behavior of exocytotic vesicles: II. Fate of the trichocyst membranes in *Paramecium* after induced trichocyst discharge, *J. Cell Biol.* **69**:313–326.

Haynes, D. H., Lansman, J., Cahill, A. L., and Morris, S. J., 1979, Kinetics of cation-induced aggregation of *Torpedo* electric organ synaptic vesicles, *Biochim. Biophys. Acta* **557**:340–353.

Haywood, A. M., 1974, Fusion of Sendai virus with model membranes, *J. Mol. Biol.* **87**:625–628.

Haywood, A. M., and Boyer, B. P., 1981, Initiation of fusion and disassembly of Sendai virus membranes into liposomes, *Biochim. Biophys. Acta* **646**:31–35.

Haywood, A. M., and Boyer, B. P., 1982, Sendai virus membrane fusion: Time course and effect of temperature, pH, calcium and receptor concentration, *Biochemistry* **24**:6041–6046.

Helenius, A., and Marsh, M., 1982, Endocytosis of enveloped animal viruses, in: *Membrane Recycling, Ciba Foundation Symposium 92*, pp. 59–76, Pitman Books, Ltd., London.

Helenius, A., Marsh, M., and White, J., 1980a, The entry of viruses into animal cells, *Trends Biochem. Sci.* **5**:104–106.

Helenius, A., Kartenbeck, J., Simons, K., and Fries, E., 1980b, On the entry of Semliki Forest virus into BHK-21 cells, *J. Cell Biol.* **84**:404–420.

Henn, F. A., and Thompson, T. E., 1969, Synthetic lipid bilayer membranes, *Ann. Rev. Biochem.* **38**:241–262.

Herman, B. A., and Fernandez, S. M., 1978, Changes in membrane dynamics associated with myogenic cell fusion, *J. Cell Physiol.* **94**:253–264.

Herrmann, A., Pratsch, L., Arnold, K., Lassmann, G., 1983, Effect of poly(ethylene glycol) on the polarity of aqueous solutions and on the structure of vesicle membranes, *Biochim. Biophys. Acta* **733**:87–94.

Heuser, J. E., and Reese, T. S., 1973, Evidence for recycling of synaptic vesicle membrane during transmitter release at the frog neuromuscular junction, *J. Cell Biol.* **57**:315–344.

Heuser, J. E., and Reese, T. S., 1981, Structural changes after transmitter release at the frog neuromuscular junction, *J. Cell Biol.* **88**:564–580.

Heuser, J. E., Reese, T. S., and Landis, D. M., 1976, Preservation of synaptic structure by rapid freezing, *Cold Spring Harbor Symp. Quant. Biol.* **40**:17–24.

Heuser, J. E., Reese, T. S. Dennis, M. J., Jan, Y., Jan, L., and Evans, L., 1979, Synaptic vesicle exocytosis captured by quick-freezing and correlated with quantal transmitter release, *J. Cell Biol.* **81**:275-300.

Hoekstra, D., 1982a, Kinetics of intermixing of lipids and mixing of aqueous contents during vesicle fusion, *Biochim. Biophys. Acta* **692**:171-175.

Hoekstra, D., 1982b, Fluorescence method for measuring the kinetics of Ca^{2+}-induced phase separations in phosphatidylserine-containing lipid vesicles, *Biochemistry* **21**:1055-1061.

Hoekstra, D., 1982c, Role of lipid phase separations and membrane hydration in phospholipid vesicle fusion, *Biochemistry* **21**:2833-2840.

Hoekstra, D., Yaron, A., Carmel, A., and Scherphof, G., 1979, Fusion of phospholipid vesicles containing a trypsin-sensitive fluorogenic substrate and trypsin, *FEBS Lett.* **106**:176-180.

Hoekstra, D., Wilschut, J., and Scherphof, G., 1983, Kinetics of calcium phosphate-induced fusion of human erythrocyte ghosts monitored by mixing of aqueous contents, *Biochim. Biophys. Acta* **732**:327-331.

Hoekstra, D., Düzgüneş, N., and Wilschut, J., 1985, Agglutination and fusion of globoside GL-4 containing phospholipid vesicles mediated by lectins and Ca^{2+}, *Biochemistry* **24** 565-572.

Hogg, R., Healy, T. W., and Fuerstenau, D. W., 1966, Mutual coagulation of colloidal dispersons, *Trans. Faraday Soc.* **62**:1638-1651.

Hokin, M. R., and Hokin, L. E., 1953, Enzyme secretion and the incorporation of P^{32} into phospholipides of pancreas slices, *J. Biol. Chem.* **203**:967-977.

Holtzman, E., Teichberg, S., Abrahams, S. J., Citkowitz, E., Crain, S. M., Kawai, N., and Peterson, E. R., 1973, Notes on synaptic vesicles and related structures, endoplasmic reticulum, lysosomes and peroxisomes in nervous tissue and the adrenal medulla, *J. Histochem. Cytochem.* **21**:349-385.

Holtzman, E., Gronowicz, G., Mercurio, A., and Masur, S. K., 1979, Notes on the heterogeneity, circulation, and modification of membranes, with emphasis on secretory cells, photoreceptors, and the toad bladder, in: *Biomembranes,* Vol. 10 (L. A. Manson, ed.), pp. 77-139, Plenum Press, New York.

Holz, R. W., and Stratford, C. A., 1979, Effects of divalent ions on vesicle–vesicle fusion studied by a new luminescence assay for fusion, *J. Memb. Biol.* **46**:331-358.

Homma, M., and Ohuchi, M., 1973, Trypsin action on the growth of Sendai virus in tissue culture cells, *J. Virol.* **12**:1457-1465.

Homma, M., Shimizu, K., Shimizu, Y. K., and Ishida, N., 1976, On the study of Sendai virus hemolysis I. Complete Sendai virus lacking in hemolytic activity, *Virology* **71**:41-47.

Honda, K., Maeda, Y., Sasakawa, S., Ohno, H., and Tsuchida, E., 1981, The components contained in polyethylene glycol of commercial grade (PEG-6,000) as cell fusogen, *Biochem. Biophys. Res. Commun.* **101**:165-171.

Hong, K., Düzgüneş, N., and Papahadjopoulos, D., 1981, Role of synexin in membrane fusion, *J. Biol. Chem.* **256**:3641-3644.

Hong, K., Düzgüneş, N., and Papahadjopoulos, D., 1982a, Modulation of membrane fusion by calcium-binding proteins, *Biophys. J.* **37**:297-305.

Hong, K., Düzgüneş, N., Ekerdt, R., and Papahadjopoulos, D., 1982b, Synexin facilitates fusion of specific phospholipid vesicles at divalent cation concentrations found intracellularly, *Proc. Natl. Acad. Sci. USA* **79**:4942-4944.

Hong, K., Düzgüneş, N., and Papahadjopoulos, D., 1983a, Clathrin-liposome interactions, *Fed. Am. Soc. Exp. Biol. Proc.* **42**:1826.

Hong, K. Schuber, F., and Papahadjopoulos, D., 1983b, Polyamines: Biological modulators of membrane fusion, *Biochim. Biophys. Acta* **732**:469-472.

Hope, M. J., and Cullis, P. R., 1979, The bilayer stability of inner monolayer lipids from the human erythrocyte, *FEBS Lett.* **107**:323-326.

Hope, M. J., Bruckdorfer, K. P., Hart, C. A., and Lucy, J. A., 1977, Membrane cholesterol and cell fusion of hen and Guinea pig erythrocytes, *Biochem. J.* **166**:255-263.

Hope, M. J., Walker, D. C., and Cullis, P. R., 1983, Calcium and pH-induced fusion of small unilamellar vesicles consisting of phosphatidylethanolamine and negatively charged phospholipids: A freeze-fracture study, *Biochem. Biophys. Res. Commun.* **110**:15–22.

Hortnagl, H., 1976, Membranes of adrenal medulla: A comparison of membranes of chromaffin granules and endoplasmic reticulum, *Neuroscience* **1**:9–18.

Horwitz, A. F., Wight, A., Ludwig, P., and Cornell, R., 1978, Interrelated lipid alterations and their influence on the proliferation and fusion of cultured myogenic cells, *J. Cell Biol.* **77**:334–357.

Horwitz, A. F., Wight, A., and Knudsen, K., 1979, A role for lipid in myoblast fusion, *Biochem. Biophys. Res. Commun.* **86**:514–521.

Horwitz, A., Neff, A., Sessions, A., and Decker, C., 1982, Cellular interactions in myogenesis, in: *Muscle Development: Molecular and Cellular Control* (M. L. Pearson and H. F. Epstein, eds.), pp. 291–300, Cold Spring Harbor Laboratory, Cold Spring Harbor, New York.

Hosaka, Y., and Shimizu, K., 1972, Artificial assembly of envelope particles of HVJ (Sendai virus) II. Lipid components for formation of the active hemolysin, *Virology* **49**:640–646.

Hosaka, Y., and Shimizu, K., 1977, Cell fusion by Sendai virus, in: *Virus Infection and the Cell Surface* (G. Poste and G. L. Nicolson, eds.), pp. 129–155, Elsevier/North-Holland Biomedical Press, Amsterdam.

Hosaka, Y., Semba, T., and Fukai, K., 1974, Artificial assembly of envelope particles of HVJ (Sendai virus). Fusion activity of envelope particles. *J. Gen. Virol.* **25**:391–404.

Hoss, W., Okumura, K., Formaniak, M., and Tanaka, R., 1980, Cation-binding sites on synaptic vesicles, *Arch. Biochem. Biophys.* **203**:647–653.

Howell, J. I., Fisher, D., Goodall, A. H., Verrinder, M., and Lucy, J. A., 1973, Interactions of membrane phospholipids with fusogenic lipids, *Biochim. Biophys. Acta* **332**:1–10.

Hsu, M.-C., Scheid, A., and Choppin, P., 1979, Reconstruction of membranes with individual paramyxovirus glycoproteins and phospholipid in cholate solution, *Virology* **95**:476–491.

Hsu, M.-C., Scheid, A., and Choppin, P. W., 1981, Activation of the Sendai virus fusion protein (F) involves a conformational change with exposure of a new hydrophobic region, *J. Biol. Chem.* **256**:3557–3563.

Huang, R. T. C., 1983a, Involvement of glycolipids in myxovirus-induced membrane fusion (haemolysis), *J. Gen. Virol.* **64**:221–224.

Huang, R. T. C., 1983b, The role of neutral glycolipids and phospholipids in myxovirus-induced membrane fusion, *Lipids* **18**:489–492.

Huang, R. T. C., Wahn, K., Klenk, H.-D., and Rott, R., 1980a, Fusion between cell membrane and liposomes containing the glycoproteins of influenza virus, *Virology* **104**:294–302.

Huang, R. T. C., Rott, R., Wahn, K., Klenk, H.-D., and Kohama, T., 1980b, The function of the neuraminidase in membrane fusion induced by myxoviruses, *Virology* **107**:313–319.

Huber, E., Konig, P., Schuler, G., Aberer, W., Plattner, H., and Winkler, H., 1979, Characterization and topography of the glycoproteins of adrenal chromaffin granules, *J. Neurochem.* **32**:35–47.

Hui, S. W., 1981, Geometry of phase separated domains in phospholipid bilayers by diffraction-contrast electron microscopy, *Biophys. J.* **34**:383–395.

Hui, S. W., and Parsons, D. F., 1975, Direct observation of domains in wet lipid bilayers, *Science* **190**:383–384.

Hui, S. W., Stewart, T. P., Boni, L. T., and Yeagle, P. L., 1981, Membrane fusion through point defects in bilayers, *Science* **212**:921–923.

Hui, S. W., Stewart, T. P., Yeagle, P. L., and Albert, A. D., 1981, Bilayer to non-bilayer transition in mixtures of phosphatidylethanolamine and phosphatidylcholine: Implications for membrane properties, *Arch. Biochem. Biophys.* **207**:227–240.

Ingolia, T. D., and Koshland, D. E., 1978, The role of calcium in fusion of artificial vesicles, *J. Biol. Chem.* **253**:3821–3829.

Israel, D., Ginsberg, D., Laster, Y., Zakai, N., Milner, Y., and Loyter, A., 1983, A possible involvement of virus-associated protease in the fusion of Sendai virus envelopes with human erythrocytes, *Biochim. Biophys. Acta* **732**:337–346.

Ito, T., and Ohnishi, S.-I., 1974, Ca^{2+}-induced lateral phase separations in phosphatidic acid-phosphatidylcholine membranes, *Biochim. Biophys. Acta* **352**:29–37.

Izumi, F., Kashimoto, T., Miyashita, T., Wada, A., and Oka, M., 1977, Involvement of membrane associated protein in ADP-induced lysis of chromaffin granules, *FEBS Lett.* **78**:177–180.

Izumi, F., Oka, M., Morita, K., and Azuma, H., 1975, Catecholamine releasing factor in bovine adrenal medulla, *FEBS Lett.* **56**:73–76.

Jacobson, K., and Papahadjopoulos, D., 1975, Phase transitions and phase separations in phospholipid membranes induced by changes in temperature, pH and concentration of bivalent cations, *Biochemistry* **14**:152–161.

Jacques, P. J., 1975, The endocytic uptake of macromolecules, in: *Pathobiology of Cell Membranes*, Vol. 1 (B. J. Trump and A. U. Arstila, eds.), pp. 255–279, Academic Press, New York.

Jendrasiak, G. L. and Hasty, J. H., 1974, The hydration of phospholipids, *Biochim. Biophys. Acta* **337**:79–91.

Jendrasiak, G. L., and Mendible, J. C., 1976, The effect of the phase transition on the hydration and electrical conductivity of phospholipids, *Biochim. Biophys. Acta* **424**:133–148.

Jockusch, B. M., Burger, M. M., daPrada, M., Richards, J. G., Chaponnier, C., and Gabbiani, G., 1977, α-actinin attached to membranes of secretory vesicles, *Nature* **270**:628–629.

Jones, C. W., Mastrangelo, I. A., Smith, H. H., Liu, H. Z., and Meck, R. A., 1976, Interkingdom fusion between human (HeLa) cells and tobacco hybrid (GGLL) protoplasts, *Science* **193**:401–403.

Judah, J. D., and Quinn, P. S., 1978, Calcium ion-dependent vesicle fusion in the conversion of proalbumin to albumin, *Nature* **271**:384–385.

Kalderon, N., 1980, Muscle cell fusion, in: *Membrane–Membrane Interactions* (N. B. Gilula, ed.), pp. 99–118, Raven Press, New York.

Kalderon, N., and Gilula, N. B., 1979, Membrane events involved in myoblast fusion, *J. Cell Biol.* **81**:411–425.

Kalderon, N., Epstein, M. L., and Gilula, N. B., 1977, Cell-to-cell communication and myogenesis, *J. Cell Biol.* **75**:788–806.

Kao, K. N., and Kichayluk, M. R., 1974, A method for high-frequency intergeneric fusion of plant protoplasts, *Planta* **115**:355–367.

Kanno, T., Cochrane, D. E. and Douglas, W. W., 1973, Exocytosis (secretory granule extrusion) induced by injection of calcium into mast cells, *Can. J. Physiol. Pharmacol.* **51**:1001–1004.

Kantor, H. L., and Prestegard, J. H., 1975, Fusion of fatty acid containing lecithin vesicles, *Biochemistry* **14**:1790–1795.

Kantor, H. L., and Prestegard, J. H., 1978, Fusion of phosphatidylcholine bilayer vesicles: Role of free fatty acid, *Biochemistry* **17**:3592–3597.

Karnovsky, M. J., Kleinfeld, A. M., Hoover, R. L., and Klausner, R. D., 1982, The concept of lipid domains in membranes, *J. Cell Biol.* **94**:1–6.

Katz, B., 1969, *The Release of Neural Transmitter Substances*, C. C. Thomas, Springfield, IL.

Kaufman, S. J., 1982, Introduction: Membrane events during myogenesis, in: *Muscle Development: Molecular and Cellular Control* (M. L. Pearson and H. F. Epstein, eds.), pp. 271–280, Cold Spring Harbor Laboratory, Cold Spring Harbor, New York.

Kaufman, S. J., and Lawless, M. L., 1980, Thiodigalactoside binding lectin and skeletal myogenesis, *Differentiation* **16**:41–48.

Kaur, H., and Sanwal, B. D., 1981, Regulation of the activity of a calcium-activated neutral protease during differentiation of skeletal myoblasts, *Can. J. Biochem.* **59**:743–747.
Kawasaki, K., Sato, S. B., and Ohnishi, S.-I., 1983, Membrane fusion activity of reconstituted vesicles of influenza virus hemagglutinin glycoproteins, *Biochim. Biophys. Acta* **733**:286–290.
Kendall, D. A., and MacDonald, R. C., 1982, A fluorescence assay to monitor vesicle fusion and lysis, *J. Biol. Chem.* **257**:13892–13895.
Kent, C., Schimmel, S. D., and Vagelos, P. R., 1974, Lipid composition of plasma membranes from developing chick muscle cells in culture, *Biochim. Biophys. Acta* **360**:312–321.
Kim, J., and Okada, Y., 1981, Morphological changes in Ehrlich ascites tumor cells during the cell fusion reaction with HVJ (Sendai virus) II. Cluster formation of intramembrane particles in the early stage of cell fusion, *Exp. Cell Res.* **132**:125–136.
Klausner, R. D., Kleinfeld, A. M., Hoover, R. L., and Karnovsky, M. J., 1980, Lipid domains in membranes: Evidence derived from structural perturbations induced by free fatty acids and lifetime heterogeneity analysis, *J. Cell Biol.* **255**:1286–1295.
Knight, D. E. and Baker, P. F., 1983, The phorbol ester TPA increases the affinity of exocytosis for calcium in "leaky" adrenal medulla cells, *FEBS Lett.* **160**:98–100.
Knudsen, K. A., and Horwitz, A. F., 1977, Tandem events in myoblast fusion, *Dev. Biol.* **58**:328–338.
Knudsen, K. A., and Horwitz, A. F., 1978, Differential inhibition of myoblast fusion, *Dev. Biol.* **66**:294–307.
Knutton, S., 1977a, Studies of membrane fusion I. Paramyxovirus-induced cell fusion, a scanning electron-microscope study, *J. Cell Sci.* **28**:179–188.
Knutton, S., 1977b, Studies of membrane fusion II. Fusion of human erythrocytes by Sendai virus, *J. Cell Sci.* **28**:189–210.
Knutton, S., 1979a, Studies of membrane fusion III. Fusion of erythrocytes with polyethylene glycol, *J. Cell Sci* **36**:61–72.
Knutton, S., 1979b, Studies of membrane fusion IV. Fusion of HeLa cells with Sendai virus, *J. Cell Sci.* **36**:73–84.
Knutton, S., 1979c, Studies of membrane fusion V. Fusion of erythrocytes with non-haemolytic Sendai virus, *J. Cell Sci.* **36**:85–96.
Knutton, S., and Pasternak, C. A., 1979, The mechanism of cell–cell fusion, *Trends. Biochem. Sci.* **4**:220–223.
Koerker, R. L., Hahn, W. E., and Schneider, F. H., 1974, Electron translucent vesicles in adrenal medulla following catecholamine depletion, *Eur. J. Pharmacol.* **28**:350–359.
Koike, H., and Meldolesi, J., 1981, Post-stimulation retrieval of luminal surface membrane in parotid acinar cells is calcium-dependent, *Exptl. Cell Res.* **134**:377–388.
Kolber, M. A. and Haynes, D. H., 1979, Evidence for a role of phosphatidylethanolamine as a modulator of membrane–membrane contact, *J. Memb. Biol.* **48**:95–114.
Konings, F., and DePotter, W., 1981, Calcium-dependent in vitro interaction between bovine adrenal medullary cell membranes and chromaffin granules as a model for exocytosis, *FEBS Lett.* **126**:103–106.
Konings, F., and DePotter, W., 1982, A role for sialic acid containing substrates in the exocytosis-like in vitro interaction between adrenal medullary plasma membranes and chromaffin granules, *Biochem. Biophys. Res. Commun.* **106**:1191–1195.
Korchak, H. M., Vienne, K., Rutherford, L. E., Wilkenfeld, C., Finkelstein, M. C. and Weissmann, G., 1984, Stimulus response coupling in the human neutrophil II. Temporal analysis of changes in cytosolic calcium and calcium efflux, *J. Biol. Chem.,* **259**:4076–4082.
Korn, E. D., Bowers, B., Batzri, S., Simmons, S. R., and Victoria, E. J., 1974, Endocytosis and exocytosis: Role of microfilaments and involvement of phospholipids in membrane fusion, *J. Supramolec. Struct.* **2**:517–528.

Koter, M., deKruijff, B., and van Deenen, L. L. M., 1978, Calcium-induced aggregation and fusion of mixed phosphatidylcholine-phosphatidic acid vesicles as studied by ^{31}P NMR, *Biochim. Biophys. Acta* **514**:255–263.
Kuroda, K., Maeda, T., and Ohnishi, S.-I., 1980, Enhancement of phospholipid transfer from Sendai virus to erythrocytes is mediated by target cell membrane, *Proc. Natl. Acad. Sci. USA* **77**:804–807.
Ladbrooke, B. D., and Chapman, D., 1969, Thermal analysis of lipids, proteins and biological membranes: A review and summary of some recent studies, *Chem. Phys. Lipids* **3**:304–356.
Lagunoff, D., 1973, Membrane fusion during mast cell secretion, *J. Cell Biol.* **57**:252–259.
Lampe, P. D., and Nelsestuen, G. L., 1982, Myelin basic protein-enhanced fusion of membranes, *Biochim. Biophys. Acta* **693**:320–325.
Lang, R. D. A., Wickenden, C., Wynne, J., and Lucy, J. A., 1984, Proteolysis of ankyrin and of band 3 protein in chemically induced cell fusion. Ca^{2+} is not mandatory for fusion, *Biochem. J.* **218**:295–305.
Lansman, J., and Haynes, D. H., 1975, Kinetics of a Ca^{2+}-triggered membrane aggregation reaction of phospholipid membranes, *Biochim. Biophys. Acta* **394**:335–347.
Larrabee, A. L., 1979, Time-dependent changes in the size distribution of distearoylphosphatidylcholine vesicles, *Biochemistry* **18**:3321–3326.
Lau, A. L. Y., and Chan, S. I., 1975, Alamethicin-mediated fusion of lecithin vesicles, *Proc. Natl. Acad. Sci. USA* **72**:2170–2174.
Lawaczeck, R., Kainosho, M., Girardet, J.-L., and Chan, S. I., 1975, Effects of structural defects in sonicated phospholipid vesicles on fusion and ion permeability, *Nature* **256**:584–586.
Lawson, D., Raff, M. C., Gomperts, B., Fewtrell, C., and Gilula, N. B., 1977, Molecular events during membrane fusion: A study of exocytosis in rat peritoneal mast cells, *J. Cell Biol.* **72**:242–259.
Lazarowitz, S. G. and Choppin, P. W., 1975, Enhancement of the infectivity of influenza A and B viruses by proteolytic cleavage of the hemagglutinin polypeptide, *Virology* **68**:440–454.
Leblond, C. P., and Bennett, G., 1977, Role of the Golgi apparatus in terminal glycosylation, in: *International Cell Biology* (B. Brinkley and K. Porter, eds.), pp. 326–336, Rockefeller University Press, New York.
Lee, A. G., Birdsall, N. J. M., Metcalfe, J. C., Toon, P. A., and Warren, G. B., 1974, Clusters in lipid bilayers and the interpretation of thermal effects in biological membranes, *Biochemistry* **13**:3699–3705.
Lefort-Tran, M., Aufderheide, K., Pouphile, M., Rossignol, M., and Beisson, J., 1981, Control of exocytotic processes: Cytologial and physiological studies of trichocyst mutants in *Paramecium tetraurelia, J. Cell Biol.* **88**:301–311.
Lenard, J., and Miller, D. K., 1981, pH-dependent hemolysis by influenza, Semliki Forest virus, and Sendai virus, *Virology* **110**:479–482.
Lentz, B. R., Hoechli, M., and Barenholz, Y., 1981, Acyl chain order and lateral domain formation in mixed phosphatidylcholine-sphingomyelin multilamellar and unilamellar vesicles, *Biochemistry* **20**:6803–6809.
LeNeveu, D. M., Rand, R. P., and Parsegian, V. A., 1976, Measurement of forces between lecithin bilayers, *Nature* **259**:601–603.
LeNeveu, D. M., Rand, R. P., Parsegian, V. A., and Gingell, D., 1977, Measurement and modification of forces between lecithin bilayers, *Biophys. J.* **18**:209–230.
Liao, M.-J., and Prestegard, J. H., 1979, Fusion of phosphatidic acid–phosphatidylcholine mixed lipid vesicles, *Biochim. Biophys. Acta* **550**:157–173.
Liao, M.-J., and Prestegard, J. H., 1980, Ion specificity in fusion of phosphatidic acid–phosphatidylcholine mixed lipid vesicles, *Biochim. Biophys. Acta* **601**:453–461.
Liberman, Y. A., and Nenashev, V. A., 1968, Study of the interaction of artificial phospholipid membranes, *Biofizika* **13**:193–196.

Liberman, Y. A., and Nenashev, V. A., 1970, Modelling of the interaction of cell membranes by artificial phospholipid membranes, *Biofizika* **15**:1014–1021.

Liberman, Y. A., and Nenashev, V. A., 1972a, Kinetics of adhesion and surface electrical conductivity of bimolecular phospholipid membranes, *Biofizika* **17**:231–238.

Liberman, Y. A., and Nenashev, V. A., 1972b, Modelling of the changes in permeability of cellular contact with bimolecular phospholipid membranes, *Biofizika* **17**:1017–1023.

Lichtenberg, D., and Schmidt, C. F., 1981, Molecular packing and stability in the gel phase of curved phosphatidylcholine vesicles, *Lipids* **16**:555–557.

Lichtenberg, D., Freire, E., Schmidt, C. F., Barenholz, Y., Felgner, P. L., and Thompson, T. E., 1981, Effect of surface curvature on stability, thermodynamic behavior, and osmotic activity of dipalmitoylphosphatidylcholine single lamellar vesicles, *Biochemistry* **20**:3462–3467.

Lienhard, G. E., 1983, Regulation of cellular membrane transport by the exocytotic insertion and endocytotic retrieval of transporters, *Trends Biochem. Sci.* **8**:125–127.

Lin, K.-C., Weis, R. M., and McConnell, H. M., 1982, Induction of helical liposomes by Ca^{2+}-mediated intermembrane binding, *Nature* **296**:164–165.

Lis, L. J., Lis, W. T., Parsegian, V. A., and Rand, R. P., 1981, Adsorption of divalent cations to a variety of phosphatidylcholine bilayers, *Biochemistry* **20**:1771–1777.

Lis, L. J., McAlister, M., Fuller, N., Rand, R. P., and Parsegian, V. A., 1982, Interactions between neutral phospholipid bilayer membranes, *Biophys. J.*, **37**:657–666.

Llinás, R., and Nicholson, C., 1975, Calcium role in depolarization-secretion coupling: An aequorin study in squid giant synapse, *Proc. Natl. Acad. Sci. USA* **72**:187–190.

Loyter, A., Lalazar, A., 1980, Induction of membrane fusion in human erythrocyte ghosts: Involvement of spectrin in the fusion process, in: *Membrane–Membrane Interactions* (N. B. Gilula, ed.), pp. 11–26, Raven Press, New York.

Lucy, J. A., 1970, The fusion of biological membranes, *Nature* **227**:814–817.

Lucy, J. A., 1977, The membrane of the hen erythrocyte as a model for studies on membrane fusion, in: *Structure of Biological Membranes, 34th Nobel Symposium* (S. Abrahamsson and I. Pascher, eds.), pp. 275–291, Plenum Press, New York.

Lucy, J. A., 1978, Mechanisms of chemically induced cell fusion, in: *Membrane Fusion* (G. Poste and G. L. Nicolson, eds.), pp. 267–304, Elsevier/North-Holland Biomedical Press, Amsterdam.

Lucy, J. A., 1984, Do hydrophobic sequences cleaved from cellular polypeptides induce membrane fusion reactions in vivo? *FEBS Lett.* **166**:223–231.

Luzzati, V., and Tardieu, A., 1974, Lipid phases: Structure and structural transitions, *Ann. Rev. Phys. Chem.* **25**:79–94.

Luzzati, V., Gulik-Krzywicki, T., and Tardieu, A., 1968, Polymorphism of lecithins, *Nature* **218**:1031–1034.

Lyles, D. S., and Landsberger, F. R., 1979, Kinetics of Sendai virus envelope fusion with erythrocytes membranes and virus-induced hemolysis, *Biochemistry* **18**:5088–5095.

MacBride, R. G. and Przybylski, R. J., 1980, Purified lectin from skeletal muscle inhibits myotube formation in vitro, *J. Cell Biol.* **85**:617–625.

Maeda, T., and Ohnishi, S.-I., 1980, Activation of influenza virus by acidic media causes hemolysis and fusion of erythrocytes, *FEBS Lett.* **122**:283–287.

Maeda, T., Kawasaki, K., and Ohnishi, S.-I., 1981, Interaction of influenza virus hemagglutinin with target membrane lipids is a key step in virus-induced hemolysis and fusion at pH 5.2, *Proc. Natl. Acad. Sci. USA* **78**:4133–4137.

Maggio, B., Ahkong, Q. F., and Lucy, J. A., 1976, Poly(ethylene glycol), surface potential and cell fusion, *Biochem. J.* **158**:647–650.

Majumdar, S., and Baker, R. F., 1980, Phosphate-calcium induced fusion of chicken erythrocytes, *Exp. Cell Res.* **126**:175–182.

Majumdar, S., Baker, R. F., and Kalra, V. K., 1980, Fusion of human erythrocytes induced by uranyl acetate and rare earth metals, *Biochim. Biophys. Acta* **598**:411–416.
Markwell, M. A. K., Svennerholm, L. and Paulson, J. C., 1981, Specific gangliosides function as host cell receptors for Sendai virus, *Proc. Natl. Acad. Sci. USA* **78**:5406–5410.
Maroudas, N. G., 1975, Polymer exclusion, cell adhesion and membrane fusion, *Nature* **254**:695–696.
Marsh, M., 1984, The entry of enveloped viruses into cells by endocytosis, *Biochem. J.,* **218**:1–10.
Marsh, M., and Helenius, A., 1980, Adsorptive endocytosis of Semliki Forest virus, *J. Mol. Biol.* **142**:439–454.
Marsh, M., Bolzau, E., and Helenius, A., 1983a, Penetration of Semliki Forest virus from acidic prelysosomal vacuoles, *Cell* **32**:931–940.
Marsh, M., Bolzau, E., White, J., and Helenius, A., 1983b, Interactions of Semliki Forest virus spike glycoprotein rosettes and vesicles with cultured cells, *J. Cell Biol.* **96**:455–461.
Martin, F. J. and MacDonald, R. C., 1976, Phospholipid exchange between bilayer membrane vesicles, *Biochemistry* **15**:321–327.
Mason, W. T., Lane, N. J., Miller, N. G. A. and Bangham, A. D., 1980, Fusion of liposome membranes by the n-alkyl bromides, *J. Memb. Biol.* **55**:69–79.
Matlin, K. S., Reggio, H., Helenius, A., and Simons, K., 1981, Infectious entry pathway of influenza virus in a canine kidney cell line, *J. Cell Biol.* **91**:601–613.
Matlin, K. S., Reggio, H., Helenius, A., and Simons, K., 1982, Pathway of vesicular stomatitis virus entry leading to infection, *J. Mol. Biol.* **156**:609–631.
Matt, H., Bilinski, M., and Plattner, H., 1978, Adenosinetriphosphate, calcium, and temperature requirements for the final step of exocytosis in *Paramecium* cells, *J. Cell Sci.* **32**:67–86.
Mayer, L. D., and Nelsestuen, G. L., 1981, Calcium- and prothrombin-induced lateral phase separation in membranes, *Biochemistry* **20**:2457–2463.
McIntosh, T. J., 1980, Differences in hydrocarbon chain tilt between hydrated phosphatidylethanolamine and phosphatidylcholine bilayers, *Biophys. J.* **29**:237–246.
McIntyre, J. A., Gilula, N. B., and Karnovsky, M. J., 1974, Cryoprotectant-induced redistribution of intramembranous particles in mouse lymphocytes, *J. Cell Biol.* **60**:192–203.
McIver, D. J. L., 1979, Control of membrane fusion by interfacial water: a model for the actions of divalent cations, *Physiol. Chem. Phys.* **11**:289–302.
McKanna, J. A., 1973, Membrane recycling: Vesiculation of the Amoeba contractile vacuole at systole, *Science* **179**:88–90.
McLaughlin, S., Mulrine, N., Gresalfi, T., Vaio, G., and McLaughlin, A., 1981, The adsorption of divalent cations to bilayer membranes containing phosphatidylserine, *J. Gen. Physiol.* **77**:445–473.
Meldolesi, J., and Ceccarelli, B., 1981, Exocytosis and membrane recycling, *Phil. Trans. R. Soc. Lond. B.* **296**:55–65.
Melikyan, G. B., Abidor, I. G., Chernomordik, L. V., and Chailakhyan, L. M., 1983, Electrostimulated fusion and fission of bilayer lipid membranes, *Biochim. Biophys. Acta* **730**:395–398.
Metcalfe, J. C., Birdsall, N. J. M., and Lee, A. G., 1972, NMR studies of dynamic features of membrane structure, in: *Mitochondria/Biomembranes, FEBS 8th Meeting* (S. G. van den Burgh, P. Borst, L. L. M. van Deenen, J. C. Riemeroma, E. C. Slater, and J. M. Tager, eds.), pp. 197–217, Elsevier/North-Holland Biomedical Press, Amsterdam.
Meyer, D. I., and Burger, M. M., 1976, The chromaffin granule surface: Localization of carbohydrate on the cytoplasmic surface of an intracellular organelle, *Biochim. Biophys. Acta* **443**:428–436.

Meyer, D. I., and Burger, M. M., 1979a, Isolation of a protein from the plasma membrane of adrenal medulla which binds to secretory vesicles, *J. Biol. Chem.* **254**:9854–9859.

Meyer, D. I., and Burger, M. M., 1979b, The chromaffin granule surface: The presence of actin and the nature of its interaction with the membrane, *FEBS Lett.* **101**:129–133.

Michaelson, D. M., Pinchasi, I., and Sokolovsky, M., 1978, Factors required for calcium dependent acetylcholine release from isolated *Torpedo* synaptic vesicles. *Biochem. Biophys. Res. Commun.* **80**:547–552.

Michell, R. H., 1975, Inositol phospholipids and cell surface receptor function, *Biochim. Biophys. Acta* **415**:81–147.

Michell, R. H., Kirk, C. J., Jones, L. M., Downes, C. P., and Creba, J. A., 1981, The stimulation of inositol lipid metabolism that accompanies calcium mobilization in stimulated cells: defined characteristics and unanswered questions, *Phil. Trans. R. Soc. Lond. Ser. B.*, **296**:123–137.

Miledi, R., 1973, Transmitter release induced by injection of calcium ions into nerve terminals, *Proc. R. Soc. London Ser. B.* **183**:421–425.

Miller, C., and Racker, E., 1976a, Ca^{2+}-induced fusion of fragmented sacroplasmic reticulum with artificial planar bilayers, *J. Memb. Biol.* **30**:283–300.

Miller, C., and Racker, E., 1976b, Fusion of phospholipid vesicles reconstituted with cytochrome c oxidase and mitochondrial hydrophobic protein, *J. Memb. Biol.* **26**:319–333.

Miller, C., Arvan, P., Telford, J. N., and Racker, E., 1976, Ca^{2+}-induced fusion of proteoliposomes: Dependence on transmembrane osmotic gradient, *J. Memb. Biol.* **30**:271–382.

Miller, D. C. and Dahl, G. P., 1982, Early events in calcium-induced liposome fusion, *Biochim. Biophys. Acta* **689**:165–169.

Miller, D. K., and Lenard, J., 1981, Antihistamines, local anesthetics and other amines as antiviral agents, *Proc. Natl. Acad. Sci. USA* **78**:3605–3609.

Miller, R. G., 1980, Do 'lipidic particles' represent intermembrane attachment sites? *Nature* **287**:166–167.

Milutinovic, A., Argent, B. E., Schulz, I., and Sachs, G., 1977, Studies on isolated subcellular components of cat pancreas. II. A Ca^{2+}-dependent interaction between membranes and zymogen granules of cat pancreas, *J. Memb. Biol.* **36**:281–295.

Miura, N., Uchida, T., and Okada, Y., 1982, HVJ (Sendai virus)-induced envelope fusion and cell fusion are blocked by monoclonal anti-HN protein antibody that does not inhibit hemaglutination activity of HVJ, *Exp. Cell Res.* **141**:409–420.

Mombers, C., vanDijck, P. W. M., vanDeenen, L. L. M., deGier, J. and Verkleij, A. J., 1977, The interaction of spectrin-actin and synthetic phospholipids, *Biochim. Biophys. Acta* **470**:152–160.

Mooney, J. J., Dalrymple, J. M., Alving, C. R., and Russell, P. K., 1975, Interaction of Sindbis virus with liposomal model membranes, *J. Virol.* **15**:225–231.

Mueller, P., Rudin, D. O., Tien, H. T., and Wescott, W. C., 1962, Reconstitution of cell membrane structure *in vitro* and its transformation into an excitable system, *Nature* **194**:979–980.

Moore, M. R., 1976, Fusion of liposomes containing conductance probes with black lipid films, *Biochim. Biophys. Acta* **426**:765–771.

Morgan, C. G., Williamson, H., Fuller, S., and Hudson, B., 1983, Mellitin induces fusion of unilamellar phospholipid vesicles, *Biochim. Biophys. Acta* **732**:668–674.

Morré, D. J., 1980, Flow-differentiation of membranes: Pathways and mechanisms, in: *Cell Compartmentation and Metabolic Channeling* (L. Nover, F. Lynen and K. Mothes, eds.), pp. 47–61, VEB Gustav Fischer Verlag, Jena and Elsevier/North-Holland Biomedical Press, Amsterdam.

Morris, G. E., Piper, M., and Cole, R., 1976, Differential effects of calcium ion concentration on cell fusion, cell division, and creatine kinase activity in muscle cell cultures, *Exp. Cell Res.* **99**:106–114.

Morris, S. J., and Schober, R., 1977, Demonstration of binding sites for divalent and trivalent ions on the outer surface of chromaffin granule membranes, *Eur. J. Biochem.* **75**:1–12.

Morris, S. J., and Hughes, J. M. X., 1979, Synexin protein is non-selective in its ability to increase Ca^{2+}-dependent aggregation of biological and artificial membranes, *Biochem. Biophys. Res. Commun.* **91**:345–350.

Morris, S. J., Hellweg, M. A., and Haynes, D. J., 1979a, Light scattering turbidity changes as a measure of the kinetics of Ca^{2+}-promoted aggregation of chromaffin granule membrane ghosts, *Biochim. Biophys. Acta* **553**:342–350.

Morris, S. J., Chiu, V. C. K., and Haynes, D. J., 1979b, Divalent cation-induced aggregation of chromaffin granule membranes, *Memb. Biochem.* **2**:163–201.

Morris, S. J., Hughes, J. M. X., and Whittaker, V. P., 1982, Purification and mode of action of synexin: A protein enhancing calcium-induced membrane aggregation, *J. Neurochem.* **39**:529–536.

Morris, S. J., Smith, P. D., Gibson, C. G., Haynes, D. H., and Blumenthal, R., 1983, Ca^{2+}-induced fusion of negatively charged small unilamellar vesicles is rapid but leaky to entrapped material; fusion of large unilamellar vesicles is rapid but non-leaky, *Biophys. J.,* **41**:28a.

Moskowitz, H., Schook, W., and Puszkin, S., 1982, Interaction of brain synaptic vesicles induced by endogenous Ca^{2+}-dependent phospholipase A_2, *Science* **216**:305–307.

Moss, M., Norris, J. S., Peck, E. J., Jr. and Schwartz, R. J., 1978, Alterations in iodinated cell surface proteins during myogenesis, *Exp. Cell Res.* **113**:445–450.

Nagasawa, J., and Douglas, W. W., 1972, Thorium dioxide uptake into adrenal medullary cells and the problem of recapture of granule membranes following exocytosis, *Brain Res.* **37**:141–145.

Nagasawa, J., Douglas, W. W., and Schulz, R. A., 1971, Micropinocytotic origin of coated and smooth microvesicles ("synaptic vesicles") in neurosecretory terminals of posterior pituitary glands demonstrated by incorporation of horseradish peroxidase, *Nature* **232**:341–342.

Nagy, A., Baker, R. R., Morris, S. J., and Whittaker, V. P., 1976, The preparation and characterization of synaptic vesicles of high purity, *Brain Res.* **109**:285–309.

Nameroff, M., Trotter, J. A., Keller, J. M., and Munar, E., 1973, Inhibition of cellular differentiation by phospholipase C. I. Effects of the enzyme on myogenesis and chondrogenesis in vitro, *J. Cell Biol.* **58**:107–118.

Nayar, R., Hope, M. J., and Cullis, P. R., 1982, Phospholipids as adjuncts for calcium ion stimulated release of chromaffin granule contents: Implications for mechanisms of exocytosis, *Biochemistry* **21**:4583–4589.

Neher, E., 1974, Asymmetric membranes resulting from the fusion of two black lipid bilayers, *Biochim. Biophys. Acta* **373**:327–336.

Nenashev, V. A., Grishin, A. F., Berestovsky, G. N., 1978, Interaction of liposomes with bimolecular lipid membranes, *Stud. Biophys.* **73**:119–126.

Neutra, M. R. and Schaeffer, S. F., 1977, Membrane interactions between adjacent mucuous secretion granules, *J. Cell Biol.* **74**:983–991.

Newton, C., Pangborn, W., Nir, S., and Papahadjopoulos, D., 1978, Specificity of Ca^{2+} and Mg^{2+} binding to phosphatidylserine vesicles and resultant phase changes of bilayer membrane structure, *Biochim. Biophys. Acta* **506**:281–287.

Nickel, E., and Potter, L. T., 1971, Synaptic vesicles in freeze-etched electric tissue of *Torpedo, Phil. Trans. Roy. Soc. Lond. Ser. B.* **261**:383–385.

Nir, S., 1977, Van der Waals interactions between surfaces of biological interest, *Prog. Surface Sci.* **8**:1–58.

Nir, S., and Bentz, J., 1978, On the forces between phospholipid bilayers, *J. Coll. Interface Sci.* **65**:399–414.

Nir, S., Newton, C., and Papahadjopoulos, D., 1978, Binding of cations to phosphatidylserine vesicles, *Bioelectrochem. Bioenerg.* **5**:116–133.

Nir, S., Bentz, J., and Wilschut, J., 1980, Mass action kinetics of phosphatidylserine vesicle fusion as monitored by coalescence of internal vesicle volumes, *Biochemistry* **19**:6030–6036.

Nir, S., Bentz, J., and Düzgüneş, N., 1981, Two modes of reversible vesicle aggregation: Particle size and the DLVO theory, *J. Colloid Interface Sci.*, **84**:266–269.

Nir, S., Wilschut, J., and Bentz, J., 1982, The rate of fusion of phospholipid vesicles and the role of bilayer curvature, *Biochim. Biophys. Acta* **688**:275–278.

Nir, S., Bentz, J., Wilschut, J., and Düzgüneş, N., 1983a, Aggregation and fusion of phospholipid vesicles, *Prog. Surface Sci.* **13**:1–124.

Nir, S., Düzgüneş, N., and Bentz, J., 1983b, Binding of monovalent cations to phosphatidylserine and modulation of Ca^{2+}- and Mg^{2+}-induced vesicle fusion, *Biochim. Biophys. Acta* **735**:160–172.

Nordmann, J. J., Dreifuss, J. J., Baker, P. F., Ravazzola, M., Malaisse-Lagae, F., and Orci, L., 1974, Secretion-dependent uptake of extracellular fluid by the rat neurohypophysis, *Nature* **250**:155–157.

Nowak, T. P., Haywood, P. L., and Barondes, S. H., 1976, Developmentally regulated lectin in embryonic chick muscle and a myogenic cell line, *Biochem. Biophys. Res. Commun.* **68**:650–657.

Oates, P. J., and Touster, O., 1976, *In vitro* fusion of *Acanthamoeba* phagolysosomes. I. Demonstration and quantitation of vacuole fusion in *Acanthamoeba* homogenates, *J. Cell Biol.* **68**:319–339.

Oates, P. J. and Touster, O., 1978, *In vitro* fusion of *Acanthamoeba* phagolysosomes. II. Quantitative characterization of *in vitro* vacuole fusion by improved electron microscope and new light microscope techniques, *J. Cell Biol.* **79**:217–234.

Odenwald, W. F. and Morris, S. J., 1983, Identification of a second synexin-like adrenal medullary and liver protein that enhances calcium-induced membrane aggregation, *Biochem. Biophys. Res. Commun.* **112**:147–154.

O'Doherty, J., Youmans, S. J., Armstrong, W. M. and Stark, R. J., 1980, Calcium regulation during stimulus-secretion coupling: Continuous measurement of intracellular calcium activities, *Science* **209**:510–513.

Ohki, S., 1982, A mechanism of divalent ion-induced phosphatidylserine membrane fusion, *Biochim. Biophys. Acta* **689**:1–11.

Ohki, S., 1984, Effects of divalent cations, temperature, osmotic pressure gradient and vesicle curvature on phosphatidylserine vesicle fusion, *J. Memb. Biol.* **77**:265–275.

Ohki, S., and Sauvé, R., 1978, Surface potential of phosphatidylserine monolayers I. Divalent ion binding effect, *Biochim. Biophys. Acta* **511**:377–387.

Ohki, S., and Kurland, R., 1981, Surface potential of phosphatidylserine monolayers II. Divalent and monovalent ion binding, *Biochim. Biophys. Acta* **645**:170–176.

Ohki, S., and Düzgüneş, N., 1979, Divalent cation induced interaction of phospholipid vesicle and monolayer membranes, *Biochim. Biophys. Acta* **552**:438–449.

Ohki, S., and Leonards, K., 1982, Effects of proteins on phospholipid vesicle aggregation and lipid vesicle-monolayer interactions, *Chem. Phys. Lipids* **31**:307–318.

Ohki, S., Düzgüneş, N., and Leonards, K., 1982, Phospholipid vesicle aggregation: Effect of monovalent and divalent ions, *Biochemistry* **21**:2127–2133.

Ohnishi, S.-I., and Ito, T., 1974, Calcium-induced phase separations in phosphatidylserine-phosphatidylcholine membranes, *Biochemistry* **13**:881-887.

Ohnishi, S.-I., and Tokutomi, S., 1981, ESR studies of calcium and proton-induced phase separations in phosphatidylserine/phosphatidylcholine mixed membranes, in: *Biological Magnetic Resonance*, Vol. 3 (L. J. Berliner and J. Reuben, eds.), pp. 121-153, Plenum Press, New York.

Ohsawa, K., Ohshima, H., and Ohki, S., 1981, Surface potential and surface charge density of the cerebral-cortex synaptic vesicle and stability of vesicle suspension, *Biochim. Biophys. Acta* **648**:206-214.

Oka, M., Ohuchi, T., Yoshida, H., and Imizumi, R., 1965, Effect of adenosine triphosphate and magnesium on the release of catecholamines from adrenal medullary granules, *Biochim. Biophys. Acta* **97**:170-171.

Oku, N., Nojima, S., and Inoue, K., 1982, Studies on the interaction HVJ (Sendai virus) with liposomal membranes. HVJ-induced permeability increase of liposomes containing glycophorin, *Virology* **116**:419-427.

Olden, K., Law, J., Hunter, V. A., Romain, R., and Parent, K., 1981, Inhibition of fusion of embryonic muscle cells in culture by tunicamycin is prevented by leupeptin, *J. Cell Biol.* **88**:199-204.

Oliver, J. M., and Berlin, R. D., 1979, Microtubules, microfilaments and the regulation of membrane functions, *Symp. Soc. Exp. Biol.* **33**:277-298.

O'Loughlin, J., Lehr, L., Havaranis, A., and Heywood, S. M., 1981, Encapsulation of "core" eIF3, regulatory components of eIF3 and mRNA into liposomes and their subsequent uptake into myogenic cells in culture, *J. Cell Biol.* **90**:160-168.

Op den Kamp, J. A. F., 1979, Lipid asymmetry in membranes, *Ann. Rev. Biochem.* **48**:47-71.

Orci, L., 1982, Macro- and micro-domains in the endocrine pancreas, *Diabetes* **31**:538-565.

Orci, L., and Malaisse, W., 1980, Single and chain release of insulin secretory granules is related to anionic transport at exocytotic sites, *Diabetes* **29**:943-944.

Orci, L., Malaisse-Lagae, F., Ravazzola, M., Amherdt, M., and Renold, A. E., 1973, Exocytosis-endocytosis coupling in the pancreatic beta cell, *Science* **181**:561-562.

Orci, L., Amherdt, M., Malaisse-Lagae, F., Rouiller, C., Renold, A. E., 1973, Insulin release by emiocytosis: Demonstration with freeze-etching technique, *Science* **179**:82-84.

Orci, L., Amherdt, M., Montesano, R., Vassalli, P., and Perrelet, A., 1981, Topology of morphologically detectable protein and cholesterol in membranes of polypeptide secreting cells, *Phil. Trans. R. Soc. Lond. Ser. B.* **296**:47-54.

Ornberg, R. L., and Reese, T. S., 1981, Beginning of exocytosis captured by rapid-freezing of *Limulus* amebocytes, *J. Cell Biol.* **90**:40-54.

Ozawa, M., and Asano, A., 1981, The preparation of cell fusion-inducing proteoliposomes from purified glycoproteins of HVJ (Sendai virus) and chemically defined lipids, *J. Biol. Chem.* **256**:5954-5956.

Palade, G. E., 1959, Functional changes in the structure of cell components, in: *Subcellular Particles* (T. Hayashi, ed.), pp. 64-80, Ronald Press, New York.

Palade, G., 1975, Intracellular aspects of the process of protein synthesis, *Science* **189**:347-358.

Palade, G. E., and Bruns, R. R., 1968, Structural modulations of plasmalemmal vesicles, *J. Cell Biol.* **37**:633-649.

Papahadjopoulos, D., 1978, Calcium-induced phase changes and fusion in natural and model membranes, in: *Membrane Fusion* (G. Poste and G. L. Nicolson, eds.), pp. 765-790, Elsevier/North-Holland Biomedical Press, Amsterdam.

Papahadjopoulos, D., and Bangham, A. D., 1966, Biophysical properties of phospholipids. II. Permeability of phosphatidylserine liquid crystals to univalent ions. *Biochim. Biophys. Acta* **126**:185-188.

Papahadjopoulos, D., and Ohki, S., 1969, Stability of asymmetric phospholipid membranes, *Science* **164**:1075–1077.

Papahadjopoulos, D., and Ohki, S., 1970, Conditions of stability for liquid-crystalline phospholipid membranes, in: *Liquid Crystals and Ordered Fluids* (J. F. Johnson and R. S. Porter, eds.), pp. 13–32, Plenum Press, New York.

Papahadjopoulos, D., and Vail, W. J., 1978, Incorporation of macromolecules within large unilamellar vesicles (LUV), *Ann. N.Y. Acad. Sci* **308**:259–267.

Papahadjopoulos, D., Poste, G., Schaeffer, B. E., and Vail, W. J., 1974, Membrane fusion and molecular segregation in phospholipid vesicles, *Biochim. Biophys. Acta* **352**:10–28.

Papahadjopoulos, D., Vail, W. J., Jacobson, K., and Poste, G., 1975, Cochleate lipid cylinders: Formation by fusion of unilamellar lipid vesicles, *Biochim. Biophys. Acta* **394**:483–491.

Papahadjopoulos, D., Hui, S., Vail, W. J., and Poste, G., 1976a, Studies on membrane fusion I. Interactions of pure phospholipid membranes and the effect of myristic acid, lysolecithin, proteins and DMSO, *Biochim. Biophys. Acta* **448**:245–264.

Papahadjopoulos, D., Vail, W. J., Pangborn, W. A., and Poste, G., 1976b, Studies on membrane fusion. II. Induction of fusion in pure phospholipid membranes by calcium ions and other divalent metals, *Biochim. Biophys. Acta* **448**:265–283.

Papahadjopoulos, D., Vail, W. J., Newton, C., Nir, S., Jacobson, K., Poste, G., and Lazo, R., 1977, Studies on membrane fusion. III. The role of calcium-induced phase changes, *Biochim. Biophys. Acta* **465**:579–598.

Papahadjopoulos, D., Portis, A., and Pangborn, W., 1978a, Calcium-induced lipid phase transitions and membrane fusion, *Ann. NY Acad. Sci.* **308**:50–66.

Papahadjopoulos, D., Portis, A., Pangborn, W., and Newton, C., 1978b, Fusion of artificial membranes with special emphasis on the role of calcium-induced lipid phase transitions, in: *Transport of Macromolecules in Cellular Systems* (S. C. Silverstein, ed.), pp. 413–430, Dahlem Konferenzen, Berlin.

Papahadjopoulos, D., Poste, G., and Vail, W. J., 1979, Studies on membrane fusion with natural and model membranes, *Methods Memb. Biol.* **10**:1–121.

Parfett, C. L. J., Jamieson, J. C., and Wright, J. A., 1981, A correlation between loss of fusion potential and defective formation of mannose-linked lipid intermediates in independent concanavalin A-resistant myoblast cell-lines, *Exp. Cell Res.* **136**:1–14.

Parfett, C. L. J., Jamieson, J. C., and Wright, J. A., 1983, Changes in cell surface glycoproteins on non-differentiating L6 rat myoblasts selected for resistance to concanavalin A, *Exp. Cell Res.* **144**:405–415.

Parsegian, V. A., 1973, Long-range physical forces in the biological milieu, *Ann. Rev. Biophys. Bioeng.* **2**:221–255.

Pastan, I. H., and Willingham, M. C., 1981, Journey to the center of the cell: Role of the receptosome, *Science* **214**:504–509.

Pasternak, C. A., and Micklem, K. J., 1973, Permeability changes during cell fusion, *J. Memb. Biol.* **14**:293–303.

Patzak, A., Bock, G., Fischer-Colbrie, R., Schauenstein, K., Schmidt, W., Lingg, G., and Winkler, H., 1984, Exocytotic exposure and retrieval of membrane antigens of chromaffin granules: Quantitative evaluation of immunofluorescence on the surface of chromaffin cells, *J. Cell Biol.* **98**:1817–1824.

Patzer, E. J., Wagner, R. R., and Dubovi, E. J., 1979, Viral membranes: Model systems for studying biological membranes, *CRC Crit. Rev. Biochem.* **6**:165–217.

Patzer, E. J., Schlossman, D. M. and Rothman, J. E., 1982, Release of clathrin from coated vesicles dependent upon nucleotide triphosphate and a cytosol fraction, *J. Cell Biol.* **93**:230–236.

Pauw, P. G., and David, J. D., 1979, Alterations in surface proteins during myogenesis of a rat myoblast cell line, *Dev. Biol.* **70**:27–38.

Pazoles, C. J., and Pollard, H. B., 1978, Evidence for stimulation of anion transport in ATP-evoked transmitter release from isolated secretory vesicles, *J. Biol. Chem.* **253**:3962–3969.

Pearse, B. M. F., and Bretscher, M. S., 1981, Membrane recycling by coated vesicles, *Ann. Rev. Biochem.* **50**:85–101.

Peixoto de Menezes, A., and Pinto da Silva, P., 1978, Freeze-fracture observations of the lactating rate mammary gland: Membrane events during milk fat secretion, *J. Cell Biol.* **76**:767–778.

Pelletier, G., 1973, Secretion and uptake of peroxidase by rat adenohypophyseal cells, *J. Ultrastruct. Res.* **43**:445–459.

Phillips, J. N., Burridge, K., Wilson, S. P., and Kirschner, N., 1983, Visualization of the exocytosis/endocytosis secretory cycle in cultured adrenal chromaffin cells, *J. Cell Biol.* **97**:1906–1917.

Phillips, M. C., and Chapman, D., 1968, Monolayer characteristics of saturated 1,2-diacyl phosphatidylcholines (lecithins) and phosphatidylethanolamines at the air–water interface, *Biochim. Biophys. Acta* **163**:301–313.

Pinto da Silva, P., and Martinez-Palomo, A., 1974, Induced redistribution of membrane particles, anionic sites and con A receptors in *Entamoeba histolytical, Nature* **249**:170–171.

Pinto da Silva, P., and Nogueira, M. L., 1977, Membrane fusion during secretion. A hypothesis based on electron microscope observation of *Phytophthora palmivora* zoospores during encystment, *J. Cell Biol.* **73**:161–181.

Pinto da Silva, P., Shimizu, K., and Parkison, C., 1980, Fusion of human erythrocytes induced by Sendai virus: Freeze-fracture aspects, *J. Cell Sci.* **43**:419–432.

Plattner, H., 1978, Fusion of cellular membranes, in: *Transport of Macromolecules in Cellular Systems* (S. C. Silverstein, ed.), pp. 465–488, Dahlem Konferenzen, Berlin.

Plattner, H., Schmitt-Fumian, W. W., and Bachmann, L., 1973, Cryofixation of single cells by spray-freezing, in: *Freeze-Etching: Techniques and Applications* (E. L. Benedetti and P. Favard, eds.), pp. 81–100, Société Française Microscopie Electronique, Paris.

Podleski, T. R., and Greenberg, I., 1980, Distribution and activity of endogenous lectin during myogenesis as measured with antilectin antibody, *Proc. Natl. Acad. Sci. USA* **77**:1054–1058.

Podleski, T. R., Greenberg, I., Schlessinger, J., and Yamada, K. M., 1979, Fibronectin delays the fusion of L_6 myoblasts, *Exp. Cell Res.* **122**:317–326.

Pohl, G. W., Stark, G., and Trissl, H.-W., 1973, Interaction of liposomes with black lipid membranes, *Biochim. Biophys. Acta* **318**:478–481.

Pollard, H. B., and Scott, J. H., 1982, Synhibin: A new calcium-dependent membrane-binding protein that inhibits synexin-induced chromaffin granule aggregation and fusion, *FEBS Lett.* **150**:201–206.

Pollard, H. B., Tack-Goldman, K. M., Pazoles, C. J., Creutz, C. E., and Shulman, H. R., 1977, Evidence for control of serotonin secretion from human platelets by hydroxyl ion transport and osmotic lysis, *Proc. Natl. Acad. Sci. USA* **74**:5295–5299.

Pollard, H. B., Pazoles, C. J., Creutz, C. E., and Zinder, O., 1979, The chromaffin granule and possible mechanisms of exocytosis, *Int. Rev. Cytol.* **58**:159–197.

Pollard, H. B., Creutz, C. E., Pazoles, C. J., and Scott, J. H., 1980, Fusion and fission processes in exocytosis: Possible role for synexin and osmotic lysis in the two events, in: *Proceedings of the Electron Microscope Society of America 38th Annual Meeting,* (C. W. Bailey, ed.), pp. 594–597, Claitor's Publishing Division, Baton Rouge.

Pollard, H. B., Pazoles, C. J., and Creutz, C. E., 1981, Mechanisms of calcium action and release of vesicle-bound hormones during exocytosis, *Recent Prog. Hormone Res.* **37**:299–332.

Pontecorvo, G., 1975, Production of mammalian somatic cell hybrids by means of polyethylene glycol treatment, *Somat. Cell Genet.* **1**:397–400.

Pontecorvo, G., Riddle, P. N., and Hales, A., 1977, Time and mode of fusion of human fibroblasts treated with polyethylene glycol (PEG), *Nature* **265**:257–258.

Poole, A. R., Howell, J. I., and Lucy, J. A., 1970, Lysolecithin and cell fusion, *Nature* **227**:810–813.

Portis, A., Newton, C., Pangborn, W., and Papahadjopoulos, D., 1979, Studies on the mechanism of membrane fusion: Evidence for an intermembrane Ca^{2+}-phospholipid complex, synergism with Mg^{2+}, and inhibition by spectrin, *Biochemistry* **18**:780–790.

Portner, A., Scroggs, R. A., Marx, P. A., and Kingsbury, D. W., 1975, A temperature sensitive mutant of Sendai virus with an altered hemagglutinin-neuraminidase polypeptide: Consequences for virus assembly and cytopathology, *Virology* **67**:179–187.

Poste, G., 1972, Mechanisms of virus-induced cell fusion, *Int. Rev. Cytol.* **33**:157–252.

Poste, G., and Allison, A. C., 1973, Membrane fusion, *Biochim. Biophys. Acta* **300**:421–465.

Poste, G., and Pasternak, C. A., 1978, Virus-induced cell fusion, in: *Membrane Fusion* (G. Poste and G. L. Nicolson, eds.), pp. 305–367, Elsevier/North-Holland Biomedical Press, Amsterdam.

Poste, G., and Nicolson, G. L., eds., 1978, *Membrane Fusion*, pp. 1–862, Elsevier/North Holland Biomedical Press, Amsterdam.

Poste, G., Lyon, N. C., Macander, P., Porter, C. W., Reeve, P., and Bachmeyer, H., 1980, Liposome-mediated transfer of integral membrane glycoproteins into the plasma membrane of cultured cells, *Exp. Cell Res.* **129**:393–408.

Prives, J., and Shinitzky, M., 1977, Increased membrane fluidity precedes fusion of muscle cells, *Nature* **268**:761–763.

Prusch, R. D., 1980, Endocytotic sucrose uptake in *Amoeba proteus* induced by the calcium ionophore A23187, *Science* **209**:691–692.

Quinn, P. S., and Judah, J. D., 1978, Calcium-dependent Golgi-vesicle fusion and cathepsin B in the conversion of proalbumin into albumin in rat liver, *Biochem. J.* **172**:301–309.

Rand, R. P., 1981, Interacting phospholipid bilayers: Measured forces and induced structural changes, *Ann. Rev. Biophys. Bioeng.* **10**:277–314.

Rand, R. P., Tinker, D. O., and Fast, P. G., 1971, Polymorphism of phosphatidylethanolamine from two natural sources, *Chem. Phys. Lipids* **6**:333–342.

Rand, R. P., Parsegian, V. A., Henry, J. A. C., Lis, L. J., and McAlister, M., 1980, The effect of cholesterol on measured interaction and compressibility of dipalmitoylphosphatidylcholine bilayers, *Can. J. Biochem.* **58**:959–968.

Rand, R. P., Reese, T. S., and Miller, R. G., 1981, Phospholipid bilayer deformations associated with interbilayer contact and fusion, *Nature* **293**:237–238.

Raz, A., and Goldman, R., 1974, Spontaneous fusion of rat liver lysosomes *in vitro*, *Nature* **247**:206–208.

Razin, M., and Ginsburg, H., 1980, Fusion of liposomes with planar lipid bilayers, *Biochim. Biophys. Acta* **598**:285–292.

Recktenwald, D. J., and McConnell, H., 1981, Phase equilibria in binary mixtures of phosphatidylcholine and cholesterol, *Biochemistry* **20**:4505–4510.

Rehfeld, S. J., Düzgüneş, N., Newton, C., Papahadjopoulos, D., and Eatough, D. J., 1981, The exothermic reaction of calcium with unilamellar phosphatidylserine vesicles: Titration microcalorimetry, *FEBS Lett.* **123**:249–251.

Reporter, M., and Norris, G., 1973, Reversible effects of lysolecithin on fusion of cultured rat muscle cells, *Differentiation* **1**:83–95.

Reiss-Husson, F., 1967, Structure des phases liquide-cristallines de différents phospholipides, monoglycérides, sphingolipides, anhydres ou en présence d'eau, *J. Molec. Biol.* **25**:363–382.

Rink, T. J., Smith, S. W., and Tsien, R. Y., 1982, Cytoplasmic free Ca^{2+} in human platelets:

Ca^{2+} thresholds and Ca-independent activation for shape change and secretion, *FEBS Lett.* **148**:21-26.
Rink, T. J., Sanchez, A., and Hallam, T. J., 1983, Diacylglycerol and phorbol ester stimulate secretion without raising cytoplasmic free calcium in human platelets, *Nature* **305**:317-319.
Rohlich, P., Anderson, P., and Uvnas, B., 1971, Electron microscope observations on compound 48/80-induced degranulation in rat mast cells: Evidence for sequential exocytosis of storage granules, *J. Cell Biol.* **51**:465-483.
Roseman, M. A., Holloway, P. W., Calabro, M. A., and Thompson, T. E., 1977, Exchange of cytochrome b_5 between phospholipid vesicles, *J. Biol. Chem.* **252**:4841-4849.
Rosenberg, J., Düzgüneş, N., and Kayalar, C., 1983, Comparison of two liposome fusion assays monitoring the intermixing of aqueous contents and of membrane components, *Biochim. Biophys. Acta* **735**:173-180.
Rothman, J. E., 1981, The Golgi apparatus: Two organelles in tandem, *Science* **213**:1212-1219.
Rubin, R. P., 1974, *Calcium and the Secretory Process,* Plenum Press, New York.
Ryan, K. J., Kalant, A., Thomas, E. L., 1971, Free-flow electrophoretic separation and electrical surface properties of subcellular particles from Guinea pig brain, *J. Cell Biol.* **49**:235-246.
Sandra, A., 1980, Interaction of phospholipid vesicles with cultured myogenic cells: Effects on cell fusion, *Exp. Cell Res.* **125**:411-419.
Sandra, A., and Pagano, R. E., 1978, Phospholipid asymmetry in LM cell plasma membrane derivatives: Polar head group and acyl chain distributions, *Biochemistry* **17**:332-338.
Sandra, A., Leon, M. A., and Przybylski, R. J., 1977, Suppression of myoblast fusion by concanavalin A: Possible involvement of membrane fluidity, *J. Cell Sci.* **28**:251-272.
Sanger, J. W., Holtzer, S., and Holtzer, H., 1971, Effects of cytochalasin B on muscle cells in tissue culture, *Nature New Biol.* **229**:121-123.
Satir, B., 1974, Membrane events during the secretory process, *Symp. Soc. Exp. Biol.* **28**:399-418.
Satir, B., 1976, Genetic control of membrane mosaicism, *J. Supramol. Struct.* **5**:381-389.
Satir, B., 1977, Dibucaine-induced synchronous mycocyst secretion in *Tetrahymena, Cell Biol. Intl. Rep.* **1**:69-73.
Satir, B., Schooley, C., and Satir, P., 1972, Membrane reorganization during secretion in *Tetrahymena, Nature,* **235**:53-54.
Satir, B., Schooley, C., and Satir, P., 1973, Membrane fusion in a model system: Mucocyst secretion in *Tetrahymena, J. Cell Biol.* **56**:153-176.
Satir, B. H., and Oberg, S. G., 1978, *Paramecium* fusion rosettes: Possible function as Ca^{2+} gates, *Science* **199**:536-538.
Scheid, A., and Choppin, P., 1974, Identification of biological activities of paramyxovirus glycoproteins. Activation of cell fusion, hemolysis and infectivity by proteolytic cleavage of an inactive precursor protein of Sendai virus, *Virology* **57**:475-490.
Scheid, A., and Choppin, P., 1976, Protease activation mutants of Sendai virus. Activation of biological properties by specific proteases, *Virology* **69**:265-277.
Scheid, A., Hsu, M.-C., and Choppin, P. W., 1980, Role of paramyxovirus glycoproteins in the interactions between viral and cell membranes, in: *Membrane-Membrane Interactions* (N. B. Gilula, ed.), pp. 119-130, Raven Press, New York.
Schlegel, R., Tralka, T. S., Willingham, M. C., and Pastan, I., 1983, Inhibition of VSV binding and infectivity by phosphatidylserine: Is phosphatidylserine a VSV-binding site? *Cell* **32**:639-646.
Schlegel, R., and Wade, M., 1984, A synthetic peptide corresponding to the NH_2 terminus of vesicular stomatitis virus glycoprotein is a pH-dependent hemolysin, *J. Biol. Chem.* **259**:4691-4694.
Schenkman, S., Araujo, P. S., Dijkman, R., Quina, F. H., and Chaimovich, H., 1981, Effects of temperature and lipid composition on the serum albumin-induced aggregation and fusion of small unilamellar vesicles, *Biochim. Biophys. Acta* **649**:633-641.

Schmidt, C. F., Lichtenberg, D., and Thompson, T. E., 1981, Vesicle–vesicle interactions in sonicated dispersions of dipalmitoylphosphatidylcholine, *Biochemistry* **20**:4792–4797.

Schober, R., Nitsch, C., Rinne, U., and Morris, S. J., 1977, Calcium-induced displacement of membrane-associated particles upon aggregation of chromaffin granules, *Science* **195**:495–497.

Schubart, U. K., Fleisher, N., and Ehrlichman, J., 1980, Ca^{2+}-dependent protein phosphorylation and insulin release in intact hamster insulinoma cells: Inhibition by trifluorperazine, *J. Biol. Chem.* **255**:11063–11066.

Schuber, F., Hong, K., Düzgüneş, N., and Papahadjopoulos, D., 1983, Polyamines as modulators of membrane fusion: Aggregation and fusion of liposomes, *Biochemistry* **22**:6134–6140.

Schubert, D., and LaCorbiere, M., 1980, Role of a 16S glycoprotein complex in cellular adhesion, *Proc. Natl. Acad. Sci. USA* **77**:4137–4144.

Schubert, D., and LaCorbiere, M., 1982, Properties of extracellular adhesion-mediating particles in myoblast clone and its adhesion-deficient variant, *J. Cell Biol.* **94**:108–114.

Schudt, C., and Pette, D., 1975, Influence of the ionophore A23187 on myogenic cell fusion, *FEBS Lett.* **59**:36–38.

Schudt, C., and Pette, D., 1976, Influence of monosaccharides, medium factors and enzymatic modification on fusion of myoblasts in vitro, *Cytobiol.* **13**:74–84.

Schudt, C., vanderBosch, J., and Pette, D., 1973, Inhibition of muscle cell fusion *in vitro* by Mg^{2+} and K^+ ions, *FEBS Lett.* **32**:296–298.

Schudt, C., Dahl, G., and Gratzl, M., 1976, Calcium-induced fusion of plasma membranes isolated from myoblasts grown in culture, *Cytobiol.* **13**:211–223.

Schullery, S. E., Schmidt, C. F., Felgner, P., Tillack, T. W., and Thompson, T. E., 1980, Fusion of dipalmitoylphosphatidylcholine vesicles, *Biochemistry* **19**:3919–3923.

Sekiguchi, K., and Asano, A., 1978, Participation of spectrin in Sendai virus-induced fusion of human erythrocyte ghosts, *Proc. Natl. Acad. Sci. USA* **75**:1740–1744.

Sekiguchi, K., Kuroda, K., Ohnishi, S.-I., and Asano, A., 1981, Virus-induced fusion of human erythrocyte ghosts I. Effects of macromolecules on the final stages of the fusion reaction, *Biochim. Biophys. Acta* **645**:211–225.

Sénéchal, H., Pichard, A. L., Delain, D., Schapira, G., and Wahrmann, J. P., 1982, Changes in plasma membrane phosphoproteins during differentiation of an established myogenic cell line and a non-fusing α-amanitin-resistant mutant, *FEBS Lett.* **139**:209–213.

Sessions, A., and Horwitz, A. F., 1981, Myoblast aminophospholipid asymmetry differs from that of fibroblasts, *FEBS Lett.* **134**:75–78.

Sessions, A., and Horwitz, A. F., 1983, Differentiation-related differences in the plasma membrane phospholipid asymmetry of myogenic and fibrogenic cells, *Biochim. Biophys. Acta* **728**:103–111.

Severs, N. J., and Robenek, H., 1983, Detection of microdomains in biomembranes: An appraisal of recent developments in freeze-fracture cytochemistry, *Biochim. Biophys. Acta* **737**:373–408.

Shainberg, A., Yagil, G., and Yaffe, D., 1969, Control of myogenesis in vitro by Ca^{2+} concentration in nutritional medium, *Exp. Cell Res.* **58**:163–167.

Shimshick, E. J., and McConnell, H. M., 1973, Lateral phase separation in phospholipid membranes, *Biochemistry* **12**:2351–2359.

Shimizu, Y. K., Shimizu, K., Ishida, N., and Homma, M., 1976, On the study of Sendai virus hemolysis II. Morphological study of envelope fusion and hemolysis, *Virology* **71**:48–60.

Shoback, D. M., Thatcher, J., Leombruno, R., and Brown, E. M., 1984, Relationship between parathyroid hormone secretion and cytosolic calcium concentration in dispersed bovine parathyroid cells, *Proc. Natl. Acad. Sci. USA* **81**:3113–3117.

Siegel, D. P., 1984, Inverted micellar structures in bilayer membranes: Formation rates and half-lives, *Biophys. J.* **45**:399–420.

Siegel, D. P., Ware, B. R., Green, D. J., and Westhead, E. W., 1978, The effects of calcium and magnesium on the electrophoretic mobility of chromaffin granules measured by electrophoretic light scattering, *Biophys. J.* **22**:341–346.

Silverstein, S. C., Steinman, R. M., and Cohn, Z. A., 1977, Endocytosis, *Ann. Rev. Biochem.* **46**:669–722.

Singer, S. J., and Nicolson, G. L., 1972, The fluid mosaic model of the structure of cell membranes, *Science* **175**:720–731.

Skehel, J. J., Bayley, P. M., Brown, E. B., Martin, S. R., Waterfield, M. D., White, J. M., Wilson, I. A., and Wiley, D. C., 1982, Changes in the conformation of influenza virus hemagglutinin at the pH optimum of virus-mediated membrane fusion, *Proc. Natl. Acad. Sci. USA* **79**:968–972.

Smith, C. L., Ahkong, Q. F., Fisher, D., and Lucy, J. A., 1982, Is purified poly(ethylene glycol) able to induce cell fusion? *Biochim. Biophys. Acta* **692**:109–114.

Smith, R., 1977, Non-covalent cross-linking of lipid bilayers by myelin basic protein: A possible role in myelin formation, *Biochim. Biophys. Acta* **470**:170–184.

Smith, U., Smith, D. S., Winkler, H., and Ryan, J. W., 1973, Exocytosis in the adrenal medulla demonstrated by freeze-etching, *Science* **179**:79–82.

Sokolov, Y. N., and Lishko, V. K., 1979, Study of fusion of planar bilayer phospholipid membranes with liposomes, *Biokhimiya* **44**:317–323.

Steer, C. J., Klausner, R. D., and Blumenthal, R., 1982, Interaction of liver clathrin coat protein with lipid model membranes, *J. Biol. Chem.* **257**:8533–8540.

Steinhardt, R. A., and Alderton, J. M., 1982, Calmodulin confers calcium sensitivity on secretory exocytosis, *Nature* **295**:154–155.

Steinman, R. M., Brodie, S. E., and Cohn, Z. A., 1976, Membrane flow during pinocytosis. A stereological analysis, *J. Cell Biol.* **68**:665–687.

Stewart, T. P., Hui, S. W., Portis, Jr., A. R., and Papahadjopoulos, D., 1979, Complex phase mixing of phosphatidylcholine and phosphatidylserine in multilamellar membrane vesicles, *Biochim Biophys. Acta* **556**:1–16.

Stockdale, F., Okazaki, K., Nameroff, M., and Holtzer, H., 1964, 5-Bromodeoxyuridine: Effect on myogenesis in vitro, *Science* **146**:535–535.

Stockem, W., 1977, Endocytosis, in: *Mammalian Cell Membranes*, Vol. 5 (G. A. Jamieson and D. M. Robinson, eds.), pp. 151–195, Butterworths, London.

Stokem, W., and Klein, H.-P., 1979, Pinocytosis and locomotion in Amoebae. XV. Demonstration of Ca^{2+}-binding sites during induced pinocytosis in *Amoeba proteus*, *Protoplasma* **100**:33–43.

Stollery, J. G. and Vail, W. J., 1977, Interaction of divalent cations or basic proteins with phosphatidylethanolamine vesicles, *Biochim. Biophys. Acta* **471**:372–390.

Stossel, T. P., Bretscher, M. S., Ceccarelli, B., Dales, S., Helenius, A., Heuser, J. E., Hubbard, A. L., Kartenbeck, J., Kinne, R., Papahadjopoulos, D., Pearse, B., Plattner, H., Pollard, T. D., Reutter, W., Satir, B. H., Schliwa, M., Schneider, Y.-J., Silverstein, S. C. and Weber, K., 1978, Membrane dynamics group report, in *Transport of Macromolecules in Cellular Systems,* (S. C. Silverstein, ed.) pp. 503–516, Dahlem Konferenzen, Berlin.

Strittmatter, W. J., Couch, C. B., and Elias, S. B., 1982, Role of specific proteases in rat myoblast fusion, in: *Muscle Development: Molecular and Cellular Control* (M. L. Pearson and H. F. Epstein, eds.) pp. 311–318, Cold Spring Harbor Laboratory, Cold Spring Harbor, New York.

Struck, D. K., Hoekstra, D., and Pagano, R. E., 1981, Use of resonance energy transfer to monitor membrane fusion, *Biochemistry* **20**:4093–4099.

Südhof, T. C., Walker, J. H., and Obrocki, J., 1982, Calelectrin self-aggregates and promotes membrane aggregation in the presence of calcium, *EMBO J.* **1**:1167–1170.

Südhof, T. C., Ebbecke, M., Walker, J. H., Fritsche, U., and Boustead, C., 1984, Isolation of mammalian calelectrins: A new class of ubiquitous Ca^{2+}-regulated proteins, *Biochemistry* **23**:1103–1109.

Sun, S. T., Hsang, C. C., Day, E. P. and Ho, J. T., 1979, Fusion of phosphatidylserine and mixed phosphatidylserine/phosphatidylcholine vesicles: Dependence on calcium concentration and temperature, *Biochim. Biophys. Acta* **557**:45–57.

Sundler, R., and Papahadjopoulos, D., 1981, Control of membrane fusion by phospholipid head groups. I. Phosphatidate/phosphatidylinositol specificity, *Biochim. Biophys. Acta* **649**:743–750.

Sundler, R., and Wijkander, J., 1983, Protein-mediated intermembrane contact specifically enhances Ca^{2+}-induced fusion of phosphatidate-containing membranes, *Biochim. Biophys. Acta* **730**:391–394.

Sundler, R., Düzgüneş, N., and Papahadjopoulos, D., 1981, Control of membrane fusion by phospholipid head groups. II. The role of phosphatidylethanolamine in mixtures with phosphatidate and phosphatidylinositol, *Biochim. Biophys. Acta* **649**:751–758.

Suurkuusk, J., Lentz, B. R., Barenholz, Y., Biltonen, R. L., and Thompson, T. E., 1976, A calorimetric and fluorescent probe study of the gel-liquid-crystalline phase transition in small, single-lamellar dipalmitoylphosphatidylcholine vesicles, *Biochemistry* **15**:1393–1401.

Takeuchi, Y., and Nikaido, H., 1981, Persistence of segregated phospholipid domains in phospholipid lipopolysaccharide mixed bilayers: Studies with spin-labelled phospholipids, *Biochemistry* **20**:523–529.

Tanaka, Y., deCamilli, P., and Meldolesi, J., 1980, Membrane interactions between secretion granules and plasmalemma in three exocrine glands, *J. Cell Biol.* **84**:438–453.

Taupin, C., and McConnell, H. M., 1972, Membrane fusion, in: *Mitochondria/Biomembranes,* FEBS 8th Meeting (S. G. van den Burgh, P. Borst, L. L. M. van Deenen, J. C. Riemeroma, E. C. Slater, and J. M. Tager, eds.) pp. 219–229, Elsevier/North-Holland Biomedical Press, Amsterdam.

Taupin, C., Dvolaitzky, M., and Sauterey, C., 1975, Osmotic pressure induced pores in phospholipid vesicles, *Biochemistry* **14**:4771–4775.

Teichberg, S., Holtzman, E., Crain, S. M., and Peterson, E. R., 1975, Circulation and turnover of synaptic vesicle membrane in cultured fetal mammalian spinal cord neurons, *J. Cell Biol.* **67**:215–230.

Theodosis, D. T., Dreifuss, J. J., Harris, M. C., and Orci, L., 1976, Secretion-related uptake of horseradish peroxidase in neurohypophysial axons, *J. Cell Biol.* **70**:294–303.

Thomas, P., Limbrick, A. R., and Allan, D., 1983, Limited breakdown of cytoskeletal proteins by an endogenous protease controls Ca^{2+}-induced membrane fusion events in chicken erythrocytes, *Biochem. Biophys. Acta* **730**:351–358.

Tilcock, C. P. S., and Cullis, P. R., 1981, The polymorphic phase behavior of mixed phosphatidylserine-phosphatidylethanolamine model systems as detected by ^{31}P NMR: Effects of divalent cations and pH, *Biochim. Biophys. Acta* **641**:189–201.

Tilcock, C. P. S., and Fisher, D., 1979, Interaction of phospholipid membranes with poly(ethylene glycol)s, *Biochim. Biophys. Acta* **577**:53–61.

Tocanne, J. F., Verveergaert, P. H. J. T., Verkleij, A. J., and van Deenen, L. L. M., 1974, A monolayer and freeze-etching study of charged phospholipids: I. Effects of ions and pH on the ionic properties of phosphatidylglycerol and lysylphosphatidylglycerol, *Chem. Phys. Lipids* **12**:201–219.

Toister, Z., and Loyter, A., 1971, Ca^{2+}-induced fusion of avian erythrocytes, *Biochim. Biophys. Acta* **241**:719–724.

Tokutomi, S., Lew, R., and Ohnishi, S.-I., 1981, Ca^{2+}-induced phase separation in phosphatidylserine, phosphatidylethanolamine and phosphatidylcholine mixed membranes, *Biochim. Biophys. Acta* **643**:276–282.

Tozawa, H., Watanabe, M., and Ishida, N., 1973, Structural components of Sendai virus: Serological and physicochemical characterization of hemagglutinin subunit associated with neuraminidase activity, *Virology* **55**:242–263.

Trifaró, J. M., Duerr, A. C., and Pinto, J. E. B., 1976, Membranes of the adrenal medulla: A comparison between the membrane of the Golgi apparatus and chromaffin granules, *Mol. Pharmacol.* **12**:536–545.

Uchida, T., Kim, J., Yamaizumi, M., Mikaje, Y., and Okada, Y., 1979, Reconstitution of lipid vesicles associated with HVJ (Sendai virus) spikes, *J. Cell Biol.* **80**:10–20.

Unanue, E. R., Ungewickell, E., and Branton, D., 1981, The binding of clathrin triskelions to membranes from coated vesicles, *Cell* **26**:439–446.

Uster, P. S., and Deamer, D. W., 1981, Fusion competence of phosphatidylserine-containing liposomes quantitatively measured by a fluorescence resonance energy transfer assay, *Arch. Biochem. Biophys.* **209**:385–395.

Väänänen, P., and Kääriäinen, L., 1980, Fusion and haemolysis of erythrocytes caused by three togaviruses: Semliki Forest, Sindbis and rubella, *J. Gen. Virol.* **46**:467–475.

Vacquier, V. D., 1975, The isolation of intact cortical granules from sea urchin eggs: Calcium ions trigger granule discharge, *Dev. Biol.* **43**:62–74.

Vail, W. J., and Stollery, J. G., 1979, Phase changes of cardiolipin vesicles mediated by divalent cations, *Biochim. Biophys. Acta* **551**:74–84.

van der Bosch, J. and McConnell, H. M., 1975, Fusion of dipalmitoyphosphatidylcholine membranes induced by concanavalin A, *Proc. Natl. Acad. Sci. USA* **72**:4409–4413.

van der Bosch, J., Schudt, C., and Pette, D., 1973, Influence of temperature, cholesterol, dipalmitoyllecithin and Ca^{2+} on the rate of muscle cell fusion, *Exp. Cell Res.* **82**:433–438.

VanderMeulen, J., and Grinstein, S., 1982, Ca^{2+}-induced lysis of platelet secretory granules, *J. Biol. Chem.* **257**:5190–5195.

Vanderwerf, P., and Ullman, E. F., 1979, Monitoring of phospholipid vesicle fusion by fluorescence energy transfer between membrane-bound dye labels, *Biochim. Biophys. Acta* **596**:302–314.

van Dijck, P. W. M., deKruijff, B., Aarts, P. A. M. M., Verkleij, A. J., and deGier, J., 1978, Phase transition in phospholipid model membranes of different curvature, *Biochim. Biophys. Acta* **506**:183–191.

Verkleij, A. J., de Kruyff, B., Verkleij, B., Tocanne, J. F., and vanDeenen, L. L. M., 1974, The influence of pH, Ca^{2+} and protein on the thermotropic behaviour of the negatively charged phospholipid, phosphatidylglycerol, *Biochem. Biophys. Acta* **339**:432–437.

Verkleij, A. J., Mombers, C., Gerritsen, W. J., Leunissen-Bijvelt, L., and Cullis, P. R., 1979, Fusion of phospholipid vesicles in association with the appearance of lipidic particles as visualized by freeze-fracturing, *Biochim. Biophys. Acta* **555**:358–361.

Verkleij, A. J., van Echteld, C. J. A., Gerritsen, W. J., Cullis, P. R., and deKruijff, B., 1980, The lipidic particle as an intermediate structure in membrane fusion processes and bilayer to hexagonal H_{II} transitions, *Biochim. Biophys. Acta* **600**:620–624.

Verleij, A. J., Leunissen-Bijvelt, J., de Kruijff, B., Hope, M. and Cullis, P. R., 1984, Non-bilayer structures in membrane fusion, in: *Cell Fusion, Ciba Foundation Symposium 103*, pp. 45–59, Pitman Books, London.

Ververgaert, P. H. J. T., deKruyff, B., Verkleij, B., Tocanne, J. F., and vanDeenen, L. L. M., 1975, Calorimetric and freeze-etch study of the influence of Mg^{2+} on the thermotropic behavior of phosphatidylglycerol, *Chem. Phys. Lipids*, **14**:97–101.

Verwey, E. J. A., and Overbeek, J. Th. G., 1948, *Theory of the Stability of Lyophobic Colloids*, Elsevier, Amsterdam.

Volsky, D. J., and Loyter, A., 1978, An efficient method for reassembly of fusogenic Sendai virus envelopes after solubilization of intact virions with Triton X-100, *FEBS Lett.* **92**:190–194.

Volsky, D. J., Cabantchik, Z. I., Beigel, M., and Loyter, A., 1979, Implantation of the isolated human erythrocyte anion channel into plasma membranes of Friend erythroleukemic cells by use of Sendai virus envelopes, *Proc. Natl. Acad. Sci. USA* **76**:5440–5444.

Vos, J., Kuriyama, K., and Roberts, E., 1968, Electrophoretic mobilities of brain subcellular particles and binding of γ-aminobutyric acid, acetylcholine, norepinephrine and 5-hydroxytryptamine, *Brain Res.* **9**:224–230.
Vos, J., Ahkong, Q. F., Botham, G. M., Quirk, S. J., and Lucy, J. A., 1976, Changes in the distribution of intramembranous particles in hen erythrocytes during cell fusion induced by the divalent cation ionophore A23187, *Biochem. J.* **158**:651–653.
Wagner, J. A., Carlson, S. S., and Kelly, R. B., 1978, Chemical and physical characterization of cholinergic synaptic vesicles, *Biochemistry* **17**:1199–1206.
Wagner, J. A., and Kelly, R. B., 1979, Topological organization of proteins in an intracellular secretory organelle: The synaptic vesicle, *Proc. Natl. Acad. Sci. USA* **76**:4126–4130.
Wallin, A., Glimelius, K., and Erikkson, T., 1974, The induction of aggregation and fusion of *Daucus carota* protoplasts by polyethylene glycol, *Z. Pfl. Physiol.* **74**:64–80.
Warren, G. B., Metcalfe, J. C., Lee, A. G., and Birdsall, N. J. M., 1975, Mg^{2+} regulates the ATPase activity of a calcium transport protein by interacting with bound phosphatidic acid, *FEBS Lett.* **50**:261–264.
Weinstein, J. N., Yoshikami, S., Henkart, P., Blumenthal, R., and Hagins, W. A., 1977, Liposome-cell interaction: Transfer and intracellular release of a trapped fluorescent marker, *Science* **195**:489–492.
Whitaker, M. J., and Baker, P. F., 1983, Calcium-dependent exocytosis in an in vitro secretory granule plasma membrane preparation from sea urchin eggs and the effects of some inhibitors of cytoskeletal function, *Proc. R. Soc. Lond. Ser B* **218**:397–413.
White, J., and Helenius, A., 1980, pH-dependent fusion between the Semliki Forest virus membrane and liposomes, *Proc. Natl. Acad. Sci. USA* **77**:3273–3277.
White, J., Kartenbeck, J., and Helenius, A., 1980, Fusion of Semliki Forest virus with the plasma membrane can be induced by low pH, *J. Cell Biol.* **87**:264–272.
White, J., Matlin, K., and Helenius, A., 1981, Cell fusion by Semliki Forest, influenza and vesicular stomatitis viruses, *J. Cell Biol.* **89**:674–679.
White, J., Kartenbeck, J., and Helenius, A., 1982a, Membrane fusion activity of influenza virus, *EMBO J.* **1**:217–222.
White, J., Helenius, A., and Gething, M.-J., 1982b, Haemagglutinin of influenza virus expressed from a cloned gene promotes membrane fusion, *Nature* **300**:658–659.
White, J., Kielian, M., and Helenius, A., 1983, Membrane fusion proteins of enveloped animal viruses, *Quart. Rev. Biophys.* **16**:151–195.
Williams, K. E., 1981, Endocytosis and exocytosis, in: *Biochemistry of Cellular Regulation* (P. Knox, ed.) pp. 189–214, CRC Press, Boca Raton, Florida.
Williams, R. J. P., 1974, Calcium ions: Their ligands and their functions, *Biochem. Soc. Symp.* **39**:133–138.
Williams, R. J. P., 1976, Calcium chemistry and its relation to biological function, *Symp. Soc. Exp. Biol.* **30**:1–17.
Willingham, M. C., Maxfield, F. R. and Pastan, I. H., 1979, α_2-Macroglobulin binding to the plasma membrane of cultured fibroblasts. Diffuse binding followed by clustering in coated regions, *J. Cell Biol.* **82**:614–625.
Willingham, M. C., and Yamada, S. S., 1978, A mechanism for the destruction of pinosomes in cultured fibroblasts. Piranhalysis, *J. Cell Biol.* **78**:480–487.
Wilschut, J., and Papahadjopoulos, D., 1979, Ca^{2+}-induced fusion of phospholipid vesicles monitored by mixing of aqueous contents, *Nature* **281**:690–692.
Wilschut, J., Düzgüneş, N., Fraley, R., and Papahadjopoulos, D. 1980, Studies on the mechanism of membrane fusion: Kinetics of calcium ion induced fusion of phosphatidylserine vesicles followed by a new assay for mixing of aqueous vesicle contents, *Biochemistry* **19**:6011–6021.
Wilschut, J., Düzgüneş, N. and Papahadjopoulos, D., 1981, Calcium/magnesium specificity in

membrane fusion: Kinetics of aggregation and fusion of phosphatidylserine vesicles and the role of bilayer curvature, *Biochemistry* **20**:3126–3133.

Wilschut, J., Holsappel, M., and Jansen, R., 1982, Ca^{2+}-induced fusion of cardiolipin/phosphatidylcholine vesicles monitored by mixing of aqueous contents, *Biochim. Biophys. Acta* **690**:297–301.

Wilschut, J., Düzgüneş, N., Hong, K., Hoekstra, D., and Papahadjopoulos, D., 1983, Retention of aqueous contents during divalent cation-induced fusion of phospholipid vesicles, *Biochim. Biophys. Acta* **734**:309–318.

Wilschut, J., Düzgüneş, N., Hoekstra, D., and Papahadjopoulos, D., 1985, Modulation of membrane fusion by membrane fluidity: Temperature dependence of divalent cation-induced fusion of phosphatidylserine vesicles, *Biochemistry* **24**: 8–14.

Wilson, S. P., and Kirshner, N., 1983, Calcium-evoked secretion from digitonin-permeabilized adrenal medullary chromaffin cells, *J. Biol. Chem.* **258**:4994–5000.

Wilson, T., Papahadjopoulos, D., and Taber, R., 1977, Biological properties of polio virus encapsulated in lipid vesicles: Antibody resistance and infectivity in virus-resistant cells, *Proc. Natl. Acad. Sci. USA* **74**:3471–3475.

Wojcieszyn, J. W., Schlegel, R. A., Lumley-Sapanski, K., and Jacobson, K. A., 1983, Studies on the mechanism of polyethylene glycol-mediated cell fusion using fluorescent membrane and cytoplasmic probes, *J. Cell Biol.* **96**:151–159.

Wong, M., and Thompson, T. E., 1982, Aggregation of dipalmitoylphosphatidylcholine vesicles, *Biochemistry* **21**:4133–4139.

Wong, M., Anthony, F. H., Tillack, T. W., and Thompson, T. E., 1982, Fusion of dipalmitoylphosphatidylcholine vesicles at 4 °C, *Biochemistry* **21**:4126–4132.

Wu, S. H. and McConnell, H. M., 1975, Phase separations in phospholipid membranes, *Biochemistry* **14**:847–854.

Wyke, A. M., Impraim, C. C., Knutton, S., and Pasternak, C. A., 1980, Components involved in virally mediated membrane fusion and permeability changes, *Biochem. J.* **190**:625–638.

Yaffe, D., and Dym, H., 1972, Gene expression during differentiation of contractile muscle fibers, *Cold Spring Harbor Symp. Quant. Biol.* **37**:543–547.

Yegorova, Y. M., Chernomordik, L. V., Abidor, I. G., and Chizmadzhev, Y. A., 1981a, Investigation of the interaction of liposomes with BLM by a potentiodynamic method, *Biofizika* **26**:363–365.

Yegorova, Y. M., Chernomordik, L. V., Abidor, I. G., and Chizmadzhev, Y. A., 1981b, Fusion of liposomes with a flat lipid membrane, *Biofizika* **26**:145–147.

Yelkin, A. P., and Berestovskii, G. N., 1974, Measurement of distance between two contacting bimolecular lipid membranes by the optical method, *Biofizika* **19**:846–849.

Zakai, N., Kulka, R. G., and Loyter, A., 1976, Fusion of human erythrocyte ghosts promoted by the combined action of calcium and phosphate ions, *Nature* **263**:696–699.

Zakai, N., Kulka, R. G., and Loyter, A., 1977, Membrane ultrastructural changes during calcium phosphate-induced fusion of human erythrocyte ghosts, *Proc. Natl. Acad. Sci. USA* **74**:2417–2421.

Zimmerberg, J., Cohen, F. S., and Finkelstein, A., 1980a, Fusion of phospholipid vesicles with planar phospholipid bilayer membranes I. Discharge of vesicular contents across the planar membrane, *J. Gen. Physiol.* **75**:241–250.

Zimmerberg, J., Cohen, F. S., and Finkelstein, A., 1980b, Micromolar Ca^{2+} stimulates fusion of lipid vesicles with planar bilayers containing a calcium-binding protein, *Science* **210**:906–908.

Zwaal, R. F. A., and Bevers, E. M., 1983, Platelet phospholipid asymmetry and its significance in hemostatis, *Subcell. Biochem.* **9**:299–334.

Index

Acrosomal reaction, 27, 29, 255
Acrosome fusion, 201
Actin
  binding sites, 10, 12, 13, 15, 19–21, 24
  filaments, 1–39
    assembly, 4–6, 20, 28–29, 31, 37, 39
    bundles, 2, 13–16, 29, 31, 32, 35, 38, 39
    cross-linking, 7–16
    length, 7, 16–23
    membrane attachments, 6, 12, 15, 16, 31
    polarity, 2, 3, 5, 6, 15
    retraction, 24, 25
    stabilization, 23, 24
    structure, 4–6, 10
  fragmentation, severing and cutting, 39
    β-actinin, 23
    gelosin, 20, 34, 35, 39
    protection, 23, 24
    villin, 15, 20, 21
  gels, 10, 16
  monomer, 4, 5, 19–21, 23, 25–27, 35, 37, 39
  networks, 2–4, 7–13, 28, 29, 31
  and secretion, 207, 208
Actin-associated proteins, 1, 6–26, 31
Actin-binding protein, 7, 8, 10, 11, 32, 34, 37, 38
Actin-depolymerizing factor, 20
Actomere, 27–29, 38
Acumentin, 17, 23, 25, 28
Acyl-CoA-cholesterol acyltransferase, 55
Adherons, 215
Adhesion plaques, vinculin, 9, 15, 16
Adenylate cyclase, 75, 187–189
  receptor coupling, 72–74
Adrenal medulla, 206, 207, 247, 250

Adrenal medullary cells, 197, 198, 204, 205
Aerobacter aerogenes, 146
Aging, and cholesterol asymmetry, 75, 76
Alkaline phosphatase, 125, 128, 130, 132, 189
α-Actinin, 9, 13–15, 32, 37, 38
Amoeba, 9, 199
Amphotericin, 82, 83
Anisotropy, fluorescence probes, 79–82
Ankyrin, 12
Anthroyl derivatives, 56, 57, 65
Arabinose binding protein, 127, 128
Arrhenius plots of biophysical probe parameters, 68–70
Artifacts and fluorescent probe use, 77–83
  fluorescence determinations, 78–82
  membrane preparation, 77
  probe location, 77, 78
  quenching, 78
Asolectin, 245, 249
Assembly of actin filaments, 4–6, 17, 22, 32, 37
  cytochalasins, 19
  and membrane movement, 27, 28, 31
  profilin, 26
ATP, 147, 148, 152, 204, 207, 209, 210
  synthesis inhibitors, 55
ATPase, 147, 253

Bacillus
  B.licheniformis, 132
  B.subtilis, 111, 155
Bacteriocins, 108, 117, 135
Bacteroides nodosus, 146
Bacteriophage, 122, 140, 144
Basement membrane, sarcolemma, 181, 192

β-Actinin, 17, 22, 23, 28
β-Galactosidase, 128, 130, 136
β-Glucosidase, 160
β-Glucosides, 160
Betaine, 156
β-Lactamase, 128, 132, 134
Biochemical properties, sarcolemma, 188–190
  calcium transport, 189
  enzymes, 188, 189
  glucose transport, 189
  receptors, 190
Boundary lipids, 55, 68
Brain
  gelosin, 20
  nonerythrocyte spectrin, 8
  tropomyosin, 23
Brevin, 20
Brush border
  membrane, 60
  microvilli, 2
BtuB protein, 116, 117, 140

Calcium, Ca^{2+}, 8, 9, 12, 13, 15, 17, 18, 20–23, 25, 27, 28, 32, 34, 37–39, 67, 72, 109, 143, 145–147, 149, 190, 207–213, 216, 217, 220–255
  and endocytosis, 199
  indicators, 203, 204
  requirements for exocytosis, 196, 197, 201, 203–209, 253
  transport, sarcolemma, 189
Ca^{2+}-ATPase, 55
Ca^{2+}-calcein complex, 226
Calmodulin, 12, 25, 38, 249
Carboxyfluorescein fluorescence, 226, 227
Cardiac muscle sarcolemma, 181, 188–191
Cardiolipin, 55, 144, 222, 228, 232–234, 240, 252
Cation specificity, membrane fusion, 243–245
Cell-cell sufion, 210–220
  erythrocytes, 210–213
  myoblasts, 213–215
  virus-cells, 215–220
Cell membrane, 1, 19, 31
Cell movement, 1, 2, 13, 19, 26–31, 35
Cell shape, 2, 31
Characterization, sarcolemma, 186–188
  enzyme markers, 187
  fluorescent markers, 188
Charge, membrane asymmetry, 56

Chemiosmotic potential, 147
Cholesterol, 213, 214, 217, 218, 222, 240, 254
  asymmetry and secretion, endocytosis, transport and aging, 75–77
  and insulin receptors, 70
  in membrane, 55, 65, 66, 69, 73, 80–82
  phospholipid ratio, sarcolemma, 187
  transbilayer asymmetry, 60–64
Cholesterol oxidase, 60, 64
Cholestratrienol, 56, 59, 61, 62
Chromaffin cells, 208, 246
  catecholamine release, 196, 197
Chromaffin granules, 246–250
Cir protein, 116
Citrobacter freundii, 112
Cleavage furrow, 2, 31
Cochleate lipid cylinders, 225
Colicins, 108, 116–119, 131, 132, 139
Collagen fibers, sarcolemma, 181
Concanavalin A, 52, 251
Conjugation, 137, 149
  divalent cations, 146
  energy dependence, 147, 148
  lipids, 144
  lipopolysaccharides, 142
  proteins, 138–140
Critical concentration, actin assembly, 4, 5, 20
Cross-linking of actin filaments, 6–16, 17, 29–32, 37–39
  actin filament bundles, 13–16, 29, 31, 32, 35, 38, 39
  actin networks, 7–13, 31, 32, 34
Cyclic AMP, 162, 163
Cytochalasins, 19, 20, 24, 27, 209, 213
Cytochrome oxidase, 55, 245
Cytochrome P-450 induction, phospholipid methylation, 75
Cytoplasm, osmoregulation, 152–156
Cytoplasmic membrane, 132, 136, 140, 143, 150, 151
  E. coli, 105, 106, 130
  gram-negative bacteria, 126, 138
Cytoskeleton, 30, 35, 56, 207, 214
  detergent insoluble, 10

Dansyl derivatives, 56, 57, 65
Dehydroergosterol, 56, 59, 61, 62
Density gradient centrifugation, sarcolemma isolation, 181–186

Destabilization of membranes during fusion, 236–241, 244, 254
Dimethylsulfoxide, 210, 211
Dimyristoylphosphatidylcholine, 229, 230, 251
Dimyristoylphosphatidylethanolamine, 237, 242
Dimyristoylphsophatidylglycerol, 251
Dipalmitoylphosphatidylcholine, 229, 230, 237, 241, 245, 251, 252
Dipalmitoylphosphatidylethanolamine, 242
Dipicolnic acid, 223, 224, 226–228
DNA
  competence, 145, 146
  transfer, 137, 140, 149
  uptake by bacteria, 137–149
    divalent cations, 146, 147
    energy dependence, 147–149
    lipids, 144–146
    lipopolysaccharides, 142–144
    proteins, 138–142
DNase, 35

EDTA, 109, 224, 226, 229
Edwardsiella tarda, 112
Eggs, 2
  sea urchin, 204, 209
  villin, 15
EGTA, 15, 17, 18, 146, 190, 225, 253
Endocytic vesicles, 215
Endocytosis, 75, 76, 195, 198–200, 218
Endoplasmic reticulum, 198
  markers, 188
  membrane flow, 199
  in sarcolemma isolation, 182, 183
Endothelial cells, actin filaments, 2, 3, 31
Enterobacter
  E.aerogenes, 112
  E.cloacae, 116
Enterobacteria common antigen, 104
Enterobacteriaceae, 111, 112, 116, 118, 137, 150, 151
Enterochelin, 119, 124
EnvZ protein, 162
Enzymes, *E. coli* outer membrane, 119
Epithelial cells, 2, 60, 120, 123
  intestinal, 2, 8, 9, 15, 17, 32
  kidney, 2, 15, 32
Ergostatrienol, 61, *See also* dehydroergosterol
Erwinia amylovora, 110, 111, 129

Erythrocyte, 20, 32, 251, 252, *See also* red blood cells
  fusion induced by chemicals, 196, 210–213
  ghosts, 240
  membranes, 53, 56, 59, 76
  spectrin, 12
Escherichia, 137
  E. coli, 124, 125, 131–136, 138, 139, 141, 143, 145–150, 152–161, 163
  K-12, 103, 109, 115, 138, 143, 148, 161
  outer membrane proteins, 103–119
  pillus genes, 122, 138, 139
Exocytic vesicles, 221
Exocytosis, 195, 196, 197, 200–205, 210, 222, 246, 249, 252–255
Exocytoskeleton, 158

F-actin, 4, 6
F-actin-associated proteins, 6
Fatty acid, 110, 111, 249
  and insulin receptors, 70
  in membranes, 55, 56, 63, 67, 72, 76
  requiring mutants, 145
FecA protein, 116
FepA protein, 116, 119
F gylcoprotein, Sendai virus, 217
FhuA protein, 116, 133, 140, 143, 146
FhuE protein, 116
Fibroblasts, 52, 54, 75, 79
  actin filaments, 2, 4, 31
  filamin, 7
  membranes, 52, 54, 55, 59, 62, 64, 66, 69, 73, 75
Filamin, 7, 8, 10, 11
Filipin, 62, 63
  and artifacts, 82, 83
Filopodia, platelets, 35, 37, 38
Fimbriae, *E. coli,* 104, 106, 107
  subunit, 135
Fimbrin, 9, 15
Flu protein, 116
Fluidity
  membrane, 52, 64, 65, 72, 83
  transbilayer gradients, 65–67, 72, 78
Fluorescence determinations and artifacts, 78–82
Fluorescence probes
  membrane structure, 51–83
  quencher pairs, 59
Fodrin, 11, 12

Gangliosides, 206, 207, 217
Gelosin, 15, 17, 20–25, 28, 32, 34, 37–39
Gene expression, regulation, 150, 153, 159
Glucose, 155, 156, 159
　transport, sarcolemma, 189
Glycerol, 210, 211, 240
Glycine betaine, 155
Glycolipids, 250
　and membrane fusion, 190, 205, 206, 217, 218
Glycophorin, 217, 251
Glycoproteins, 205–207, 216
　and membrane fusion, 215, 218–220, 246
Golgi apparatus, 209
　membrane flow, 199
Gonococci, 120–124
G-protein, vesicular stomatitis virus, 219
Gram-negative bacteria, 108, 119, 124, 126, 135–137, 141, 146, 147, 149, 151, 155, 157, 159
Gram-negative cell surfaces, functional aspects, 103–163
Gram-positive bacteria, 111, 131, 141, 150, 151, 155

Haemophilus, 146
　H. influenzae, 141, 145
　H. parainfluenzae, 145
Heavy meromyosin, 2, 23, 25, 35
Hemagglutinin (HA), 219
Hemolysin, 135, 137
HN glycoprotein, Sendai virus, 217
Homogenization, muscle cells, 182–186

Influenza virus, 218, 219
Inner membrane, 135, 136
　proteins, 130–132, 134
Insulin receptors, 70, 71, 75
Iron (III) complexes, 113, 119, 126
　requirements, 123
　uptake, receptor proteins, 116, 125
Intermediate filaments, 1, 28
Intestinal epithelial cells, 8, 9, 15, 17, 32

Kdp system, K$^+$ uptake, 152, 153
Kidney epithelial cells, 2, 9, 15, 32
Klebsiella
　K. aerogenes, 116
　K. pneumoniae, 155
K proteins, 113, 114

Lactoferrin, 123, 124
LamB protein, 109, 114–116, 127, 129, 130–134, 136, 140, 144
　-LacZ hybrid proteins, 127, 136
Lateral diffusion, lipids, 69
Lateral reorganization of membranes, 236–241, 254
Lectin, 213, 215, 250, 251
Leucylnaphthylamidase, 187–189
Leukocytes, 4, 7, 8, 10, 20, 23, 27, 28, 32–35, 123
Lipids
　boundary, 55–68
　and conjugation, 144
　DNA transfer, 149
　fluidity, 71
　in membranes, 51, 53–55, 70, 81
　peroxidation, 76
　phase separations, 76
　transbilayer distribution, 56–60, 71
　and transformation, 145, 146
　and viral infections, 144, 145
Lipopolysaccharides (LPS), 104–106, 113, 117, 119, 160
　and conjugation, 138, 139, 142, 146
　and DNA uptake, 138, 149
　in transformation, 143, 144
　and viral infections, 142, 143
Lipoproteins, 210
　E. coli, 108–111, 132
　inner membrane, 131
　outer membrane, 112, 127, 129, 130, 133
Liposomes, 196, 206, 207, 216–218, 221–223, 229–232, 234, 241, 245, 249, 253
Lithium bromide extraction, sarcolemma isolation, 184, 185, 192
Liver, plasma membranes, 68–70, 72, 73
Lysolecithin, 210, 213, 230, 249, 252
Lysosomes, 34, 196, 199, 209, 210, 254

Macrophages, 8, 9, 20, 23, 33, 35
MalE protein, 114, 115, 131–133
Maltodextrins, 113, 115
　pore, 140
Maltose, 114, 115
Maltose-binding protein (MalE), 114, 128, 130, 134
Mast cells, exocytosis, 197, 201, 203, 204

Mechanisms of membrane fusion, 200–255
  electron microscopy, 201–203
  erythrocyte fusion, 210–213
  isolated secretory vesicles, 204–210
  mixed phopholipid vesicles, 230–234
  molecular mechanisms, 234–255
  myoblast fusion, 213–215
  permeabilized cells, 203, 204
  phospholipid bilayer fusion, 220–223
  pure phospholipid vesicles, 223–230
  virus-cell fusion, 215–220
Membrane, 1, 19, 26–31, 34
  actin attachment, 6, 12, 15, 16, 31
  charge asymmetry, 56
  fluorescent probes, 51–83
  lipids, 53–55
  lipid transbilayer distribution, 56–60
  physiological functions, 70–77
  potential artifacts, 77–83
  protein asymmetry, 52, 53
Membrane depth gradients, 65
Membrane derived oligosaccharides (MDO) and periplasmic space osmolarity, 158–161
Membrane flow, 199, 200
Membrane fluidity, 52, 64–67, 72, 75, 83, 211, 214
Membrane fusion, 195–255
  mechanisms, 200–234
  molecular mechanisms, 234–255
  subcellular membranes, 196–200
    endocytosis, 198–200
    exocytosis, 196, 197
Membrane potential, 56, 147–149
Membrane preparation and artifacts, 77
Membrane proteins, 236, 238, 254; see also proteins
  inner, 130–131
  and membrane fusion, 246–253
  outer
    E. coli, 103–119
    Neisseria, 119–124
    Pseudomonas, 124–146
Membrane surface charge, 56
Membrane trigger hypothesis, 126, 127, 133, 135
Meningococci, 123, 124
Magnesium (Mg^{2+}), 205–207, 210, 213, 221, 225, 227–229, 232–235, 237, 238, 242–248, 250, 253, 254

Microfilaments, 35, 37
  disrupting agents, 55
  function in endocytosis, 198
Microspikes, 26, 28
Microtubules, 1, 38
  disrupting agents, 55
  function in endocytosis, 198
Microvilli, 1, 2, 15, 30
  membrane, 70
Model membrane systems, 195, 221, 224, 234
Molecular mechanisms, membrane fusion, 234–255
  cation specificity, 243–245
  close approach of membranes, 234–236
  destabilization, 236–241
  lateral reorganization, 236–241
  osmotic effects, 245, 246
  phospholipid specificity, 241–243
  protein, polypeptides, polyamines, 246–253
  subcellular fusion, 253, 254
Moraxella
  M. morganii, 110, 111
  M. nonliquefaciens, 120
Murein, 105, 108–110, 112, 113, 152
Muscle, 13, 20, 30
  skeletal, 9, 13, 22–24
  smooth, 7, 17, 24
Mycoplasma membranes, 55, 63
Myelin, 60, 251
Myelin basic protein, 55, 251
Myoblast fusion, 196, 213–215, 252
Myosin, 23–25, 27, 30, 31, 34, 35, 39, 208

Neisseria, 108, 124
  N. gonorrhoeae, 111–124, 141
  N. meningitidis, 120, 123, 124, 146
Neurotransmitter release, 195–197, 203
Nucleoside transport, 113, 115

OmpA protein, 109, 112, 117, 118, 129, 131, 133, 136, 139, 140, 142, 144
OmpC protein, 109, 112–115, 118, 129, 140, 143, 161, 162
OmpF protein, 109, 112–115, 118, 124, 129, 133, 140, 161–163
OmpR protein, 161, 162
Oocytes, 8, 15, 18, 26
Osmoregulation, 150–156
  cytoplasm, 152–156
  periplasmic space, 157–161

Osmotic 'downshock', 152, 156
Osmotic effects, membrane fusion, 245, 246, 249, 253
Osmotic 'upshock', 152–154
Outer membrane, bacteria
  and DNA uptake, 137–149
  E. coli, 103–119, 134, 159
  Neisseria, 119–124
  osmoregulation, 150–163
  Pseudomonas, 124–126
Outer membrane proteins, 133, 135, 149
  components for DNA uptake, 137–149
  and conjugation, 138–140
  E. coli, 103–119
  Neisseria, 119–124
  osmolarity and composition, 161–163
  Pseudomonas, 124–126
  role of mature sequence, 131
  signal sequences, 127, 129–130
  and viral infections, 140, 141

Pancreatic cells, 246
  exocytosis, 198, 201, 254
p-Nitrophenylphosphatase (PNPPase), sarcolemma, 187–191
Parinaric acid, probes, 67–70, 77, 79, 80
Parotid acinar cells, 197, 198, 201, 204
Pathogenicity
  E. coli, 103, 104, 108
  Neisseria, 108, 119, 120
Peptidoglycan, 105, 108, 114, 133, 150, 158, 161
Periplasm, 105, 106, 108, 109, 114, 119, 126, 131, 134–136, 157–159
Periplasmic proteins, 128, 130, 131, 133
Periplasmic space, osmoregulation, 151, 157–161
pH
  cytoplasmic, 154
  and osmoregulation, 154, 155
  secretion, 207
  transmembranous gradient, 56, 147–149
  and vesicle fusion, 221, 228, 229
  and virus-cell fusion, 218, 219
Phages, 103, 108, 113, 116–118, 125, 137, 139, 142–144, 146, 148, 149
  phage lamda, 114, 140, 144, 162
Phagocytosis, 52, 75, 198
Phagoliposomes, 209
Phagosomes, 34, 67, 196, 199
  membranes, 59, 66, 77

Phase separations of lipids, 73
  independent monolayers, 69, 70
  lateral rearrangements, 67, 68
PhoA protein, 132, 133
PhoE protein, 114, 115, 125, 129
Phosphatidate, 228, 232–234, 237, 240, 241, 243, 247, 248, 250, 252–254
Phosphatidylcholine, 53–55, 59–61, 65, 71, 74, 76, 210, 212, 218–223, 228–237, 239–242, 244–249, 251–254
Phosphatidylethanolamine, 53, 59, 61, 69, 74, 76, 77, 144, 160, 205, 214, 218, 222, 223, 229, 232–234, 236, 239–242, 244, 247–250, 252, 254
Phosphatidylglycerol, 77, 144, 160, 228, 234, 236, 241
Phosphatidylinosital, 53, 61, 221, 229, 234, 241–243, 248, 254
Phosphaticylinosital response, 243
Phosphatidylserine, 53, 55, 59, 61, 65, 69, 71, 76, 145, 187, 207, 210, 214, 218, 219, 221–223, 225–228, 230–254
Phosphodiesterase, 187–189
Phosphoethanolamine, 159, 160, 207
Phosphoglycerol, 159, 160
Phospholipase A, 119
Phospholipase C, 125, 212–214
Phospholipids, 158–160, 205, 207, 210–212, 214, 216–220, 224, 230, 234–238, 240, 241, 246–248, 252, 254
  E. coli, 105, 106
  liposomes, 113, 144
  in membrane, 53, 54, 56, 57, 61, 63, 64, 66, 67, 76
  methylation and receptor function, 74, 75
  model systems, 69
  Neisseria, 119
Phospholipid bilayers, 220–223, 237, 238, 254
Phospholipid vesicles, 243, 245, 246, 250, 252
  mixed, 230–234, 236–241
  pure, 223–230
Pili
  f-pilus, 138, 142, 146
  Neisseria, 120–123
  plasmids, 138
  sex pilus, 138, 144
  subunits, 120, 121
Pilin, 138
Pinocytosis, 198, 199
Pinosomes, 199

Index

Plasma membrane, sarcolemma, 181, 182
   ultrastructure during secretion, 201
Plasmid, 108, 134, 138, 142, 147–149
   F-plasmid, 138, 155
Plasmid-coded mating function, 138, 139
Platelets, 2, 3, 8–10, 17, 20, 23, 27, 35–39, 55, 65, 71, 246, 249
   fusion, 204
   microspikes, 26
   secretory granules, 246
Polarity, actin filaments, 2, 3, 5, 6, 15
Polyamines and membrane fusion, 246, 252, 253
Polyene-sterol interactions, 63, 82
Poly(ethylene glycol), 210, 211
Polymorphonuclear leukocytes, 3, 8, 123
Polypeptides and membrane fusion, 246, 252, 253
Polysaccharides, 104, 106, 113
Pore-forming proteins, 115, 116
Porin, 136, 140, 143, 163, 246
   E. coli, 105, 112–114, 116, 159, 161
   Neisseria, 123
   Pseudomonas, 124, 125
Porin-type proteins, 119, 120
Potassium transport, 152–154, 162
Probe location and artifacts, 77, 78
Probe molecules
   membrane impermeant, 58
   quencher pairs, 59
Profilactin, 26–28, 32, 38
Profilin, 25–29, 37, 39
Protein
   in membrane, 51–53, 80, 81
   processing, 133–135
   translocation, 135, 136
Protein export, 126–137
   cellular components, 132, 133
   models, 126, 127
   processing, 133–135
   role of sequence of mature protein, 130–132
   signal sequence role, 127
   signal sequence structure, 127–129
   translocation to outer membrane, 135, 136
Protein kinase C, 204, 254
Protein P (porin), 125
Proteus, 107
   P. mirabilis, 111, 112, 116
Pseudomonas aeruginosa, 112, 120, 124–126, 138, 151

Pseudopods, 2, 26, 28, 32, 34, 38
Pyrene derivatives, 56, 57, 65

Quenching and artifacts, 78
Quenching agents, 59, 66, 67
Quin, 2, 204, 253, 254

Receptors
   modulation and membrane asymmetry, 70–72
   sarcolemma, 190
Receptor-adenylate cyclase coupling, 72–74
Receptor function, 74, 75
Receptor protein, iron uptake, 116
Red blood cells, 83, 123; *see also* erythrocytes
   ghosts, 77
   membrane, 55, 56, 60, 62–67, 69, 76, 77
Retinal rods, 30
   outer segment membranes, 68, 69
Retinol, 210
Rhodopseudomonas sphaeroides, 146, 147
Rhodopsin, 55
Rotational rates, fluorescence probes, 80–82

Salmonella, 106, 122, 125, 141, 146, 152
   S. minnesota, 142
   S. paratyphi B, 124
   S. phage, 142
   S. typhimurium, 112, 113, 116–118, 124, 135, 155, 157, 158
Salt extraction, sarcolemma isolation, 182, 185, 191
Sarcolemma biochemistry, 181–192
   biochemical properties, 188–190
   characterization, 186–188
   isolation, 182–186
   vesicles, 190, 191
Sarcoplasmic reticulum, 65
Secretory granules, 201, 246, 253, 254
Secretory processes
   electron microscopy, 201–203
   fusion of vesicles, 204–210
Secretory vesicles, 196, 200, 201, 203, 220, 236, 238, 246, 250, 253, 255
   fusion, 202, 204–210, 243
   release, 196
Semlike virus, 218, 219
Sendai virus, 216–218
Serratia, 111, 118
   S. marcescnes, 110, 129

Shigella, 116, 144
S. dysenteriae, 111
S. flexneri, 108
Siderophores, 116, 125, 126
Signal hypothesis, 126, 127, 132, 133, 135
Signal recognition particle, 136
Signal sequence (peptide), 126, 127, 131, 132, 135, 136
Simian virus, 40, 219
Sindbus virus, 218
SITS dye, 188
Skeletal muscle, 9, 13, 22–24
  sarcolemma, 181, 183, 188, 189, 191
Smooth muscle, 7, 17
  sarcolemma, 181–183, 188–192
Solutes, and osmoregulation, 156
Spectrin, 11, 12, 216, 251
  nonerythrocyte, 8, 9, 11, 12, 31
Sperm, 27–29, 38, 201, 254
Spermidine, 154, 252
Spermine, 252
Spermatozoa membrane, 63
Sphingomyelin, 53, 54, 59, 61, 64, 71, 76, 187, 210, 218
Steady-state fluorescence polarization measurements, 79–82
Sterols
  asymmetry, 75
  in membrane, 51, 56, 64, 76, 77
  nitroxide-labelled, 60, 65
  -polyene interactions, 63, 82, 83
Stimulus-secretion coupling, 195, 207, 253
Stress fibers, 2–4, 7, 8, 16, 30, 31
Subcellular membrane fusion, 253, 254
Sucrose, 156, 157, 161, 190, 225
Surface potential, 222
Synaptic vesicles, 208, 250, 253
Synaptosome membranes, 55
Synexin, 205, 207, 208, 243, 244, 247–250, 254, 255

Tb-dipicolinic acid fusion assay, 223, 224, 226–228, 232
Temperature and membrane fusion, 221, 228, 239, 242
Terminal web, 8
Tetrahymena, secretion, 197, 202, 203

Tight junction, 60
TolC protein, 118, 129
Toluene-lithium bromide, sarcolemma isolation, 185, 192
TonB protein, 116
TraJ, TraT protein, 139
Transbilayer structure, 64–67
  cholesterol asymmetry, 60–64
  coupling, 64, 65
  lipid distribution, 56–60
Transduction, 137
Transferrin, 123, 124
Transformation, 137, 149
  and divalent cations, 146, 147
  energy dependence, 149
  and lipids, 145, 146
  and lipopolysaccharides, 143, 144
  and proteins, 141, 142
Transmembrane pH gradients, 56, 147–149
Transport system proteins, 113–117
Trk system, K^+ uptake, 152–154
Trypomyosin, 23, 24, 38
Tsx protein, 115, 140

Uranyl acetate, 211
Uterine smooth muscle, 190
  sarcolemma, 182, 183

Vesicles, 190, 191
Vesicle aggregation, 234, 235, 238, 239, 243, 244, 247, 248, 250–252
Vesicular stomatitis virus, 218, 219
Vibrio, 137
Villin, 9, 15, 17, 18, 20–22
Vinculin, 9, 15–17, 31
Virosomes, 217–219
Virus, 60, 64, 66, 137, 236, 252
  -cell fusion, 215, 220
  membrane, 55, 77
Virus infections, 149
  divalent cations, 146
  energy dependence, 148, 149
  and lipids, 144, 145
  and lipopolysaccharides, 142, 143
  and protein, 140, 141
Vitamin B_{12} transport, 113, 117, 120